Electrochemical Phenomena in the Cathode Impedance Spectrum of PEM Fuel Cells

AF784214

Electrochemical Phenomena in the Cathode Impedance Spectrum of PEM Fuel Cells:

Fundamentals, Modelling and Applications

Samuel Cruz-Manzo
Research Fellow, University of Lincoln, United Kingdom

Paul Greenwood
Engineer, Abastecedora Electrica Tehuacan, Mexico

ELSEVIER

Elsevier
Radarweg 29, PO Box 211, 1000 AE Amsterdam, Netherlands
The Boulevard, Langford Lane, Kidlington, Oxford OX5 1GB, United Kingdom
50 Hampshire Street, 5th Floor, Cambridge, MA 02139, United States

Copyright © 2022 Elsevier Inc. All rights reserved.

No part of this publication may be reproduced or transmitted in any form or by any means, electronic or mechanical, including photocopying, recording, or any information storage and retrieval system, without permission in writing from the publisher. Details on how to seek permission, further information about the Publisher's permissions policies and our arrangements with organizations such as the Copyright Clearance Center and the Copyright Licensing Agency, can be found at our website: www.elsevier.com/permissions.

This book and the individual contributions contained in it are protected under copyright by the Publisher (other than as may be noted herein).

Notices
Knowledge and best practice in this field are constantly changing. As new research and experience broaden our understanding, changes in research methods, professional practices, or medical treatment may become necessary.

Practitioners and researchers must always rely on their own experience and knowledge in evaluating and using any information, methods, compounds, or experiments described herein. In using such information or methods they should be mindful of their own safety and the safety of others, including parties for whom they have a professional responsibility.

To the fullest extent of the law, neither the Publisher nor the authors, contributors, or editors, assume any liability for any injury and/or damage to persons or property as a matter of products liability, negligence or otherwise, or from any use or operation of any methods, products, instructions, or ideas contained in the material herein.

ISBN: 978-0-323-90607-4

For information on all Elsevier publications
visit our website at https://www.elsevier.com/books-and-journals

Publisher: Susan Dennis
Editorial Project Manager: Veronica III Santos
Production Project Manager: Kumar Anbazhagan
Cover Designer: Christian J. Bilbow

Typeset by STRAIVE, India

ELSEVIER Book Aid International
Working together to grow libraries in developing countries
www.elsevier.com • www.bookaid.org

Dedication

To my lovely wife Mariel Arellano-Huerta who afforded me the time and freedom and looked after me and the house during the preparation of this book.

Samuel Cruz-Manzo

Contents

2. Fundamentals of PEM fuel cells

3. Electrochemical impedance spectroscopy in PEM fuel cells

Part II
Modelling

4. Impedance model of the cathode catalyst layer

5. Impedance model of the gas diffusion layer and air channel

Part III
Modelling application

6. Electrochemical phenomena in the cathodic impedance spectrum

Preface

Electrochemical impedance spectroscopy (EIS) is an experimental technique that can be applied in situ to reveal the electrochemical phenomena that occur at different rates within electrochemical power systems. The use of sophisticated apparatus and precise experimental equipment to carry out EIS measurements has simplified the application of the experimental EIS technique in different electrochemical power systems, e.g. polymer electrolyte membrane fuel cells (PEMFCs) and modern batteries. Mathematical models based on electrochemistry and diffusion theories are required for the effective interpretation of EIS measurements carried out in electrochemical systems operated at different operating conditions. However, the development of such models requires a high level of mathematical skills, which are sometimes not encountered among the electrochemical community. Hence, the measured results from the EIS technique are widely interpreted through equivalent electrical circuits (EECs) using dedicated software. However, the analysis provided by the EECs on the phenomenological processes of the electrochemical system studied can be obscure and ambiguous because different EEC configurations can analyse the same set of EIS measurements.

In this book, we provide a backbone for understanding how the electrochemical phenomena occurring across the different layers comprising the cathodic side of the PEMFC are represented in EIS measurements using impedance models derived from fundamental electrode theory. Students new to the field of EIS-PEMFC technology as well as engineers from the fuel cell industry who aim to understand how electrochemistry, thermodynamics, kinetics, and mass transport phenomena of PEMFCs are related to EIS measurements will gain benefit from this book. A basic knowledge of complex number theory as well as Laplace transform theory will be of help to understand the technical content of this book. A basic knowledge of MATLAB® programming can also be of benefit to execute the MATLAB script examples presented in the book.

This book is a result of the authors' years of experience working on EIS modelling in PEMFCs. The technical information presented in this book establishes a wider scope to relate the EIS technique to the fundamental theory of PEMFCs. The chapters have been classified into three parts: the first part is focused on the fundamentals of EIS and PEMFCs and comprises Chapters 1, 2, and 3; in the second part, the modelling framework is developed and comprises Chapters 4 and 5; and the third part describes the application of the modelling framework and comprises Chapters 6 and 7.

The authors met during PhD studies at Loughborough University, United Kingdom, and have since maintained a fruitful collaboration through the publication of several research manuscripts related to the field of EIS-PEMFC modelling. We hope that this book will provide a fundamental understanding of EIS modelling and will encourage new researchers in finding new ideas for the modelling of EIS in PEMFCs. We are grateful to the people who provided us support to complete this book during a very difficult period dominated by the coronavirus disease COVID-19.

Samuel Cruz-Manzo
Paul Greenwood

List of symbols

Abbreviations

AC	alternating current
ACL	anode catalyst layer
CCL	cathode catalyst layer
CO	carbon monoxide
CPE	constant phase element
DC	direct current
DHE	dynamic hydrogen electrode
EEC	equivalent electrical circuit
EIS	electrochemical impedance spectroscopy
EPIS	electrochemical pressure impedance spectroscopy
FFP	flow field plate
FFT	fast Fourier transform
FRA	frequency response analyser
GDE	gas diffusion electrode
GDL	gas diffusion layer
HOR	hydrogen oxidation reaction
K–K	Kramers–Kronig
LFL	low frequency limit
LP	lumped parameters
LST	linear system theory
MATLAB	matrix laboratory
MEA	membrane electrode assembly
MPL	microporous layer
OCV	open circuit voltage
ORR	oxygen reduction reaction
PEM	polymer electrolyte membrane
PEMFC	polymer electrolyte membrane fuel cell
PTFE	polytetrafluoroethylene
Pt	platinum
PtO	platinum oxide
RH	relative humidity
STP	standard temperature and pressure
TLM	transmission-line model

Symbol

b	Tafel slope (V/dec)
b_c	Tafel slope for cathodic reaction (V)
$b_{H_2O_2}$	Tafel slope for hydrogen peroxide reaction (V)
b_{O_2}	absolute value of Tafel slope from cathodic reaction (V)
$b_{O_2}^{n}$	Tafel slope estimated at operating conditions of EIS-PEMFC (V)
C	capacitance (F)
C_{dl}	double-layer capacitance in the CCL (F/cm^2)
C_{ext}	capacitance attributed to carbon/Nafion and carbon/water interface in the CCL (F/cm^2)
C_{Wv}	water vapour capacitance in the CCL (F/cm^2)
$c_{bOx,bRed}$	bulk concentration of reactants in the PEMFC (mol/cm^3)
c_{bO_2}	bulk concentration of oxygen in the FFP channel (mol/cm^3)
c_{ch}	oxygen concentration in the channel (mol/m^3)
$\tilde{c}_{ch}$	oscillating oxygen concentration in the channel (mol/m^3)
c_{ch_in}	oxygen concentration at inlet of the channel (mol/m^3)
c_{GDL}	oxygen concentration in the GDL (mol/m^3)
$\tilde{c}_{GDL}$	oscillating oxygen concentration in the GDL (mol/m^3)
$c_{Ox,Red}$	local reactant concentration in the CCL (mol/cm^3)
$\tilde{c}_{Ox}$	oscillating reactant concentration in the CCL (mol/cm^3)
c_{O_2}	local oxygen concentration in the CCL (mol/cm^3)
$c^{*}_{Ox,Red}$	reactant bulk concentration in the CCL (mol/cm^3)
$\tilde{c}^{*}_{Ox}$	oscillating reactant bulk concentration in the CCL (mol/cm^3)
$c^{*}_{O_2}$	oxygen concentration in the GDL–CCL interface (mol/cm^3)
c'_{O_2}	oxygen concentration in the CCL–PEM interface (mol/cm^3)
$\tilde{c}'_{O_2}$	oscillating oxygen concentration in the CCL–PEM interface (mol/cm^3)
$\tilde{c}^{*}_{O_2}$	oscillating oxygen concentration in the GDL–CCL interface (mol/cm^3)
$c_{sOx,sRed}$	reactant concentration on the catalyst layer surface (mol/cm^3)
c_{sO_2}	oxygen concentration on the CCL surface (mol/cm^3)
D_{CCL}	oxygen diffusion coefficient in the CCL (m^2/s)
D_{GDL}	oxygen diffusion coefficient in the GDL (m^2/s)
D_{H_2O}	water vapour diffusion coefficient in the CCL (m^2/s)

E	electrode potential (V)
$\tilde{E}$	oscillating potential (V)
E^0	potential at standard conditions (V)
E_{eq}	equilibrium potential in the electrode (V)
E_{Nernst}	reversible potential in the PEMFC (V)
E_R	reversible potential at different temperatures and pressures (V)
E_{SS}	steady-state potential (V)
E_Φ	potential in the solution phase of the CCL (V)
E_ϕ	potential in the solid phase of the CCL (V)
F	Faraday constant (sA/mol)
f	frequency (Hz)
$f(t)$	time-dependent function
G	Gibbs free energy (J/mol)
G^0	Gibbs free energy at standard conditions (J/mol)
H	humidity molar ratio (dimensionless)
H^0	heat of formation at standard conditions (J/mol)
H_T	heat of formation of the electrochemical reaction (J/mol)
h_c	depth of the channel (m)
I	current (A)
$\tilde{I}$	oscillating current (A)
I_{AC}	amplitude of sinusoidal current (A)
I_{cell}	current of the PEMFC (A)
I_j	imaginary component of complex current (A)
I_r	real component of complex current (A)
i	current density in the electrode (A/cm^2)
$\tilde{i}$	oscillating current density in electrode (A/cm^2)
i_{AC}	amplitude of oscillating contribution of sinusoidal current density (A/cm^2)
$i_{C_{dl}}$	current density in the charge double-layer (A/cm^2)
i_{cell}	current density in the PEMFC (A/cm^2)
i_e	electron current density in the CCL (A/cm^2)
i_F	faradaic current density (A/cm^2)
$\tilde{i}_F$	oscillating faradaic current density (A/cm^2)
i_{F_in}	current density at inlet of the air channel (A/cm^2)
i_L	limit current density (A/cm^2)
i_m	proton current density in the membrane (A/cm^2)
i_n	internal current density (A/cm^2)
i_p	proton current density in CCL (A/cm^2)
i_{SS}	steady-state current density (A/cm^2)
i_0	exchange current density (A/cm^2)
j	imaginary component of a complex number (dimensionless)

k	reaction rate coefficient (cm/s)
k_σ	ratio between electron and proton conductivities (dimensionless)
L	inductance (H)
ℓ_c	length of the air channel (m)
ℓ_{CCL}	thickness of the CCL (m)
ℓ_{GDL}	thickness of the GDL (m)
ℓ_P	pore length (cm)
n	number of electrons travelling from anode to cathode (dimensionless)
P	parameter of the constant phase element (dimensionless)
p	pressure of the reactant (Pascal)
p_a	pressure of air entering the channel (Pascal)
p_{vs}	saturation pressure of vapour (Pascal)
p^0	pressure at standard conditions (Pascal)
R	gas constant (J/mol K)
R_{ad}	resistance of adsorption of intermediate species (Ωcm^2)
R_C	charge transfer resistance during ORR (Ωcm^2)
R_{C_LP}	charge transfer resistance considering lumped parameters (Ωcm^2)
R_{C,H_2O_2}	charge transfer resistance for the hydrogen peroxide reaction (Ωcm^2)
R_{C,O_2}	charge transfer resistance for the oxygen reaction in two-step ORR (Ωcm^2)
R_e	electron resistance in the CCL (Ωcm^2)
R_F	faradaic resistance (Ω)
R_{GDL-ch}	oxygen transport resistance in the GDL and channel (Ωcm^2)
R_{GW}	resistance of generalised Warburg impedance (Ωcm^2)
R_m	mass transport resistance (Ωcm^2)
$R_{O_2\_CCL}$	oxygen transport resistance in the CCL (Ωcm^2)
R_{ohm}	ohmic resistance (Ωcm^2)
R_P	proton resistance in CCL (Ωcm^2)
R_{PtO}	resistance for platinum oxide formation (Ωcm^2)
R_{pol}	polarisation resistance (Ωcm^2)
R_W	resistance of the Warburg diffusion impedance (Ω or Ωcm^2)
R_{W_CCL}	Warburg resistance accounting for oxygen transport resistance in CCL (Ωcm^2)
R_{W_GDL-ch}	Warburg resistance accounting for oxygen transport resistance in GDL-channel (Ωcm^2)
$R_{W\_O_2}$	Warburg resistance accounting for oxygen transport resistance in GDL (Ωcm^2)
S_{O_2}	oxygen stoichiometry (dimensionless)

S_T	entropy (J/mol K)
S^0	entropy at standard conditions (J/mol K)
s	Laplace transform (dimensionless)
T	temperature (K)
T	period of sinusoidal signal, see Section 1.2.2 (s)
T^0	temperature at standard conditions (K)
T_a	temperature of air entering the channel (K)
T_{cell}	temperature of the PEMFC (K)
t	time (s)
u_{ch}	velocity of oxygen in the channel (m/s)
V	voltage (V)
$\tilde{V}$	oscillating voltage (V)
V_{AC}	amplitude of sinusoidal voltage (V)
V_j	imaginary component of complex voltage (V)
V_m	molar volume of reactant gas (m^3/mol)
V_{PEMFC}	output voltage in PEMFC (V)
V_r	real component of complex voltage (V)
v	flux of the chemical species (mol/scm^2)
w_s	water saturation in GDL (dimensionless)
Y	parameter of constant phase element ($s^P/\Omega cm^2$)
Z	impedance (Ω)
Z_{ad}	impedance of side reactions with intermediate species (Ωcm^2)
Z_{CCL}	impedance of the CCL (Ωcm^2)
Z_{CCL_LP}	impedance of the CCL considering lumped parameters (Ωcm^2)
$Z_{C_{ext}}$	impedance of the external capacitance (Ωcm^2)
Z_{e^-}	electron resistivity (Ωcm)
Z_F	faradaic impedance in the transmission-line model (Ωcm^3)
Z_{Fw}	faradaic impedance for the binary oxygen–water vapour (Ωcm^2)
Z_{GDL-ch}	local impedance of the GDL and channel (Ωcm^2)
Z_{GDL-ch_LP}	impedance of the GDL-channel considering lumped parameters (Ωcm^2)
Z_{GW}	generalised finite-length Warburg impedance (Ωcm^2)
Z_{H^+}	proton resistivity (Ωcm)
Z_{int}	interfacial impedance in pore (Ωcm)
Z_j	imaginary component of impedance (Ω)
Z_{K_GDL-ch}	impedance of GDL-channel considering charge transfer of the CCL (Ωcm^2)
Z_r	real component of impedance (Ω)
Z_{O_2}	oxygen diffusion impedance (Ωcm^2)
$Z_{O_2\_CCL}$	oxygen diffusion impedance in the CCL (Ωcm^2)
$Z_{PEMFC-cathode}$	cathodic impedance of the PEMFC (Ωcm^2)

$Z_{\mathrm{PEMFC-cathode\_LP}}$	cathodic impedance of PEMFC considering lumped parameters (Ωcm^2)
$Z_{\mathrm{T\_GDL-ch}}$	total impedance of the GDL and channel (Ωcm^2)
Z_{TLM}	impedance of the transmission-line model (Ωcm^2)
Z_{Voigt}	impedance of the Voigt circuit (Ω)
Z_{W}	finite-length Warburg impedance (Ω or Ωcm^2)
$Z_{\mathrm{W\_CCL}}$	Warburg impedance accounting for oxygen diffusion impedance in the CCL (Ωcm^2)
$Z_{\mathrm{W\_GDL-ch}}$	Warburg impedance accounting for GDL-channel impedance (Ωcm^2)
$Z_{\mathrm{W\_O_2}}$	Warburg impedance accounting for oxygen diffusion impedance in the GDL (Ωcm^2)
Z_{Wv}	impedance for water vapour diffusion in the CCL (Ωcm^2)
z	number of electrons in the redox reaction (dimensionless)

A bar over a letter represents the Laplace transform s of the variable, e.g. $\overline{c}_{\mathrm{ch}}(y,s)$, $\overline{V}(s)$, $\overline{i_{\mathrm{F}}}(y,s)$.

Greek symbol

α	charge transfer coefficient (dimensionless)
$\hat{\alpha}$	deviation factor of charge transfer coefficient (dimensionless)
$\gamma_{\mathrm{H_2O_2}}$	fractional surface coverage of hydrogen peroxide (dimensionless)
γ_{PtO}	fractional surface coverage of platinum oxide (dimensionless)
$\gamma_{\mathrm{X^+}}$	fractional surface coverage of the intermediate species (dimensionless)
δ	diffusion distance (m)
δ_{CCL}	oxygen diffusion distance in the CCL (m)
δ_{GDL}	oxygen diffusion distance in the GDL (m)
δ_{Wv}	diffusion distance for water vapour in the CCL (m)
ε	radius of frequency domain contour, see Fig. 1.5 (rad/s)
ε	porosity of the GDL, see Section 6.5 (dimensionless)
$\in$	water molecules during electro-osmotic drag (dimensionless)
η	overpotential (V)
η_{act}	activation overpotential (V)
η_{con}	concentration overpotential (V)
η_{F}	faradaic overpotential (V)
η_{ohm}	ohmic overpotential (V)
η_{SS}	steady-state overpotential (V)
θ	total phase shift (radians or degrees)
$\tilde{\vartheta}_{\mathrm{O_2}}$	oscillating oxygen concentration gradient between the CCL–PEM interface and GDL–CCL interface (dimensionless)

ξ	parameter to reduce oxygen concentration in the GDL/CCL interface (dimensionless)
σ_{e^-}	electron conductivity (S/cm)
σ_{H^+}	proton conductivity (S/cm)
$\tilde{\varsigma}_{OX}$	oscillating reactant concentration gradient between the CCL surface and bulk (dimensionless)
τ_F	electrochemical time constant (s)
τ_{GW}	time constant of the generalised Warburg diffusion impedance (s)
$\tau_{H_2O_2}$	time constant for hydrogen peroxide formation (s)
$\tau_{O_2\_CCL}$	oxygen diffusion time constant in the CCL (s)
$\tau_{O_2\_GDL}$	oxygen diffusion time constant in the GDL (s)
τ_{PtO}	time constant for platinum oxide formation (s)
τ_W	time constant of the Warburg diffusion impedance (s)
τ_{W_CCL}	oxygen diffusion time constant in CCL with lumped parameters (s)
τ_{W_GDL}	oxygen diffusion time constant in GDL with lumped parameters (s)
τ_{W_v}	time constant for water vapour diffusion in the CCL (s)
ϕ_I	phase shift of sinusoidal current (radians)
ϕ_V	phase shift of sinusoidal voltage (radians)
φ	relative humidity (dimensionless)
φ_{ch}	molar oxygen concentration in the channel (dimensionless)
φ_{GW}	phase constant of the generalised Warburg impedance (dimensionless)
ω	angular frequency (rad/s)
ω_o	reference angular frequency for Hilbert Transform (rad/s)
ω^*	angular frequency at the minimum imaginary component (rad/s)

Part I

Fundamentals of PEM fuel cells and EIS

Chapter 1

Introduction to electrochemical impedance spectroscopy

1.1 Fundamental concepts of electrochemical impedance spectroscopy

Olivier Heaviside is considered the founder of the theoretical principles of electrochemical impedance spectroscopy (EIS) [1]. Heaviside defined the term impedance as the relation between voltage and current in the Laplace domain through the study of electrical circuit theory. Among his scientific contributions, the relation between the Laplace space and Fourier space is well recognised through the following transformation:

$$s = j\omega \tag{1.1}$$

where s is the Laplace domain, j is the imaginary component of a complex number, $\omega = 2\pi f$ is the angular frequency in radians/second, and f is the frequency in Hertz (Hz). This transformation from Laplace to Fourier and vice versa is known as Heaviside Transform.

Emil Warburg [2] derived an impedance equation that still bears his name (Warburg impedance) and relates diffusion of electroactive species through an electrode surface. The contribution of the Warburg impedance to study diffusive processes allowed extending the study of impedance on electrochemical systems. EIS measurements have been carried out over the past 70 years [3,4] with the invention of the potentiostat and have been applied to investigate the physical processes in electrochemical systems when applying a sinusoidal perturbation as a function of the frequency. One of the earliest approaches for EIS measurements has been applied to characterise the physical processes in electrodes of electrochemical systems. de Levie [5] studied the electrochemical processes in porous and planar electrodes through EIS measurements. The conclusions of his work state that for a planar electrode the electrical current lines flow in the normal plane to the electrode surface but for a porous electrode the current lines flow effectively parallel to the electrode surface. The total impedance of a planar electrode is the sum of the individual impedances of the

Electrochemical Phenomena in the Cathode Impedance Spectrum of PEM Fuel Cells.
https://doi.org/10.1016/B978-0-323-90607-4.00007-0
Copyright © 2022 Elsevier Inc. All rights reserved.

components representing physical processes in the electrode, but the impedance of a porous electrode is the geometric mean of those components.

EIS is an experimental technique to measure the opposition to the flow of an AC current within an electrochemical system, which contains elements that dissipate energy and store energy. The opposition to the AC load, or impedance, can be measured over a range of frequencies thereby revealing the frequency response of the electrochemical system.

1.1.1 Impedance response in the frequency domain

The impedance is the opposition to the flow of an AC current that an electrochemical system presents when an AC voltage is applied. The impedance response of an electrochemical system can be measured over a range of frequencies. The term 'frequency response' corresponds to the response of an electrochemical system subjected to a sinusoidal input. When an electrochemical system is subjected to a sinusoidal input (e.g. sinusoidal current), a sinusoidal output (voltage) will be produced at the same frequency as the input. However, the amplitude and phase of the output will be different from those of the input. To characterise an electrochemical system in the frequency domain, it is necessary to know the amplitude ratio between the sinusoidal output (e.g. voltage) and the sinusoidal input (current), and the phase shift between the sinusoidal output and the sinusoidal input as a function of the frequency. In electrical engineering, it is very common to represent sinusoidal functions in electrical networks comprising the same frequency with different amplitude and phases in terms of complex numbers. Fig. 1.1 shows a graphical representation of the relation between a trigonometric function with the exponential complex function defined in the Euler's formula.

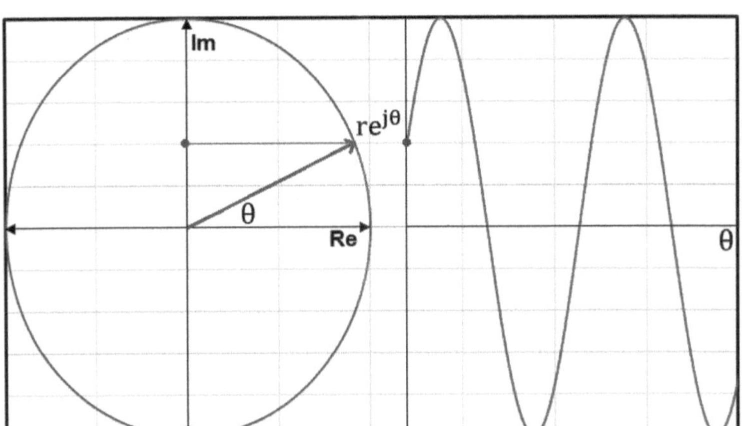

FIG. 1.1 Graphical representation between a trigonometric function and Euler's formula.

The impedance response of an electrochemical system can be expressed in terms of complex numbers by considering the general Euler's formula $re^{j\theta}=r[\cos(\theta)+j\sin(\theta)]$ as such:

$$Z(\omega) = \frac{r_V e^{j[\omega t+\phi_V]}}{r_I e^{j[\omega t+\phi_I]}} = \frac{r_V}{r_I} e^{j\theta} = |Z|[\cos(\theta) + j\sin(\theta)] \quad (1.2)$$

where $|Z|=r_V/r_I$ is the magnitude or amplitude ratio between the AC voltage and the AC current. The term $|Z|$ is also coined as the modulus of the impedance response. $\theta=\phi_V-\phi_I$ is the total phase shift or the difference between the AC voltage and AC current phase shift.

Eq. (1.2) can be expressed through the following relation:

$$Z(\omega) = Z_r + jZ_j \quad (1.3)$$

with

$$Z_r = |Z| \cos(\theta) \quad (1.4)$$

represented as the real component of the impedance response and

$$Z_j = |Z| \sin(\theta) \quad (1.5)$$

represented as the imaginary component of the impedance response. The phase shift can be related with the real and imaginary component of the impedance response as follows:

$$\theta = \tan^{-1}\left(\frac{Z_j}{Z_r}\right) \quad (1.6)$$

1.2 Methods for analysis of impedance measurements

One key advantage of the EIS technique is that it is noninvasive and can be applied in situ. Another advantage is that the frequency response tests are simple to carry out and can be easily tuned for greater accuracy by using readily available sinusoidal generators and precise measuring equipment. Impedance experiments involve the conversion of time-domain input and output signals into a complex quantity that is a function of frequency. A sinusoidal current signal of amplitude I_{AC} (amps) as shown in Fig. 1.2 can be defined as:

$$I(t) = I_{AC} \cos(\omega t + \phi_I) \quad (1.7)$$

where I_{AC} is the amplitude of the sinusoidal current, t is time and ϕ_I is the phase shift. The sinusoidal voltage signal obtained from the electrochemical system can be defined as:

$$V(t) = V_{AC} \cos(\omega t) \quad (1.8)$$

where V_{AC} is the amplitude of the sinusoidal voltage.

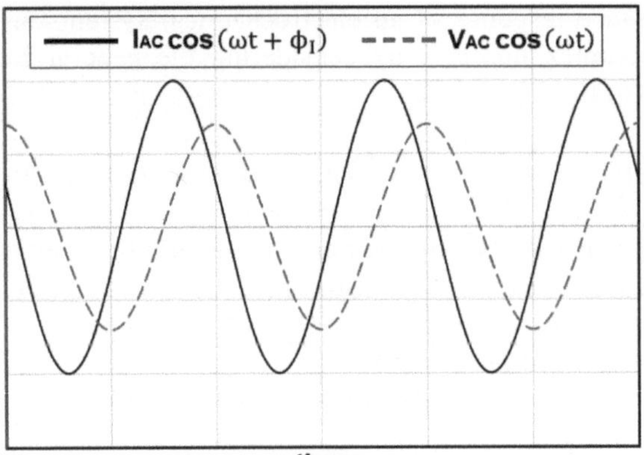

FIG. 1.2 Sinusoidal current and voltage with different amplitude and phase.

1.2.1 Lissajous analysis

Prior to the use of frequency response analysers (FRAs), which convert time domain signals into frequency domain signals through a Fourier analysis, an analysis of Lissajous figures on oscilloscope screens was the accepted method for impedance measurements. Fig. 1.3 shows that the ellipse represented in the Lissajous plot provides information about the magnitude and phase shift θ between the sinusoidal current $I(t)$ and the sinusoidal voltage $V(t)$.

The magnitude of the impedance response represented in the Lissajous plot shown in Fig. 1.3 is calculated as [6]:

$$|Z| = \frac{OB}{OC} \tag{1.9}$$

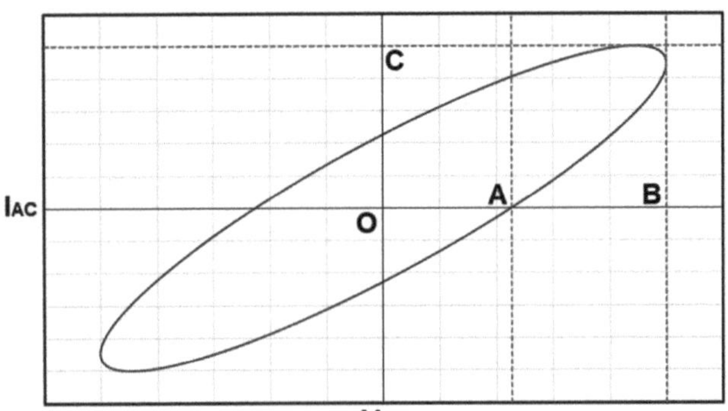

FIG. 1.3 Amplitudes of the sinusoidal current I_{AC} and sinusoidal voltage V_{AC} represented in the Lissajous plot.

The phase shift θ between sinusoidal current $I(t)$ and the sinusoidal voltage $V(t)$ represented in the Lissajous plot shown in Fig. 1.3 is calculated as:

$$\theta = \sin^{-1}\left(-\frac{OA}{OB}\right) \tag{1.10}$$

The shape and direction of the ellipse represented in the Lissajous plot of Fig. 1.3 changes at different frequencies.

1.2.2 Frequency Fourier analysis

Currently, FRAs have replaced the Lissajous analysis through mathematical transformations in Fourier analysis. FRAs use an orthogonal system that involves sines and cosines to determine the complex impedance of a single-frequency input signal. A periodic function of time $f(t)$ that is integrated on $[-\pi,\pi]$ can be expressed as a Fourier series:

$$f(t) = a_0 + \sum_{n=1}^{\infty}[a_n \cos(n\omega t) + b_n \sin(n\omega t)] \tag{1.11}$$

where

$$a_0 = \frac{1}{2\pi}\int_{-\pi}^{\pi} f(t)dt$$

$$a_n = \frac{1}{\pi}\int_{-\pi}^{\pi} f(t)\cos(n\omega t)dt$$

$$b_n = \frac{1}{\pi}\int_{-\pi}^{\pi} f(t)\sin(n\omega t)dt$$

are the Fourier coefficients of the periodic function $f(t)$. Eq. (1.11) can be represented in terms of exponentials through Euler's formula, as such:

$$f(t) = \sum\nolimits_{n=-\infty}^{\infty}\left[c_n e^{jn\omega t}\right] \tag{1.12}$$

and the Fourier coefficients a_0, a_n, b_n are related to

$$c_n = \begin{matrix} [a_n - jb_n]/2 & n > 0 \\ a_0 & n = 0 \\ [a_{-n} + jb_{-n}]/2 & n < 0 \end{matrix}$$

The coefficient c_n from Eq. (1.12) can be evaluated from:

$$c_n = \frac{1}{T}\int_0^T f(t)e^{-jn\omega t}dt \tag{1.13}$$

where T is the period of the signal. Eq. (1.13) can be expressed in terms of trigonometric relations within the Euler's formula:

$$c_n = \frac{1}{T}\int_0^T f(t)[\cos(n\omega t) - j\sin(n\omega t)]dt \tag{1.14}$$

The input current signal in the frequency domain can be expressed as:

$$I(\omega) = \frac{1}{T}\int_0^T I(t)[\cos(n\omega t) - j\sin(n\omega t)]dt \tag{1.15}$$

and the output voltage signal in the frequency domain can be expressed as:

$$V(\omega) = \frac{1}{T}\int_0^T V(t)[\cos(n\omega t) - j\sin(n\omega t)]dt \tag{1.16}$$

The real components of the current and voltage signals in the frequency domain are defined as:

$$I_r(\omega) = \frac{1}{T}\int_0^T I(t)\cos(n\omega t)dt \tag{1.17}$$

$$V_r(\omega) = \frac{1}{T}\int_0^T V(t)\cos(n\omega t)dt \tag{1.18}$$

The imaginary components of the current and voltage signals in the frequency domain are defined as:

$$I_j(\omega) = -\frac{1}{T}\int_0^T I(t)\sin(n\omega t)dt \tag{1.19}$$

$$V_j(\omega) = -\frac{1}{T}\int_0^T V(t)\sin(n\omega t)dt \tag{1.20}$$

The impedance is the ratio between the complex output signal and the complex input signal, as such:

$$Z(\omega) = \frac{V_r(\omega) + jV_j(\omega)}{I_r(\omega) + jI_j(\omega)} \tag{1.21}$$

MATLAB exercise. Demonstrate that a sinusoidal signal for voltage and current in the frequency domain can be constructed from the Fourier coefficients and demonstrate that the ratio between the sinusoidal voltage and current in the frequency domain yields a representation of the impedance response.

Solution. The script shown in Fig. 1.4 has been written in MATLAB software.

The following signals were defined in the MATLAB script shown in Fig. 1.4:

$$V(t) = V_{AC}e^{j[\omega t + \phi_V]} \tag{1.22}$$

```
Fourier_Transform_Signal.m
% Definition of symbolic parameters
syms t theta phi f T Iac Vac w
% Period of the signal
T=1/f;
% Coefficients of the Fourier transform
n=[-1 0 1];
% Angular frequency
w=2*pi*f;
% V(t) signal with phase phi
Vt=Vac*exp(1i*(w*t+phi));
% I(t) signal with phase theta
It=Iac*exp(1i*(w*t+theta));
%%% Sum of Fourier coefficients yields the signal in frequency domain
% Voltage signal in frequency domain
Vw=sum((1/T)*int((Vt.*exp(-n*1i*w*t)),t,[0 T]))
% Current signal in frequency domain
Iw=sum((1/T)*int((It.*exp(-n*1i*w*t)),t,[0 T]))
% Impedance
Zw=Vw/Iw
```

```
Command Window
>> Fourier_Transform_Signal

Vw =

Vac*exp(phi*1i)


Iw =

Iac*exp(theta*1i)


Zw =

(Vac*exp(phi*1i)*exp(-theta*1i))/Iac

fx >>
```

FIG. 1.4 Script in MATLAB to mathematically represent the impedance response from a sinusoidal voltage and sinusoidal current in the frequency domain. Note, 1i in MATLAB means the imaginary component *j*.

$$I(t) = I_{AC} e^{j[\omega t + \phi_I]} \tag{1.23}$$

Eqs (1.22) and (1.23) were substituted into Eq. (1.13) and the coefficients c_n with $n = -1, 0, 1$ were calculated. The sum of the resulting coefficients c_n yielded the sinusoidal signal in the frequency domain.

$$V(\omega) = V_{AC} e^{j\phi_V} \tag{1.24}$$

$$I(\omega) = I_{AC} e^{j\phi_I} \tag{1.25}$$

The impedance response resulted in

$$Z(\omega) = \frac{V_{AC} e^{j[\phi_V - \phi_I]}}{I_{AC}} = \frac{V_{AC}}{I_{AC}} e^{j\theta} \tag{1.26}$$

Eq. (1.26) is similar to Eq. (1.2) and can be represented in real and imaginary components as demonstrated in Eqs (1.4) and (1.5).

1.3 Validity of EIS measurements

The impedance response of an electrochemical system is commonly analysed through an equivalent electrical circuit. The use of physics-based impedance models comprised of differential equations describing the phenomenological processes of the electrochemical system can also be considered for the analysis of impedance measurements. But before trying to relate physical phenomena represented in the impedance response of the electrochemical system with electrical components or through mathematical differential equations, an evaluation of the validity or correctness of the electrochemical impedance response has to be carried out.

The impedance response is constrained by the Linear System Theory (LST). Four properties are imposed on electrochemical systems by the LST [1]:

1. The response of the electrochemical system must comply with the superposition principle for linear systems.
2. The response of the electrochemical system must be at steady state.
3. The response of the electrochemical system must be causal, there is no change in response of the system before the AC perturbation is applied.
4. The impedance response generated by the electrochemical system must be finite for the range of frequencies applied.

The aforementioned properties are considered to evaluate the validity or correctness of the measured impedance response of an electrochemical system. This evaluation can be carried out by data transformation of impedance measurements using Kramers–Kronig (K–K) relations or using the Hilbert transformation.

1.3.1 Kramers–Kronig relations

The impedance response resulting from the Fourier analysis (Eq. 1.21) has to be consistent with K–K relations. These relations are mathematical properties that relate the real and imaginary components of the impedance response resulting from the electrochemical system studied. K–K relations are applied to systems, which are linear, causal, and stable and have been applied in electrochemical impedance data over the past 30 years [7–9]. In K–K the condition of stationarity implies that the causality condition is satisfied, and conversely causality condition implies stationarity of the corresponding electrochemical system. The derivation of these mathematical relations begins with an application of Cauchy's Integral Theorem [6], which defines the integral around a closed contour and evaluates the real or imaginary component of the impedance at a particular frequency with a pole ω located on the real axis of the frequency domain.

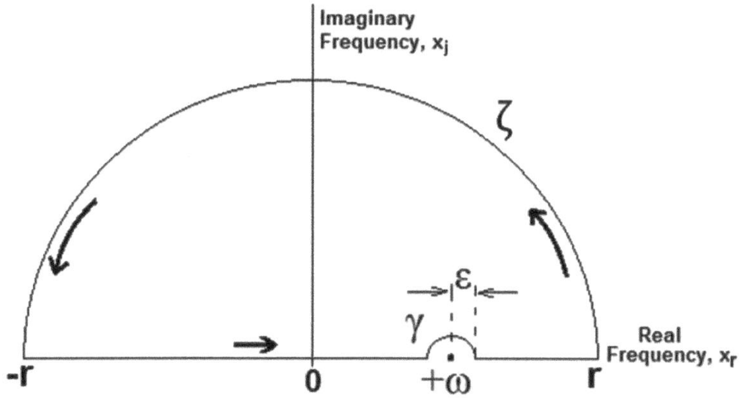

FIG. 1.5 Frequency domain integral contour for deriving Kramers–Kronig relations.

The contour integral of the frequency domain for EIS measurements considering Cauchy's Theorem and a singularity in the right half plane is shown in Fig. 1.5.

Cauchy's Theorem states that if a function $Z(x)$ is solvable in a simple connected domain (in this case, frequency domain), the integral around the closed contour of the domain is:

$$\oint Z(x)dx \tag{1.27}$$

where x is an independent and continuous variable that represents the complex frequency.

Fig. 1.5 shows that the function $Z(x)$ has a singularity on the real frequency axis x_r; therefore, for the function $Z(x)$ to be solvable at point ω on the boundary of the frequency domain, it is necessary to apply Cauchy's Residue Theorem, as such:

$$\oint \frac{Z(x)}{x-\omega}dx = 0 \tag{1.28}$$

Eq. (1.28) can be expanded to define the closed contour shown in Fig. 1.5, as:

$$\oint \frac{Z(x)}{x-\omega}dx = \mathrm{P}\int_{-\infty}^{\infty} \frac{Z(x)}{x-\omega}dx + \int_{\gamma} \frac{Z(x)}{x-\omega}dx + \int_{\zeta} \frac{Z(x)}{x-\omega}dx = 0 \tag{1.29}$$

where $\mathrm{P}\int_{-\infty}^{\infty} f(x)dx$ represents the Cauchy principal value of the integral [6]. The second term on the right-hand side of Eq. (1.29) represents the integral around the pole ω from the closed contour shown in Fig. 1.5 and can be expressed with $x=\omega+\varepsilon e^{j\theta}$, $dx=j\varepsilon e^{j\theta}d\theta$, $0\leq\theta\leq\pi$ as:

$$\begin{aligned}\int_{\gamma} \frac{Z(x)}{x-\omega}dx &= \int_{\pi}^{0} \frac{Z(x)}{x-\omega}dx = \int_{\pi}^{0} \frac{Z\left(\omega+\varepsilon e^{j\theta}\right)}{\omega+\varepsilon e^{j\theta}-\omega} j\varepsilon e^{j\theta}d\theta \\ &= \int_{\pi}^{0} Z\left(\omega+\varepsilon e^{j\theta}\right) jd\theta \end{aligned} \tag{1.30}$$

if $\omega \to \infty$, ε shrinks to zero $\varepsilon \approx 0$ and evaluating the integral in Eq. (1.30) yields:

$$\int_{\gamma} \frac{Z(x)}{x-\omega} dx = \int_{\pi}^{0} jZ(\omega) d\theta = -j\pi Z(\omega) \tag{1.31}$$

The third term on the right-hand side of Eq. (1.29) represents the integral around the contour ζ from the closed contour shown in Fig. 1.5 and can be expressed with $x = re^{j\theta}$, $dx = jre^{j\theta}d\theta$, $0 \le \theta \le \pi$ as:

$$\int_{\zeta} \frac{Z(x)}{x-\omega} dx = \int_{0}^{\pi} \frac{Z(x)}{x-\omega} dx = \int_{0}^{\pi} \frac{Z(re^{j\theta})}{re^{j\theta}-\omega} jre^{j\theta} d\theta = \int_{0}^{\pi} \frac{Z(re^{j\theta})}{1-\frac{\omega}{re^{j\theta}}} jd\theta \tag{1.32}$$

for large $r \to \infty$ and evaluating the integral in Eq. (1.32) yields:

$$\int_{\zeta} \frac{Z(x)}{x-\omega} dx = \int_{0}^{\pi} \frac{Z_{r,\infty}}{1-\frac{\omega}{\infty}} jd\theta = j\pi Z_{r,\infty} \tag{1.33}$$

Substituting Eqs (1.31) and (1.33), into Eq. (1.29) yields: $\mathrm{P}\int_{-\infty}^{\infty} \frac{Z(x)}{x-\omega} dx - j\pi Z(\omega) + j\pi Z_{r,\infty} = 0$, which can be arranged as:

$$Z(\omega) = Z_{r,\infty} + \frac{1}{j\pi} \mathrm{P} \int_{-\infty}^{\infty} \frac{Z(x)}{x-\omega} dx \tag{1.34}$$

To comply with the causality property, the residual expressed in Eq. (1.34) of the singularity has to be zero. Therefore, the function $Z(x)$ in the integrand of Eq. (1.34) has to be combined with another function as such the result vanishes as ω tends to infinity. Eq. (1.34) takes the form:

$$Z(\omega) = Z_{r,\infty} + \frac{1}{j\pi} \mathrm{P} \int_{-\infty}^{\infty} \frac{Z(x) - Z^{*}(\omega)}{x-\omega} dx \tag{1.35}$$

The term $Z^{*}(\omega)$ in Eq. (1.35) indicates that the real and imaginary values of the complex number at frequency ω, are changed from real to imaginary and vice versa. Eq. (1.35) can be split into its real and imaginary parts, as such:

$$Z(\omega) = Z_{r,\infty} + \frac{1}{j\pi} \mathrm{P} \int_{-\infty}^{\infty} \frac{Z_r(x) + jZ_j(x) - Z_r(\omega) - jZ_j(\omega)}{x-\omega} dx$$

$$Z(\omega) = Z_{r,\infty} + \frac{1}{j\pi} \mathrm{P} \int_{-\infty}^{\infty} \frac{Z_r(x) - Z_r(\omega)}{x-\omega} dx + \frac{1}{j\pi} \mathrm{P} \int_{-\infty}^{\infty} \frac{jZ_j(x) - jZ_j(\omega)}{x-\omega} dx$$

$$Z(\omega) = Z_{r,\infty} + \frac{1}{\pi} \mathrm{P} \int_{-\infty}^{\infty} \frac{Z_j(x) - Z_j(\omega)}{x-\omega} dx - \frac{j}{\pi} \mathrm{P} \int_{-\infty}^{\infty} \frac{Z_r(x) - Z_r(\omega)}{x-\omega} dx \tag{1.36}$$

1.3.1.1 Transformation from real to imaginary component

Transforming the integral of the imaginary part (third term on the right-hand side) of Eq. (1.36) into one of definite parity by multiplying the numerator and denominator of the integrand by $x+\omega$ and separating yields:

$$Z_j(\omega) = -\frac{1}{\pi} P \int_{-\infty}^{\infty} \frac{Z_r(x) - Z_r(\omega)}{x - \omega} dx$$
$$= -\frac{1}{\pi} P \int_{-\infty}^{\infty} \frac{[Z_r(x) - Z_r(\omega)][x + \omega]}{[x - \omega][x + \omega]} dx$$
$$Z_j(\omega) = -\frac{1}{\pi} P \int_{-\infty}^{\infty} \frac{xZ_r(x) + \omega Z_r(x) - xZ_r(\omega) - \omega Z_r(\omega)}{x^2 - \omega^2} dx \tag{1.37}$$

Applying the properties for even and odd functions to each term of the integrand in Eq. (1.37), and using symmetry arguments $\int_{-a}^{a} \mathrm{Even}(x)dx = 2\int_0^a \mathrm{Even}(x)dx$ and $\int_{-a}^{a} \mathrm{Odd}(x)dx = 0$ yields:

$$Z_j(\omega) = -\frac{2\omega}{\pi} P \int_0^{\infty} \frac{Z_r(x) - Z_r(\omega)}{x^2 - \omega^2} dx \tag{1.38}$$

1.3.1.2 *Transformation from imaginary to real component*

Transforming the integral of the real part (first and second terms on the right-hand side) of Eq. (1.36) into one of definite parity by multiplying the numerator and denominator of the integrand by $x+\omega$ and separating yields:

$$Z_r(\omega) = Z_{r,\infty} + \frac{1}{\pi} P \int_{-\infty}^{\infty} \frac{Z_j(x) - Z_j(\omega)}{x - \omega} dx$$
$$= Z_{r,\infty} + \frac{1}{\pi} P \int_{-\infty}^{\infty} \frac{[Z_j(x) - Z_j(\omega)][x + \omega]}{[x - \omega][x + \omega]} dx$$
$$Z_r(\omega) = Z_{r,\infty} + \frac{1}{\pi} P \int_{-\infty}^{\infty} \frac{xZ_j(x) + \omega Z_j(x) - xZ_j(\omega) - \omega Z_j(\omega)}{x^2 - \omega^2} dx \tag{1.39}$$

Applying the properties for even and odd functions to each term of the integrand in Eq. (1.39), and using symmetry arguments $\int_{-a}^{a} \mathrm{Even}(x)dx = 2\int_0^a \mathrm{Even}(x)dx$ and $\int_{-a}^{a} \mathrm{Odd}(x)dx = 0$ yields:

$$Z_r(\omega) = Z_{r,\infty} + \frac{2}{\pi} P \int_0^{\infty} \frac{xZ_j(x) - \omega Z_j(\omega)}{x^2 - \omega^2} dx \tag{1.40}$$

If the measured imaginary part does not coincide with the transformed imaginary part and the measured real part with the transformed real part; the data are not obtained under a causal system (the measured response is due only to the AC perturbation applied), linear system (for small AC perturbations) or stable system (the system returns to its original state after the perturbation is removed and does not change with time). The K–K relations cannot be used to determine if the measurements were corrupted by instrument artefacts. The application of the K–K transformations expressed in Eqs (1.38) and (1.40) requires integration of the frequency from zero to infinity, but a finite and reduced range of frequencies is commonly considered during EIS measurements. Therefore, the use of electrical circuits which comply with K–K transformations is an accepted approach to evaluate validity of impedance measurements from electrochemical systems [10]. Another approach to evaluate K–K in EIS measurements consists of extrapolating measured data to extend the frequency domain. This is commonly carried out through commercial software such as ZMAN (WonATech).

1.3.2 Hilbert transformation

Experimental EIS measurements are commonly carried out using FRAs. The resulting impedance response is calculated through mathematical transformations of Fourier transform (FT) analysis. The discrete FT (DFT) changes the time domain discrete data into frequency domain discrete data. The Fast Fourier Transform (FFT) is referred to as a fast-computerised implementation of the DFT. VanderNoot [11] reported that it is possible to calculate the imaginary component from the measured real component and vice versa and evaluate the validity of EIS measurements for steady, causal, and linear systems using the FFT and the Hilbert transformation of impedance data.

The application of the Hilbert transform considers a new frequency variable because the impedance data are usually collected on a log-spaced scale with no available data at the low frequency limit of zero. The new range of frequency variable ρ is calculated from a reference frequency ω_o and the experimental frequency ω as such [12]:

$$\rho = 2\tan^{-1}(\omega/\omega_o) \tag{1.41}$$

$$\omega = \omega_o \tan(\rho/2) \tag{1.42}$$

The following steps reported in the study of VanderNoot [11] allowed the Hilbert transformation of the impedance data:

1. Doubling the impedance data to create an even or odd function with respect to frequency.
2. Data interpolation using FFT.
3. Scaling data at frequency variable ρ.
4. FFT of scaled data at frequency variable ρ.
5. Change of FFT coefficients.
6. Inverse of FFT.
7. Conversion of impedance data to original log-spaced scale frequency ω.

Fig. 1.6 shows the resulting data transformation using the aforementioned steps for the Hilbert transformation of the impedance data using FFT. The steps shown in Fig. 1.6 were carried out using the simulated impedance data from a resistor–capacitor electrical circuit with parallel–series configuration.

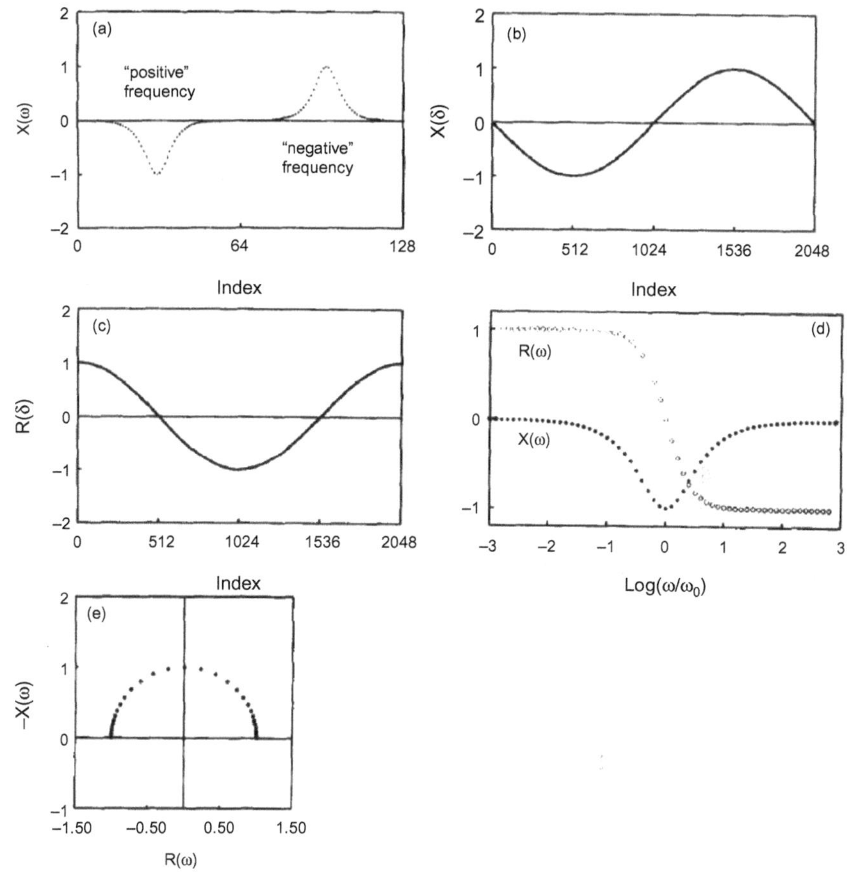

FIG. 1.6 Steps for the Hilbert transform reported by VanderNoot [11]. The use of the Fast Fourier Transform allowed the Hilbert transform of the impedance data. (a) The imaginary impedance (reactance), has been 'doubled' to create an odd function, (b) Interpolation of data, (c) Hilbert transform, (d) transformed real $R(\omega)$ and imaginary $X(\omega)$ from the Hilbert Transform as a function of frequency, and (e) transformed real $R(\omega)$, and imaginary $X(\omega)$ from the Hilbert Transform in the complex plane. *(Reproduced from T.J. VanderNoot, Hilbert transformation of immittance data using the fast Fourier transform, J. Electroanal. Chem. 322 (1992) 15, with permission of Elsevier. Copyright Elsevier.)*

In the next section, the impedance of electrical circuits will be analysed and discussed. Fig. 1.6e shows that it is possible to convert the real component to an imaginary component and vice versa considering a reference frequency ω_o of the impedance data to be transformed.

FFT provides a fast and efficient algorithm of computation of the frequency domain. Nowadays, the FFT algorithm used to compute the DFT has been developed and implemented in commercial software such as MATLAB. A computational algorithm based on FFT and the Hilbert transformation method reported by VanderNoot [11] could assist in the evaluation of correctness in measured EIS data of electrochemical systems. The computational algorithm based on the Hilbert transformation method reported by VanderNoot [11] can be developed as a script in MATLAB software. A major advantage of developing a FFT algorithm for Hilbert transformation in MATLAB software is that it can be used within low-cost FRAs [13] for EIS data evaluation. In the next section, the procedure to construct a computational algorithm based on the FFT theory and the Hilbert transformation of impedance data in a script from MATLAB software is demonstrated. The MATLAB script allows the evaluation of data transformation (real to imaginary $Z_r \rightarrow Z_j$ and imaginary to real $Z_j \rightarrow Z_r$) based on the Hilbert transformation method reported by VanderNoot [11]. The complete script written in MATLAB software is shown in Appendix A. The construction of the algorithm in MATLAB software described in the next section has also been reported in the study by Cruz-Manzo et al. [14].

1.3.2.1 Imaginary to real transformation $Z_j \rightarrow Z_r$

A new frequency range ρ is calculated from the experimentally measured frequency and through Eq. (1.41). Thereafter, the imaginary components Z_j of the impedance response to be evaluated are interpolated with a frequency range from zero to π, with 1024 equal points, as the number of data points for FFT analysis has to be a power of two. The causality property for the frequency response of an electrochemical system implies that the frequency response to an AC perturbation cannot precede the input AC perturbation. In the frequency response of a causal system, the real component is an even function of the frequency and the imaginary component is an odd function of the frequency, as such $Z(\omega)=Z_r+Z_j$, $Z(-\omega)=Z_r-Z_j$. The following procedure is developed in a script in MATLAB to calculate the real components from imaginary components ($Z_j \rightarrow Z_r$ transformation).

1. In FFT analysis, both positive and negative frequencies have to be considered. Due to the fact that the imaginary component is an odd function of the frequency to comply with the causality property, the imaginary components at negative frequencies are calculated and concatenated with the ones at positive frequencies as shown in Fig. 1.7a.
2. The FFT is applied to the imaginary interpolated data shown in Fig. 1.7a using the command *fft* in MATLAB and without considering the imaginary unit *j*; as a result, complex numbers are calculated.
3. The resulting data at positive frequencies are set to be zero.
4. The complex conjugates of the data at negative frequencies resulting from step 2 are calculated, doubled, and concatenated with the data gathered in step 3.
5. The inverse FFT *ifft(j,x,dim)* in MATLAB, where *j* is the imaginary unit, *x* is the vector of data to be transformed and *dim* is the dimension of the transformation and is equal to 2048, is applied to the set of data gathered in step 4

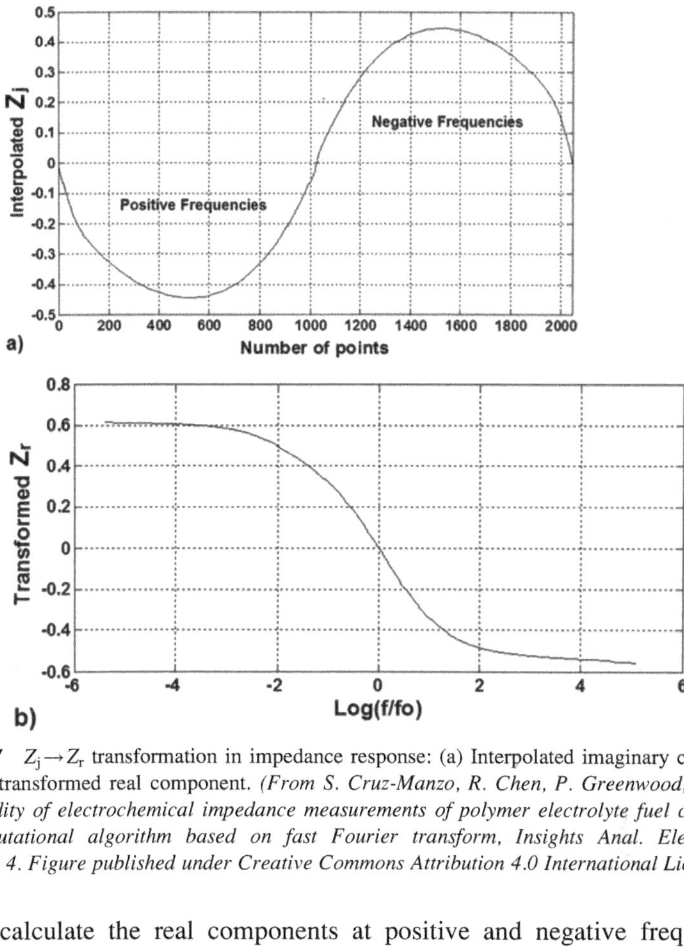

FIG. 1.7 $Z_j \rightarrow Z_r$ transformation in impedance response: (a) Interpolated imaginary component and (b) transformed real component. *(From S. Cruz-Manzo, R. Chen, P. Greenwood, Evaluate the validity of electrochemical impedance measurements of polymer electrolyte fuel cells using a computational algorithm based on fast Fourier transform, Insights Anal. Electrochem. 1 (2015) 4. Figure published under Creative Commons Attribution 4.0 International License.)*

to calculate the real components at positive and negative frequencies. Fig. 1.7b represents the transformed real components at positive frequencies at the logarithm scale f/f_O.

1.3.2.2 Real to imaginary transformation $Z_r \rightarrow Z_j$

The following procedure is developed in a script in MATLAB to calculate the imaginary components from real components ($Z_r \rightarrow Z_j$ transformation).

1. The real components Z_r of the impedance response to be evaluated are interpolated with a frequency range from zero to π, with 1024 equal points. The interpolated real components at negative frequencies are calculated and concatenated with the ones at positive frequencies due to the fact that the real component is an even function of the frequency, as shown in Fig. 1.8a. Note that Fig. 1.8a is also the result of step 5 as discussed in the previous $Z_j \rightarrow Z_r$ transformation section.

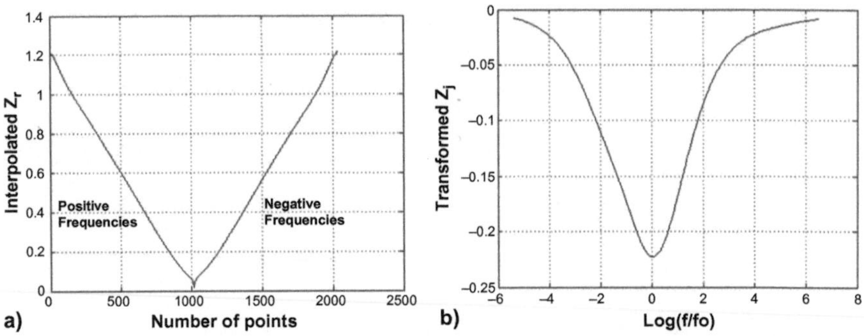

FIG. 1.8 $Z_r \rightarrow Z_j$ transformation in impedance response: (a) Interpolated real component and (b) transformed imaginary component. *(From S. Cruz-Manzo, R. Chen, P. Greenwood, Evaluate the validity of electrochemical impedance measurements of polymer electrolyte fuel cells using a computational algorithm based on fast Fourier transform, Insights Anal. Electrochem. 1 (2015) 5. Figure published under license of Creative Commons Attribution 4.0 International License.)*

2. The FFT is applied to the data shown in Fig. 1.8a using the command *fft* in MATLAB and complex numbers are generated.
3. The data at negative frequencies are set to be zero.
4. The complex conjugates of the data at positive frequencies resulting from step 2 are calculated, doubled, and concatenated with the data gathered in step 3.
5. The inverse FFT *ifft(j,x,dim)* in MATLAB, where *j* is the imaginary unit, *x* is the vector of data to be transformed, and *dim* is the dimension of the transformation and is equal to 2048, is applied to the data gathered in step 4 to calculate the imaginary components at positive and negative frequencies. This results in the imaginary components shown in Fig. 1.7a. Fig. 1.8b represents the calculated imaginary components at positive frequencies plotted at the logarithm scale f/f_O.

The results shown in Figs 1.7 and 1.8 have been obtained by implementing the steps reported in the study of VanderNoot [11] for the Hilbert transformation of the impedance data in a MATLAB script. The MATLAB script that executes the Hilbert transformation of the impedance data is shown in Appendix A. Note that the results shown in Figs 1.7b and 1.8b are similar to the results shown in Fig. 1.6d and reported in the study by VanderNoot [11].

1.4 Equivalent electrical circuits for EIS analysis

Mathematical models based on electrochemistry theory are required for the effective interpretation of EIS measurements in electrochemical systems such as polymer electrolyte membrane fuel cells (PEMFCs). However, the development of such models requires a high level of mathematical skill, which sometimes is not encountered among the electrochemical community. Hence, the measured results from EIS are widely interpreted through various equivalent electrical circuits. The use of electrical circuits with the experimental EIS technique is a well-established methodology to characterise processes in electrochemical systems. Electrical circuits to interpret the mechanisms related to the impedance response of an electrochemical system have been used for over 40 years [15]. An electrical circuit can represent an identical impedance response to that obtained from the electrochemical system studied. Each electrical component in the electrical circuit describes a physical process that takes place in the electrochemical system. Nevertheless, different electrical circuit configurations can represent the same impedance response of the electrochemical system and can create ambiguities when relating electrochemical phenomena of the electrochemical system to the electrical components. The most common electrical components to represent the impedance response in an electrochemical system are as follows:

Resistor. This component represents energy losses, electronic and ionic conductance in solid and aqueous mediums. This element does not depend upon the frequency range applied.

Inductor. This component relates to a magnetic field generated in electrical conductors of the measurement system and in the electrodes of an electrochemical system studied.

Capacitor. This component reflects the accumulation of electrostatic energy between two dissimilar materials.

Warburg element. This element represents the opposition to diffusion of chemical species in a finite or semiinfinite planar medium in the frequency domain.

Constant phase element. This element reflects the distribution of time constants in an electrode-electrolyte interface of an electrochemical system.

The values of the electrical components in an electrical circuit are commonly fitted to the measured EIS response of an electrochemical system using a nonlinear regression strategy, e.g. Levenberg–Marquardt algorithm. The fitting analysis can be carried out using commercial software, e.g. ZView (Scribner Associates, Inc.), MATLAB. The resulting values of the electrical components are commonly related to the different phenomenological processes occurring in the electrochemical system studied.

1.4.1 Impedance response of a capacitor

The current flowing through a capacitor relates its capacitance to the rate of change of voltage across its terminals and can be expressed as:

$$I(t) = C\frac{dV(t)}{dt} \tag{1.43}$$

where C is the capacitance with Farad as the unit, $I(t)$ and $V(t)$ are time-domain current and voltage signals, respectively.

The equation that represents the impedance response of the capacitor can be derived through two different approaches:

Considering sinusoidal signals. The oscillating time-dependent signals of the current and voltage are commonly considered by chemical engineers to develop impedance models that represent the impedance of electrochemical systems [6].

The oscillating time-dependent current and voltage can be represented as:

$$I(t) = Re\left\{\widetilde{I}e^{j\omega t}\right\} \tag{1.44}$$

$$V(t) = Re\left\{\widetilde{V}e^{j\omega t}\right\} \tag{1.45}$$

The term on the right-hand side of Eqs (1.44) and (1.45) represents the real component or the cosine contribution of the general Euler's formula $re^{j\theta} = r[\cos(\theta) + j\sin(\theta)]$ and considering the exponential functions $\widetilde{I} = I_{AC}e^{j\phi_I}$ and $\widetilde{V} = V_{AC}e^{j\phi_V}$ where ϕ_I and ϕ_V are the phase shift of the oscillating current and voltage. The derivative of the voltage in Eq. (1.45) can be expressed as:

$$\frac{dV(t)}{dt} = Re\left\{j\omega\widetilde{V}e^{j\omega t}\right\} \tag{1.46}$$

Substituting Eqs (1.44) and (1.46) into Eq. (1.43) yields:

$$Re\left\{\widetilde{I}e^{j\omega t}\right\} = CRe\left\{j\omega\widetilde{V}e^{j\omega t}\right\} \tag{1.47}$$

The ratio between the oscillating voltage $\widetilde{V}$ and oscillating current $\widetilde{I}$ expressed in Eq. (1.48) yields the impedance response of the capacitor as such:

$$\frac{\widetilde{V}}{\widetilde{I}} = \frac{1}{Cj\omega} \tag{1.48}$$

Further information of the aforementioned approach can be found in the book published by Orazem and Tribollet [6].

Considering Laplace domain. Considering the Laplace domain s of the time-domain current and voltage signals and converting the transfer function or impedance from Laplace space s into frequency domain with $s = j\omega$ (Eq. 1.1) is commonly considered by electrical and control engineers. Under this approach, it is also possible to convert the transfer function of the system from the frequency domain to the time domain via inverse Laplace transform and simulate the output dynamic response of the system at a given input signal [16].

The Laplace transform of a derivative of a function $f(t)$ using the Real Differentiation Theorem [16] is given by:

$$\mathcal{L}\left[\frac{d}{dt}f(t)\right] = s\bar{f}(s) - f(0) \tag{1.49}$$

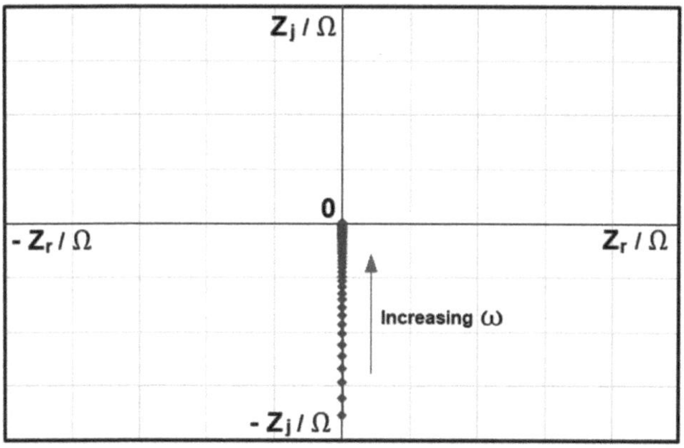

FIG. 1.9 Representation of the impedance response of a capacitor in the complex-impedance plot.

where the bar located over a letter represents the Laplace transform of a variable. Applying Eq. (1.49) into the Laplace transform of Eq. (1.43) with initial condition at $t=0$, $V(0)=0$ yields:

$$\bar{I}(s) = Cs\bar{V}(s) \tag{1.50}$$

The transfer function of a linear, time-invariant system represented in the Laplace space s can be translated into the frequency domain by considering the Laplace space in the frequency domain with $s=j\omega$ (Eq. 1.1) [16]. The transfer function of Eq. (1.50) in the frequency domain can be represented as:

$$\frac{\tilde{V}(j\omega)}{\tilde{I}(j\omega)} = \frac{1}{Cj\omega} \tag{1.51}$$

In this book, the second approach related to conversion of the time-domain signals into the Laplace domain will be considered for further derivation of the impedance of equivalent circuits and impedance of the cathode of the PEMFC presented in the next chapters.

The impedance of a capacitor represented in Eq. (1.51) is only comprised of imaginary components. Fig. 1.9 shows the impedance response of a capacitor. Noting that the terms Z_r and Z_j represent the real and imaginary parts of the complex-impedance plot (Nyquist plot). The impedance response of the capacitor shown in Fig. 1.9 is comprised of only imaginary components and becomes more negative with decreasing frequency. Noting that $\omega = 2\pi f$ is the angular frequency.

1.4.2 Impedance response of an inductor

The voltage of an inductor relates its inductance with the rate of change of the current with time and can be expressed as:

$$V(t) = L\frac{dI(t)}{dt} \tag{1.52}$$

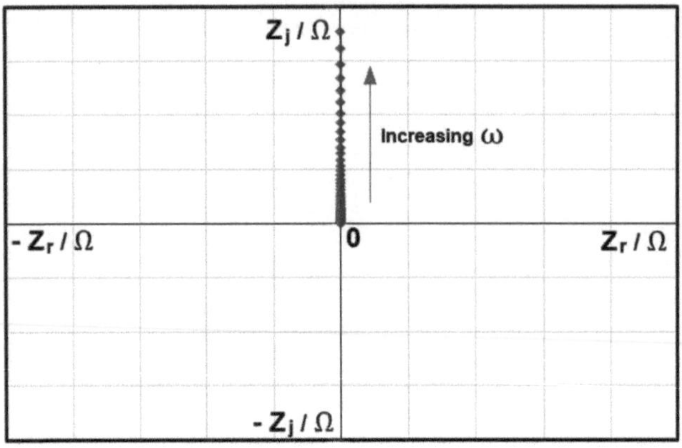

FIG. 1.10 Representation of the impedance response of an inductor in the complex-impedance plot.

where L is the inductance with Henry as the unit. The Laplace transform of Eq. (1.52) using the Real Differentiation Theorem as expressed in Eq. (1.49) with initial condition at $t=0$, $I(0)=0$ yields:

$$\bar{V}(s) = Ls\bar{I}(s) \tag{1.53}$$

The frequency response of the transfer function of Eq. (1.53) can be represented with $s=j\omega$ (Eq. 1.1) as:

$$\frac{\tilde{V}(j\omega)}{\tilde{I}(j\omega)} = Lj\omega \tag{1.54}$$

Eq. (1.54) represents the impedance response of an inductor in the frequency domain.

Fig. 1.10 demonstrates that the impedance response of an inductor, which is generated through Eq. (1.54), is comprised of only imaginary components and becomes more positive with increasing frequency.

1.4.3 Impedance response of a Warburg element considering finite diffusion

This element characterises diffusion of electroactive species through an electrode surface [1]. The name of this element is attributed to Emil Warburg [2] who was able to describe diffusional processes through an impedance equation as a function of frequency. The Warburg element can model diffusion of electroactive species through a diffusion layer with infinite or finite thickness. Finite diffusion describes diffusion of chemical species through a coating or a passive film. Finite diffusion of electroactive species is considered in electrodes of PEMFCs. Infinite diffusion is another form and describes diffusion of electroactive material in batteries or super capacitors.

In Chapter 4, it will be demonstrated that the Warburg impedance considering finite-diffusion is derived from the general theory of experimental controlled-current methods [17]. This theory solves Fick's Second Law of diffusion and relates its solution with the flux of chemical species and Faraday's Law to account for the change in concentration of chemical species during an electrochemical reaction at a fixed current.

The Warburg impedance to account for the diffusion layer with a finite thickness is represented as:

$$Z_W = \frac{R_W}{\sqrt{\tau_W j\omega}} \tanh\left(\sqrt{\tau_W j\omega}\right) \tag{1.55}$$

where R_W is the resistance for diffusion of electrochemical species with Ω as the unit, τ_W is the time constant for the diffusion process.

Eq. (1.55) is commonly applied to characterise diffusive processes in PEMFCs through EIS. In Chapters 4 and 5, a mathematical treatment based on fundamental electrochemical theory of PEMFCs will be developed to demonstrate the derivation of Eq. (1.55).

Fig. 1.11 shows the impedance response of the finite-length diffusion Warburg impedance predicted by Eq. (1.55). A 45-degree straight line at high frequencies is observed. Under the high frequency range, it is possible to relate that the period of the sinusoidal signal during EIS measurements is shorter than the time required for an electroactive species to diffuse through the diffusive layer. Therefore, an infinite diffusion process is commonly related to the high frequency range of the impedance response of the finite-length diffusion Warburg impedance (45-degree straight line shown in Fig. 1.11). The frequency ω^* at which the imaginary component reaches its minimum value at $-Z_j$ as shown in Fig. 1.11 is calculated through the following relation:

$$\omega^* = \frac{2.53}{\tau_W} \tag{1.56}$$

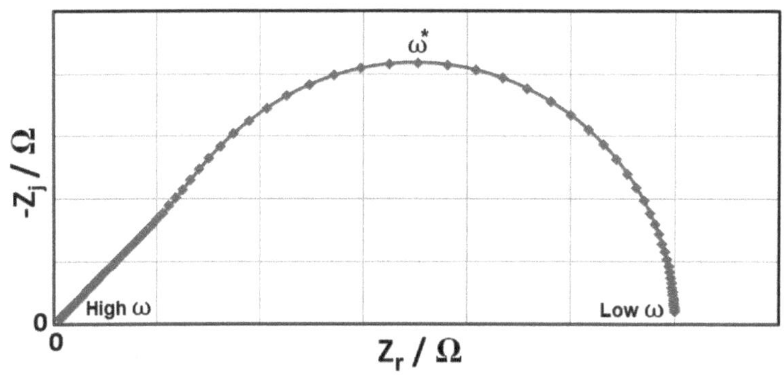

FIG. 1.11 Impedance response of the finite-length Warburg impedance.

where the factor 2.53 has been estimated by series expansion of the finite-length diffusion Warburg impedance [18]. Boukamp [19] reported that the time constant where the imaginary component of the finite-length diffusion Warburg impedance reaches its minimum value at $-Z_j$ (the highest point of the impedance spectrum) in the Nyquist plot is a factor 2.53 smaller than the characteristic time constant τ_W represented in the finite-length Warburg impedance equation. This was demonstrated through a Newton–Raphson iteration in the finite-length diffusion Warburg impedance (Eq. 1.55) as reported in the study of Boukamp [19].

1.4.4 Impedance response of a resistor–capacitor (parallel–series configuration)

The electrical circuit shown in Fig. 1.12 is commonly applied to characterise the impedance response of electrodes in electrochemical systems neglecting a mass transport process. R_1 is the resistance of electroactive species in the electrolyte, R_2 is the charge transfer resistance during the redox reaction and C is the capacitance between the dissimilar materials (electrode/electrolyte) in the electrochemical system.

The Laplace domain of the total voltage of the electrical circuit shown in Fig. 1.12 and represented through the Kirchhoff's law of voltage is:

$$\bar{V}(s) = \bar{V}_1(s) + \bar{V}_{R_2\|C}(s) \tag{1.57}$$

Eq. (1.57) can be represented in terms of the electrical current flowing through the circuit as such:

$$\bar{V}(s) = \bar{I}(s)R_1 + \bar{I}(s)\bar{Z}_{R_2\|C}(s) \tag{1.58}$$

where $\bar{Z}_{R_2\|C}(s)$ represents the impedance of the parallel configuration between the resistor R_2 and the capacitor C in the Laplace domain. The impedance of $\bar{Z}_{R_2\|C}(s)$ can be derived from the Kirchhoff's law of current:

$$\bar{I}(s) = \frac{\bar{V}_{R_2\|C}(s)}{R_2} + Cs\bar{V}_{R_2\|C}(s) \tag{1.59}$$

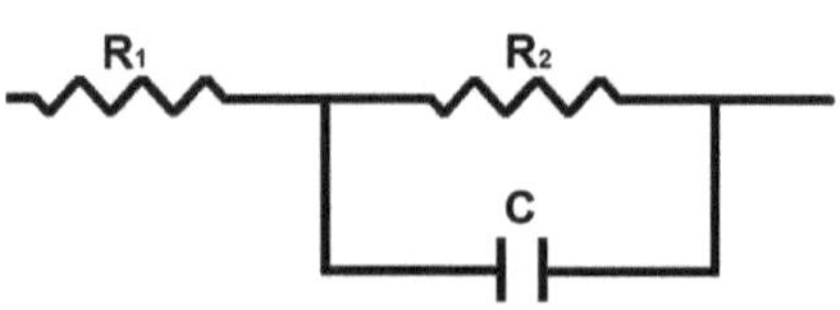

FIG. 1.12 Electrical circuit for EIS analysis in electrodes neglecting mass transport limitations.

where the second term on the right-hand side of Eq. (1.59) represents the Laplace transform of the derivative of the voltage as expressed in Eq. (1.50). Eq. (1.59) can be rearranged in terms of impedance as:

$$\bar{Z}_{\mathrm{R_2\|C}}(s) = \frac{\bar{V}_{\mathrm{R_2\|C}}(s)}{\bar{I}(s)} = \frac{R_2}{1 + R_2Cs} \tag{1.60}$$

Substituting Eq. (1.60) into Eq. (1.58) yields:

$$\bar{V}(s) = \bar{I}(s)R_1 + \bar{I}(s)\frac{R_2}{1 + R_2Cs} \tag{1.61}$$

The transfer function of Eq. (1.61) in the frequency domain with $s = j\omega$ can be expressed as:

$$\widetilde{Z}_{\mathrm{RC}}(j\omega) = \frac{\widetilde{V}(j\omega)}{\widetilde{I}(j\omega)} = R_1 + \frac{R_2}{1 + R_2Cj\omega} \tag{1.62}$$

Eq. (1.62) represents the impedance of the electrical circuit shown in Fig. 1.12. The impedance response generated by Eq. (1.62) is represented in the Nyquist plot shown in Fig. 1.13. The simulation was carried out in MATLAB software considering a range of frequencies from 100 kHz down to 0.01 Hz. The values of the resistors are $R_1 = 1\ \Omega$, $R_2 = 3\ \Omega$, and for the capacitor $C = 0.009$ F. The diameter of the semicircle shown in Fig. 1.13 is equal to the value of $R_2 = 3\ \Omega$. The semicircle is shifted towards the increasing real component Z_r of the complex-impedance plot by the value of $R_1 = 1\ \Omega$.

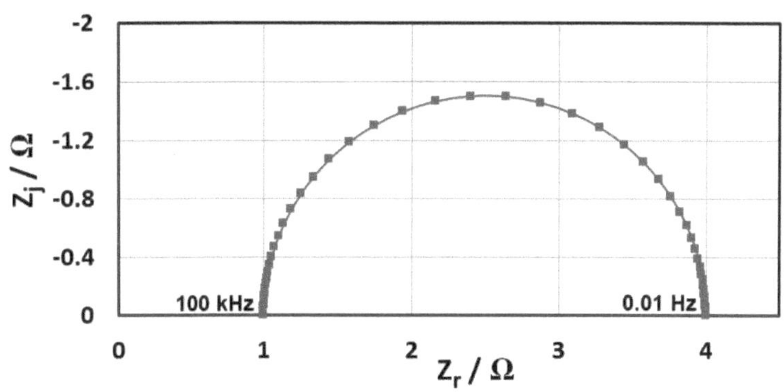

FIG. 1.13 Impedance response generated by a resistor–capacitor (parallel–series configuration) circuit and considering values of the resistors and capacitor as $R_1 = 1\ \Omega$, $R_2 = 3\ \Omega$, $C = 0.009$ F.

1.4.5 Impedance response of a constant-phase element

A constant phase element (CPE) is used in studies of corroding systems to describe the frequency dependence of a nonideal capacitive behaviour [20]. A CPE improves the quality of fitting between an equivalent electrical circuit and the impedance response of an electrochemical system. The second edition of the book published by Orazem and Tribollet [6] has a full chapter to describe the physical meaning of the CPE when applying it to the impedance data of electrochemical systems with time-constant dispersion.

The impedance of a CPE is defined as:

$$Z_{\mathrm{CPE}} = \frac{1}{Y[j\omega]^P} \tag{1.63}$$

where P has a value commonly less than 1 and Y is a parameter related to CPE with units s^P/Ω. Although Y has also been defined as the prefactor of the CPE [21]. If parameter P is equal to 1, Eq. (1.63) is analogous to the impedance of an ideal capacitor C as shown in Eq. (1.51). Nevertheless, Orazem et al. [22] demonstrated that it is not appropriate to calculate the capacitance from the parameters of the CPE by considering $P=1$ and considering $Y=C$. The capacitance can be calculated from the parameters Y and P of the CPE through different mathematical approaches as reported by Orazem and Tribollet [6].

The impedance response of a CPE can be best reflected if the CPE is connected in parallel with a resistor as shown in Fig. 1.14 and simulating the frequency response of the circuit. Note that the electrical circuit shown in Fig. 1.14 is analogous to the electrical circuit shown in Fig. 1.12 but considering the CPE instead.

The impedance response of the electrical circuit shown in Fig. 1.14 considering a CPE can be represented through the following equation:

$$\tilde{Z}_{\mathrm{R,CPE}}(j\omega) = R_1 + \frac{R_2}{1 + R_2 Y[j\omega]^P} \tag{1.64}$$

Eq. (1.64) can be derived following the same procedure from the previous Section 1.4.4 (Eqs 1.57–1.62) and considering the impedance of the CPE expressed in Eq. (1.63). If the value of the exponent P in Eq. (1.64) is equal to 1, Eq. (1.64) becomes similar to Eq. (1.62).

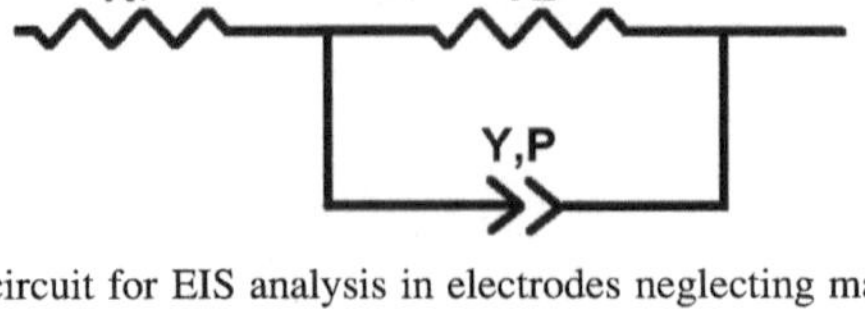

FIG. 1.14 Electrical circuit for EIS analysis in electrodes neglecting mass transport limitations and considering a constant phase element instead of a capacitor.

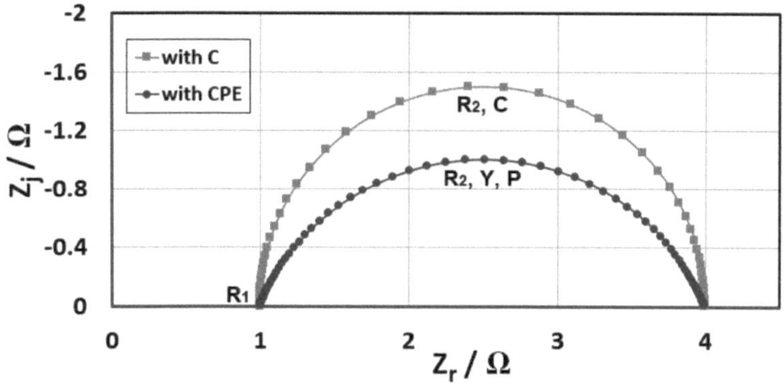

FIG. 1.15 Comparison in Nyquist plot between impedance response generated by a parallel–series RC circuit considering a constant phase element (parameters Y, P) and considering a capacitor C. The considered value of the parameters are $R_1=1\ \Omega$, $R_2=3\ \Omega$, $C=0.009$ F, $Y=0.009\ s^P/\Omega$, $P=0.75$.

In Fig. 1.15, the impedance response generated by the R||C circuit considering a capacitor (Eq. 1.62) is compared with the impedance response generated by the circuit considering a CPE (Eq. 1.64) at different frequencies from 100 kHz to 0.01 Hz. The value of the resistors ($R_1=1\ \Omega$, $R_2=3\ \Omega$) is the same for both circuit configurations represented in Eqs (1.62) and (1.64) and the value of Y for CPE was the same as the value for the capacitor C ($C=0.009$ F and $Y=0.009\ s^P/\Omega$). The considered value of the exponent P of the CPE in Eq. (1.64) was $P=0.75$. The resistors are not frequency dependant; therefore, their contribution is expressed in the real component Z_r of the Nyquist plot. The diameter of the semicircle shown in Fig. 1.15 represents the value of resistor R_2. Resistor R_1 shifts the semicircle across the real axis Z_r of the complex impedance plot. Fig. 1.15 shows that the value of the imaginary components Z_j in the impedance response generated by the circuit shown in Fig. 1.12 considering a capacitor is more negative than the impedance response generated by the circuit shown in Fig. 1.14 considering a CPE.

Fig. 1.15 has demonstrated that the electrical circuit shown in Fig. 1.14 and represented by Eq. (1.64) can simulate a depressed semicircle when the value of the parameter P in the CPE impedance is less than 1, $P<1$. An electrical circuit comprising a CPE can improve the quality of fitting between the impedance response of an electrical circuit with the impedance response of an electrochemical system. The extraction of the true value of the capacitance C from the parameters Y and P of the CPE can be possible through different mathematical approaches as reported by Orazem and Tribollet [6].

EIS measurements can also be represented in a Bode plot. EIS data are expressed as complex numbers as a function of frequency. The magnitude or modulus of the complex numbers as a function of frequency and the phase as a function of frequency are commonly plotted to represent EIS data in the

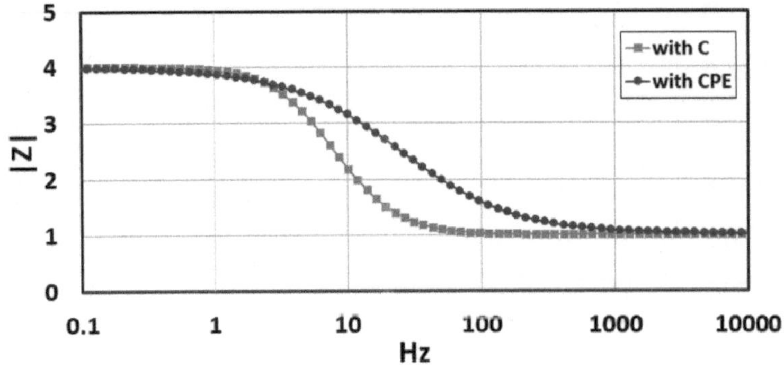

FIG. 1.16 Comparison between the modulus as a function of frequency generated by a parallel–series RC circuit considering a constant phase element (CPE) and considering a capacitor *C*.

Bode plot. Fig. 1.16 shows the modulus of the complex numbers of the impedance response generated by the circuits shown in Figs 1.12 and 1.14 and represented by Eqs (1.62) and (1.64). The difference in the slope between the modulus of the impedance response generated by the circuit considering a capacitor *C* (Fig. 1.12) and a CPE (Fig. 1.14) is observed in Fig. 1.16. The slope of the modulus increases with the impedance response of the circuit considering a capacitor rather than a CPE.

Fig. 1.17 shows the phase angle of the impedance response generated by the circuits shown in Figs 1.12 and 1.14 and through Eqs (1.62) and (1.64). The phase angle provides information about the difference between the distribution of charge between a capacitor *C* and a CPE. The phase angle is calculated through the relation between the real component and imaginary component of the complex number as represented in Eq. (1.6). For an ideal-capacitor *C* the phase angle becomes more negative, as shown in Fig. 1.17.

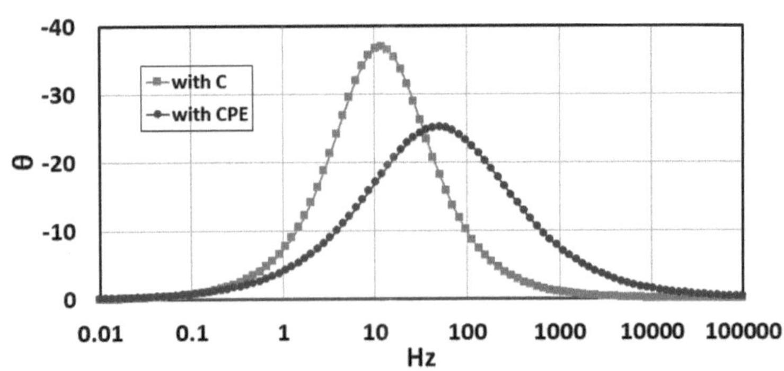

FIG. 1.17 Comparison between phase angle as a function of frequency generated by a parallel–series RC circuit considering a constant phase element (CPE) and considering a capacitor *C*.

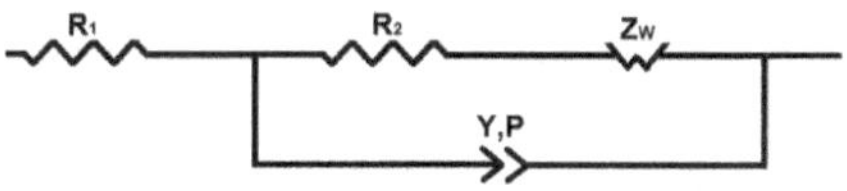

FIG. 1.18 Randles circuit for EIS analysis in electrodes with diffusion of electroactive species.

1.4.6 Impedance response of the Randles circuit

The Randles circuit shown in Fig. 1.18 has been the most common electrical circuit to characterise electrochemical phenomena at the electrochemical interface between the electrode and electrolyte of electrochemical systems. The application of the Randles circuit shown in Fig. 1.18 with EIS measurements carried out in PEMFCs will be discussed in Chapter 3. The Randles circuit comprises a parallel resistor/CPE configuration and a finite-length diffusion Warburg impedance as represented through Eq. (1.55) to account for diffusion of electroactive species during the electrochemical reaction.

The equation representing the impedance response of the Randles circuit can be expressed as:

$$Z_{\text{Randles}} = R_1 + \frac{R_2 + Z_W}{1 + [R_2 + Z_W]Y[j\omega]^P} \tag{1.65}$$

Note that Eq. (1.65) can be analogous to the impedance equation represented in Eq. (1.64), if the Warburg impedance is neglected $Z_W=0$. The impedance response generated by Eq. (1.65) is shown in Fig. 1.19. The values of the resistors are $R_1=1\ \Omega$, $R_2=3\ \Omega$, and the value of the parameters Y and P of the CPE are $Y=0.009\ s^P/\Omega$ and $P=0.75$. The parameters of the Warburg impedance represented in Eq. (1.55) are $R_W=1.5\ \Omega$ and $\tau_W=0.7$ s. The simulation of the impedance response generated by Eq. (1.65) was carried out in MATLAB software considering a range of frequencies from 100 kHz down to 0.01 Hz. An additional semicircle at low frequencies is shown in Fig. 1.19 compared with the

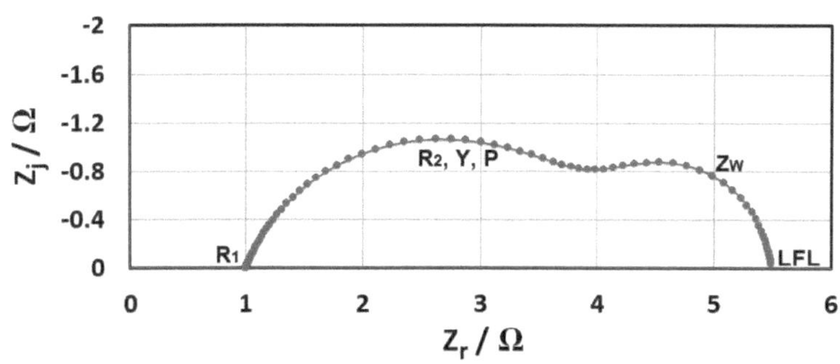

FIG. 1.19 Impedance response of the Randles circuit comprising a constant phase element and Warburg impedance. The parameters of the circuit are $R_1=1\ \Omega$, $R_2=3\ \Omega$, $Y=0.009\ s^P/\Omega$, $P=0.75$, $R_W=1.5\ \Omega$, $\tau_W=0.7$ s.

impedance response shown in Fig. 1.15 when the CPE (parameters Y, P) is considered in the electrical circuit. The additional semicircle at low frequencies is attributed to the Warburg impedance represented in Eq. (1.55).

The Randles circuit is reported in the study of Orazem et al. [23] to demonstrate a graphical representation of the impedance response of reactive electrodes in the Nyquist and Bode plots. The impedance response of the Randles circuit is represented in the complex-impedance plot as shown in Fig. 1.19 where the asymptotic limit of the real part Z_r at high frequency corresponds to resistance $R_1 = 1\ \Omega$ and the value of the asymptotic limit of the real part Z_r in the low frequency limit (LFL) corresponds to the sum of resistance R_1, resistance R_2 and resistance from the Warburg impedance R_W as such LFL $= 1\ \Omega + 3\ \Omega + 1.5\ \Omega$. If multiple loops are represented in the complex-impedance plot, it is a clear indication that more than one-time constant is required to describe the electrochemical processes. Graphical interpretations of Nyquist and Bode plots provide guidance for the development of appropriate physical models of electrochemical systems.

1.4.7 Impedance response of the Randles circuit featuring an inductive loop

An electrical circuit as shown in Fig. 1.20 considering an inductor connected in series with a resistor and these two components connected in parallel with the rest of the components of the Randles circuit (Fig. 1.18) can be considered to simulate positive imaginary components Z_j at low frequencies and represented in the complex-impedance plot. Positive imaginary components at low frequencies are commonly coined inductive loops. It is also possible to simulate inductive loops by considering a negative capacitance [24] instead of the inductor element in the circuit shown in Fig. 1.20.

The impedance response of the electrical circuit shown in Fig. 1.20 can be derived from Kirchoff's law of voltage and current and can be represented through the following equation.

$$Z_{\text{Randles\_Inductor}} = R_1 + \frac{[R_2 + Z_W][R_3 + Lj\omega]}{[R_3 + Lj\omega]\left[1 + [R_2 + Z_W]Y[j\omega]^P\right] + R_2 + Z_W} \tag{1.66}$$

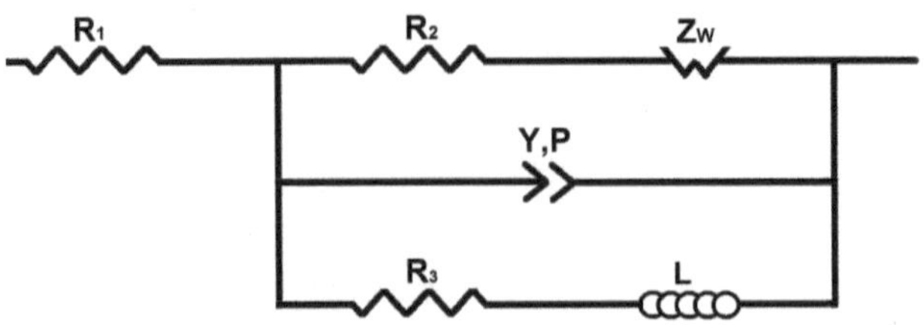

FIG. 1.20 Randles circuit with resistor–inductor impedance.

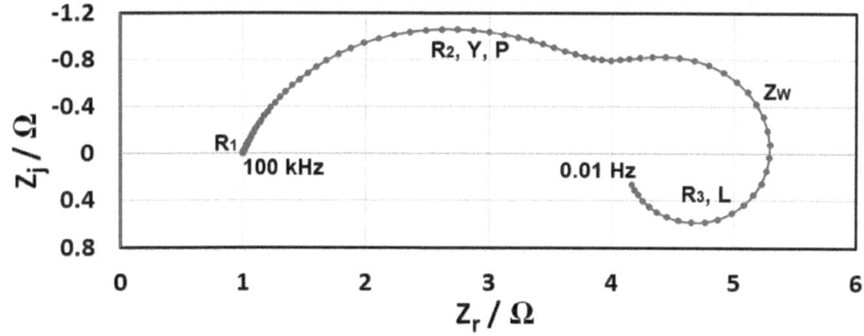

FIG. 1.21 Impedance response of the Randles circuit with resistor–inductor impedance. The parameters of the circuit are $R_1=1\ \Omega$, $R_2=3\ \Omega$, $Y=0.009\ s^P/\Omega$, $P=0.75$, $R_W=1.5\ \Omega$, $\tau_W=0.7$ s, $R_3=10\ \Omega$, $L=50$ H.

Fig. 1.21 shows the impedance response of the circuit shown in Fig. 1.20 and generated through Eq. (1.66) and considering parameters $R_1=1\ \Omega$, $R_2=3\ \Omega$, $Y=0.009\ s^P/\Omega$, $P=0.75$, $R_W=1.5\ \Omega$, $\tau_W=0.7$ s, $R_3=10\ \Omega$, and $L=50$ H. The simulation of the impedance response generated by Eq. (1.66) was carried out in MATLAB software considering a range of frequencies from 100 kHz down to 0.01 Hz. An inductive loop at low frequencies is present in the impedance response shown in Fig. 1.21. At first glance, it could be assumed that the impedance spectrum shown in Fig. 1.21 with negative imaginary components $-Z_j$ resembles the impedance spectrum of the Randles circuit shown in Fig. 1.19. Nevertheless, the inductive loop changes the value where the spectrum intercepts the real component Z_r of the impedance complex plot at low frequencies. The impedance spectrum shown in Fig. 1.19 converged close to the real component at $Z_r=5.49\ \Omega$ at low frequency. The impedance spectrum shown in Fig. 1.21 intercepted the real component at $Z_r=5.29\ \Omega$ at low frequency. The inductive loop of the impedance spectrum shown in Fig. 1.21 is mainly attributed to the impedance response of the resistor R_3 connected in series with the inductor L in the circuit shown in Fig. 1.20.

1.5 Application of Kramers–Kronig relations and Hilbert transformation

The methods described in this section allow the evaluation of EIS data to ensure consistency with linearity and stability properties of K–K relations. The impedance responses of the Randles circuit shown in Figs 1.19 and 1.21 are considered to demonstrate the application of the Voigt circuit for K–K analysis and application of Hilbert Transformation [11] in MATLAB software.

1.5.1 Kramers–Kronig analysis using the Voigt circuit in MATLAB software

The use of equivalent electrical circuits that comply with K–K is an accepted approach to evaluate validity of impedance measurements from

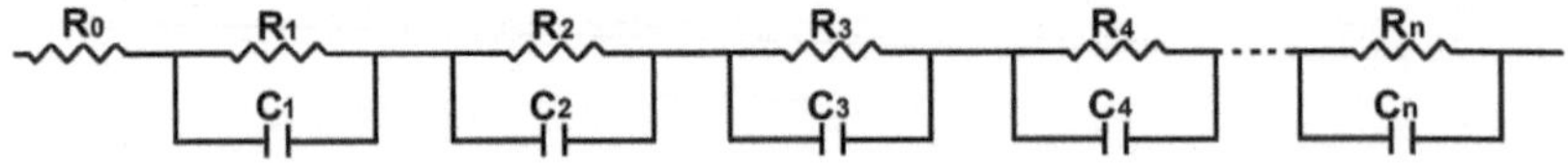

FIG. 1.22 Voigt circuit.

electrochemical systems. Agarwal and Orazem [10] and later Boukamp [25] reported that it is possible to evaluate K–K relations by fitting the Voigt circuit (a resistor–capacitor parallel configuration repeated a finite number of times) with EIS measurements. Fig. 1.22 shows a representation of the Voigt circuit.

The impedance response by the Voigt circuit can be represented through the following equation:

$$Z_{\text{Voigt}} = R_0 + \sum_{k=1}^{n} \frac{R_k}{1 + j\omega R_k C_k} \tag{1.67}$$

Eq. (1.67) is analogous to Eq. (1.62) if only a resistor–capacitor parallel configuration $n=1$ is considered. Boukamp [25] also reported that if the EIS data cannot be well approximated by the Voigt model containing a reasonable number of R-C pairs, it means that data are not transformable and do not represent the physical processes of the electrochemical system studied. ZView software (Scribner Associates, Inc.) is a tool to fit electrical circuits with EIS data. The K–K analysis of EIS data can be carried out in ZView software by constructing the Voigt circuit shown in Fig. 1.22 and fitting it with the EIS data to be evaluated. Further information about K–K evaluation in ZView software can be found in the technical report from Cruz-Manzo and Greenwood [26]. It is also possible to fit the impedance response of the Voigt circuit with EIS data using a nonlinear regression method such as the Levenberg-Marquardt algorithm in MATLAB software.

The script written in MATLAB software for fitting the impedance response of the Voigt circuit with EIS data can be found in Appendix A.

Fig. 1.23 shows a comparison between the impedance response generated by the Randles circuit previously shown in Fig. 1.19 and the fitted data generated by the Voigt circuit. A combination of six resistor–capacitor parallel configurations were considered in the Voigt circuit. There is an agreement between the impedance responses of the Randles circuit and the Voigt circuit because the Randles circuit satisfies linearity properties for K–K transformation. The Randles circuit considered a CPE to simulate a depressed semicircle at high-low frequencies. Different values of resistor and capacitors in the Voigt circuit allowed the simulation of the depressed semicircle at high-low frequencies as shown in Fig. 1.23. A time constant dispersion, which is represented through the different values of resistor and capacitors $\tau_k = R_k C_k$ in the Voigt circuit can simulate the impedance response represented by a depressed semicircle. As discussed in Section 1.4.5, a CPE not only improves the quality of fitting of an electrical circuit with EIS data, but also represents heterogenous physical

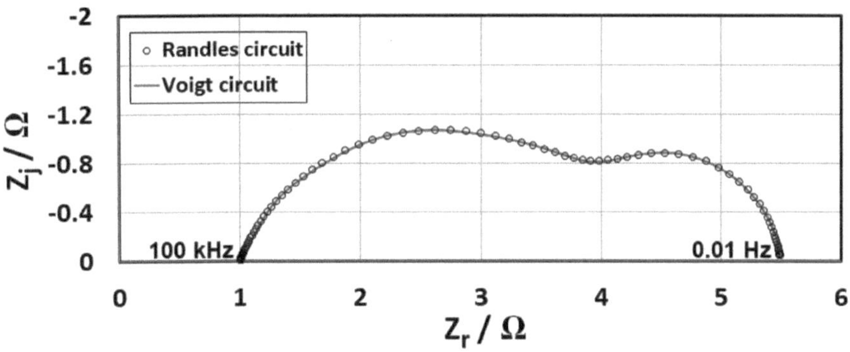

FIG. 1.23 Comparison between impedance response from Randles circuit and Voigt circuit. The value of the components of the Randles circuit are $R_1 = 1\ \Omega$, $R_2 = 3\ \Omega$, $Y = 0.009\ s^P/\Omega$, $P = 0.75$, $R_W = 1.5\ \Omega$, $\tau_W = 0.7$ s. The value of the components of the Voigt circuit are $R_0 = 1.0086\ \Omega$, $R_1 = 1.3762\ \Omega$, $C_1 = 0.0052$ F, $R_2 = 0.1233\ \Omega$, $C_2 = 9.1337$ F, $R_3 = 0.1184\ \Omega$, $C_3 = 0.0015$ F, $R_4 = 0.8878\ \Omega$, $C_4 = 0.0332$ F, $R_5 = 1.4250\ \Omega$, $C_5 = 0.1979$ F, $R_6 = 0.5470\ \Omega$, $C_6 = 0.0030$ F.

properties in electrodes of electrochemical systems. The Voigt circuit can assist the evaluation of correctness of EIS measurements in electrochemical systems.

The Voigt circuit can also be fitted to the impedance response of the Randles circuit considering the inductive loop at low frequencies previously shown in Fig. 1.21. Fig. 1.24 shows that the Voigt circuit can accurately predict the impedance response of the Randles circuit with the inductive loop at low frequencies. The prediction of the inductive loop was possible by considering a negative value in the last resistor R_6 and capacitor C_6 in the Voigt circuit configuration, but their associated time constant $\tau_6 = R_6 C_6$ was positive and was

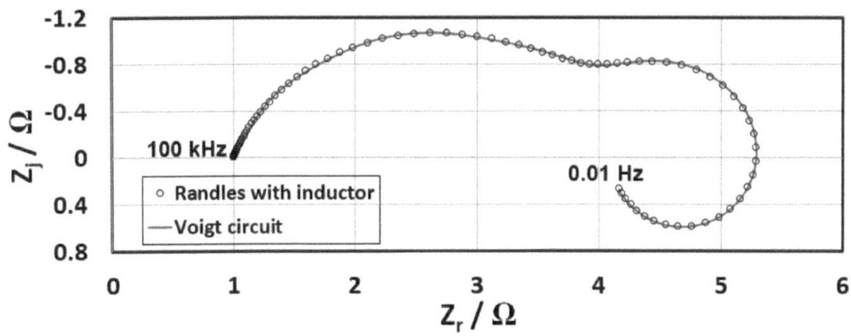

FIG. 1.24 Comparison between impedance response from Randles circuit with inductor and Voigt circuit. The value of the components of the Randles circuit are $R_1 = 1\ \Omega$, $R_2 = 3\ \Omega$, $Y = 0.009\ s^P/\Omega$, $P = 0.75$, $R_W = 1.5\ \Omega$, $\tau_W = 0.7$ s, $R_3 = 10\ \Omega$, $L = 50$ H. The value of the components of the Voigt circuit are $R_0 = 1.0089\ \Omega$, $R_1 = 0.5917\ \Omega$, $C_1 = 0.0030$ F, $R_2 = 1.4019\ \Omega$, $C_2 = 0.0055$ F, $R_3 = 1.5573\ \Omega$, $C_3 = 0.1948$ F, $R_4 = 0.8418\ \Omega$, $C_4 = 0.0387$ F, $R_5 = 0.1270\ \Omega$, $C_5 = 0.0015$ F, $R_6 = -1.4345\ \Omega$, $C_6 = -2.4756$ F.

higher than the rest of the time constants of the Voigt circuit. The electrical components associated with a high time constant predict low-frequency inductive loops represented in the impedance-complex plot.

1.5.2 Hilbert transformation through fast Fourier transform

The Hilbert transformation using the FFT introduced in Section 1.3.2 and reported by VanderNoot [11] has been considered to evaluate the impedance response generated by the Randles circuit shown in Fig. 1.19. In Section 1.3.2, the steps for the application of the Hilbert transformation with EIS data have been implemented in a script in MATLAB. The script is shown in Appendix A and allows the transformation from real to imaginary and imaginary to real components of the EIS data through the Hilbert transformation. Fig. 1.25 shows that it is possible to transform the impedance response of the Randles circuit through the Hilbert transformation. The Hilbert transformation was carried around a centre frequency ω_o of the impedance data to be transformed which is related to the highest point of the impedance spectrum as shown in Fig. 1.25. A transformation of data around a centre frequency is also observed in the results shown Fig. 1.6e and reported by VanderNoot [11]. The impedance response of the Randles circuit generated through Eq. (1.65) was shifted along the real component Z_r of the complex-impedance plot for the comparison shown in Fig. 1.25. A correct Hilbert transformation of the impedance response of the Randles circuit is shown in Fig. 1.25 because the Randles circuit complies with linearity properties for K–K and Hilbert transformation.

1.5.3 Kramers–Kronig analysis through ZView software

ZView software (Scribner Associates, Inc.) is a tool to fit any electrical circuit with EIS data. The electrical components defined in ZView software such as resistor, capacitor, inductor do not provide sufficient information about the physical processes of the electrochemical system studied. For instance, the

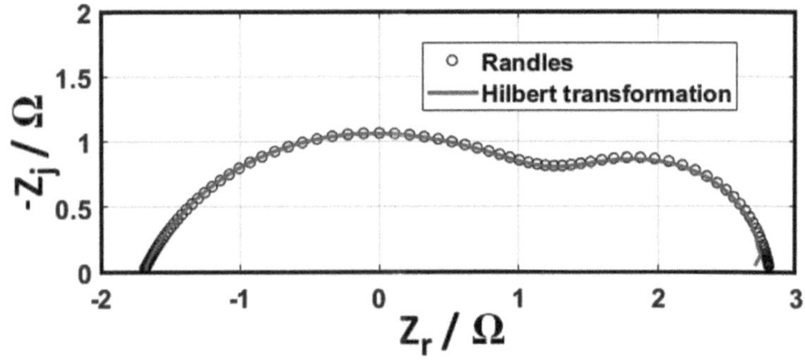

FIG. 1.25 Comparison between impedance response from Randles circuit and transformed by Hilbert transformation using MATLAB software.

Randles circuit constructed in ZView software can represent the impedance response from an electron-transfer reaction between an electrode/electrolyte interface, but the resistors do not provide enough information about electrochemical mechanisms of the electron-transfer reaction (e.g. Tafel slope, exchange current). Nevertheless, the ZView software allows the evaluation of K–K consistency of experimental EIS measurements. The Voigt circuit can be constructed in ZView software to evaluate K–K consistency in EIS data. To demonstrate the analysis of K–K transform in ZView software, the impedance spectrum generated by Eq. (1.66) (Randles circuit with resistor–inductor impedance) and shown in Fig. 1.21 will be considered. As a first step, the Voigt circuit is constructed in ZView software and fitted to the impedance spectrum as shown in Fig. 1.26. Six parallel resistor–capacitors were considered. Each parameter in the circuit must have an initial condition value different from zero. The circuit is fitted to the impedance spectrum to be analysed until no changes in the fitted values result. A disagreement between the impedance spectrum to be analysed and the impedance response of the Voigt circuit will likely result because the initial conditions defined in each parameter of the Voigt circuit during the fitting process are not appropriate as shown in Fig. 1.26.

In the second step, the K–K analysis is carried out after fitting the Voigt circuit with the impedance spectrum to be analysed. A good agreement between EIS data and transformed data as shown in Fig. 1.27 demonstrates that the EIS data comply with K–K transform properties for linear, causal, and stable electrochemical systems. More detailed information about the application of the K–K transform in EIS data using ZView software can be found in the technical report from Cruz-Manzo and Greenwood [26].

1.5.4 Application with real-world EIS measurements

The aforementioned methods can be applied with real-world EIS measurements to evaluate the validity of the EIS measurements. This is very important before trying to estimate electro-diffusion mechanisms of the electrochemical system studied. Fig. 1.28 shows the evaluation of the K–K analysis in real-world EIS measurements of a PEMFC using ZView software. The application of K–K in EIS data using ZView software has been discussed in Section 1.5.3 and EIS measurements in PEMFCs will be discussed later in Chapter 3. The PEMFC was operated with hydrogen and oxygen diluted in helium with 100% relative humidity in the gases supplied. More information about the EIS measurements of the PEMFC shown in Fig. 1.28 can be found in the study from Cruz-Manzo and Chen [27]. The results shown in Fig. 1.28 demonstrate that the experimental EIS measurements satisfy real to imaginary $Z_r \rightarrow Z_j$ and imaginary to real $Z_j \rightarrow Z_r$ data transformation.

In addition, the Hilbert transformation using the FFT discussed in Section 1.3.2 is applied to the same EIS measurements shown in Fig. 1.28. The scattered transformed data at the lowest frequency shown in Fig. 1.29 can be related to the low frequency limit considered in the EIS measurements

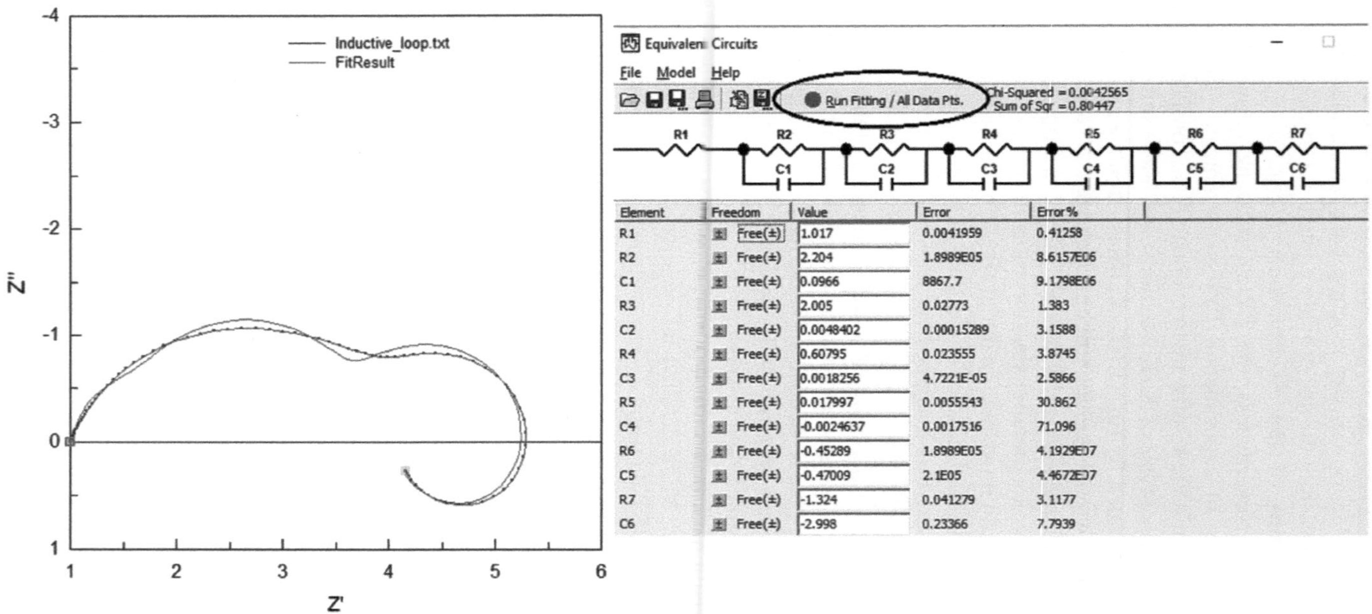

Element	Freedom	Value	Error	Error%
R1	Free(±)	1.017	0.0041959	0.41258
R2	Free(±)	2.204	1.8989E05	8.6157E06
C1	Free(±)	0.0966	8867.7	9.1798E06
R3	Free(±)	2.005	0.02773	1.383
C2	Free(±)	0.0048402	0.00015289	3.1588
R4	Free(±)	0.60795	0.023555	3.8745
C3	Free(±)	0.0018256	4.7221E-05	2.5866
R5	Free(±)	0.017997	0.0055543	30.862
C4	Free(±)	-0.0024637	0.0017516	71.096
R6	Free(±)	-0.45289	1.8989E05	4.1929E07
C5	Free(±)	-0.47009	2.1E05	4.4672E07
R7	Free(±)	-1.324	0.041279	3.1177
C6	Free(±)	-2.998	0.23366	7.7939

FIG. 1.26 Fitting Voigt circuit with the impedance spectrum generated by the Randles circuit with resistor–inductor impedance using ZView software.

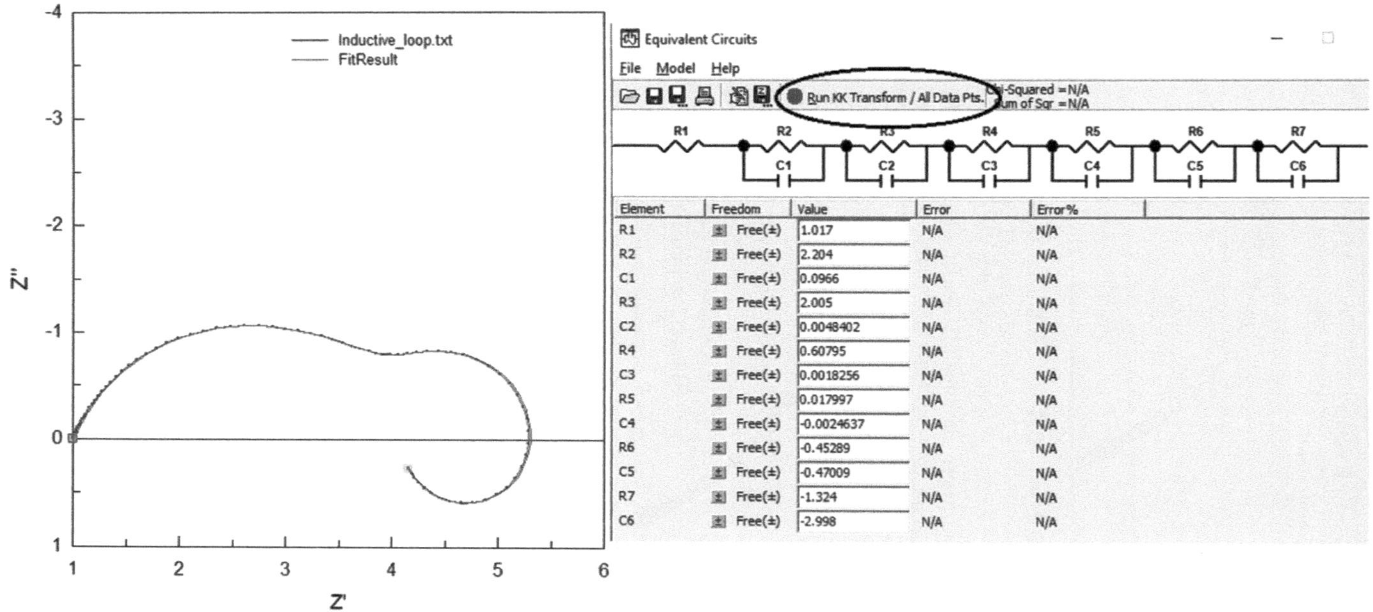

FIG. 1.27 Kramers–Kronig evaluation in the impedance response generated by the Randles circuit with resistor–inductor impedance using ZView software.

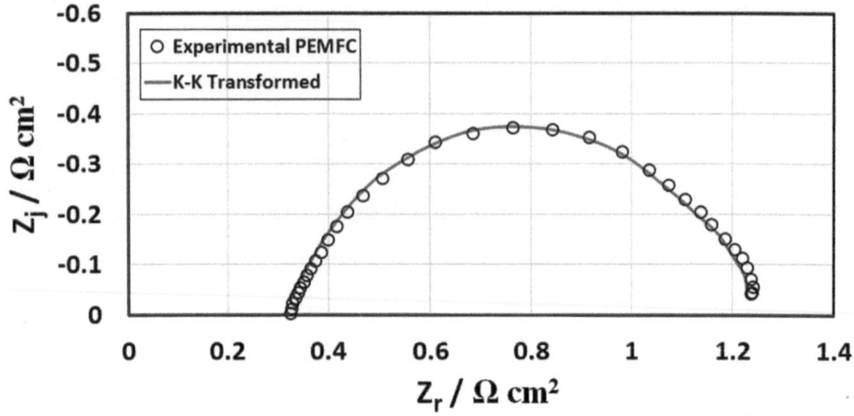

FIG. 1.28 Comparison between measured EIS data of a PEMFC and Kramers–Kronig (K–K) transformed data using ZView software. More information about the PEMFC-EIS measurements can be found in the study from Cruz-Manzo and Chen [27].

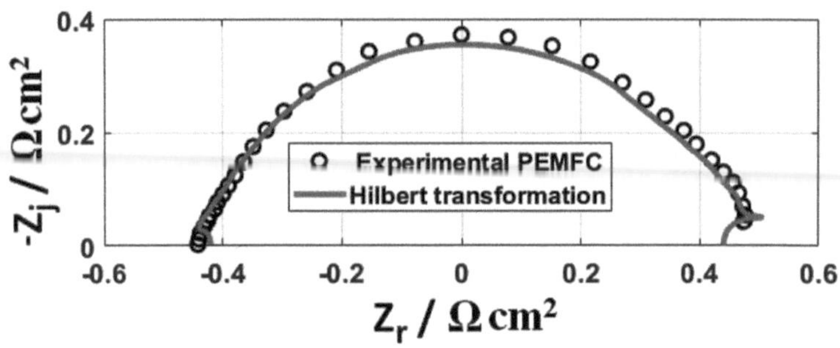

FIG. 1.29 Comparison between measured EIS data and transformed data through Hilbert transformation using MATLAB. More information about the PEMFC-EIS measurements can be found in the study by Cruz-Manzo and Chen [27].

(1 Hz). The application of the Hilbert transformation requires a lower frequency limit (e.g. 0.001 Hz) as FFT requires linearly spaced frequencies. The two aforementioned methods K–K in ZView and Hilbert transformation in MATLAB can be applied to evaluate correctness of the real-world EIS measurements to ensure that the measurements truly represent the impedance response of the electrochemical system studied. In Chapter 3, the application of the K–K and Hilbert transformation in EIS measurements of PEMFCs operated at different operating conditions will be demonstrated.

References

[1] D.D. Macdonald, Reflections on the history of electrochemical impedance spectroscopy, Electrochim. Acta 51 (2006) 1376–1388.

[2] E. Warburg, Ueber das Verhalten Sogenannter Unpolarisirbarer Elektroden gegen Wechselstrom, Ann. Phys. 303 (1899) 493–499.

[3] B.V. Ershler, Investigation of electrode reactions by the method of charging-curves and with the aid of alternating currents, Discuss. Faraday Soc. 1 (1947) 269–277.

[4] J.E.B. Randles, Kinetics of rapid electrode reactions, Discuss. Faraday Soc. 1 (1947) 11–19.

[5] R. de Levie, in: Delahay, Tobias (Eds.), Advances in Electrochemistry and Electrochemical Engineering, vol. 6, 1967, p. 329.

[6] M. Orazem, B. Tribollet, Electrochemical Impedance Spectroscopy, second ed., Wiley, Hoboken, NJ, 2017.

[7] D. Macdonald, M. Urquidi-Macdonald, Application of Kramers-Kronig transforms in the analysis of electrochemical systems I. polarization resistance, J. Electrochem. Soc. 132 (1985) 2316–2319.

[8] M. Urquidi-Macdonald, S. Real, D. Macdonald, Application of Kramers-Kronig transforms in the analysis of electrochemical impedance data II. Transformations in the complex plane, J. Electrochem. Soc. 133 (1986) 2018–2024.

[9] M. Urquidi-Macdonald, S. Real, D. Macdonald, Application of Kramers-Kronig transforms in the analysis of electrochemical impedance data III. Stability and linearity, Electrochim. Acta 35 (1990) 1559–1566.

[10] P. Agarwal, M.E. Orazem, Measurement models for electrochemical impedance spectroscopy I. demonstration of applicability, J. Electrochem. Soc. 139 (1992) 1917–1927.

[11] T.J. VanderNoot, Hilbert transformation of immittance data using the fast Fourier transform, J. Electroanal. Chem. 322 (1992) 9–24.

[12] A. Papoulis, The Fourier Integral and its Applications, McGraw-Hill, New York, 1962.

[13] Q. Meyer, S. Barass, O. Curnick, T. Reisch, D.J.L. Brett, A multichannel frequency response analyser for impedance spectroscopy on power sources, J. Electrochem. Sci. Eng. 3 (2013) 107–114.

[14] S. Cruz-Manzo, R. Chen, P. Greenwood, Evaluate the validity of electrochemical impedance measurements of polymer electrolyte fuel cells using a computational algorithm based on fast Fourier transform, Insights Anal. Electrochem. 1 (2015) 1–13.

[15] N.A. Hampson, S.A.G.R. Karunathilaka, R. Leek, The impedance of electrical storage cells, J. Appl. Electrochem. 10 (1980) 3–11.

[16] K. Ogata, Modern Control Engineering, fourth ed., Pearson, United States, 2001.

[17] A.J. Bard, L.R. Faulkner, Electrochemical Methods, Fundamentals and Applications, Wiley Interscience, New York, 1980, pp. 305–330.

[18] J.R. Macdonald, Impedance Spectroscopy, John Wiley & Sons, Inc., New York, 1987, p. 88.

[19] B.A. Boukamp, Derivation of a distribution function of relaxation times for the (fractal) finite length Warburg, Electrochim. Acta 252 (2017) 154–163.

[20] C.H. Hsu, F. Mansfeld, Concerning the conversion of the constant phase element parameter Y0 into a capacitance, Corrosion 57 (2001) 747–748.

[21] M.R.S. Abouzari, F. Berkemeier, G. Schimitz, D. Wilmer, On the physical interpretation of constant phase elements, Solid State Ion. 180 (2009) 922–927.

[22] M.E. Orazem, I. Frateur, B. Tribollet, V. Vivier, S. Marcelin, N. Pébère, A.L. Bunge, E.A. White, D.P. Riemer, M. Musiani, Dielectric properties of materials showing constant-phase-element (CPE) impedance response, J. Electrochem. Soc. 160 (2013) C215–C225.

[23] M.E. Orazem, N. Pebere, B. Tribollet, Enhanced graphical representation of electrochemical impedance data, J. Electrochem. Soc. 153 (2006) B129–B136.

[24] D. Klotz, Negative capacitance or inductive loop? – a general assessment of a common low frequency impedance feature, Electrochem. Commun. 98 (2019) 58–62.

[25] B.A. Boukamp, A linear Kronig-Kramers Transform test for immittance data validation, J. Electrochem. Soc. 142 (1995) 1885–1894.

[26] S. Cruz-Manzo, Kramers-Kronig Analysis in ZView Software for Evaluation of EIS Measurements in Polymer Electrolyte Fuel Cells, 2015, https://doi.org/10.13140/rg.2.1.1094.0887/1. last access June 2022.

[27] S. Cruz-Manzo, R. Chen, Electrochemical impedance study on estimating the mass transport resistance in the polymer electrolyte fuel cell cathode catalyst layer, J. Electroanal. Chem. 702 (2013) 45–48.

Chapter 2

Fundamentals of PEM fuel cells

2.1 The structure of the PEM fuel cell

Polymer electrolyte membrane (PEM) fuel cells or abbreviated to PEMFCs are the best candidates to be considered for future sources of power for stationary and mobile applications because they are able to produce high power density at low temperatures and are not bound by the Carnot cycle. A PEMFC converts the chemical energy of a fuel such as a hydrogen gas into electrical and thermal energies. Two electrodes (anode and cathode) separated by a PEM are the main components in the PEMFC. Each electrode consists of a gas diffusion layer (GDL) and a catalyst layer as a minimum. The active area is the extent of the electrochemical surface on the PEMFC's electrodes where the electrochemical reaction takes place.

The catalyst layer is a layer made of a matrix of carbon grains providing the electron conductivity, and platinum Pt supported on carbon as the catalyst to drive the hydrogen-oxidation or oxygen-reduction processes. The GDL is a layer made of porous carbon cloth and its main function is to distribute reactant gases to the active zones of the catalyst layer. The GDL typically contains a wetproofing agent such as polytetrafluoroethylene (PTFE) or fluorinated ethylene propylene (FEP) to facilitate the movement of liquid water through the cell. An additional layer that can be placed between the GDL and catalyst layer is the microporous layer (MPL), which is similar in composition to the GDL but treated to a higher level of wetproofing agent to force the flow of liquid water toward certain directions within the cell. A primary hardware component of the PEMFC is the flow field plate (FFP), which provides structural rigidity to the cell and provides conductive pathways that enable the energy-carrying electrons that are lost in the anode catalyst layer (ACL) due to a hydrogen oxidation reaction (HOR) and transferred to the cathode catalyst layer (CCL) for an oxygen reduction reaction (ORR). Electrons e^- are sent to the cathode by an external circuit and travel through a load to produce an electrical potential difference. Fig. 2.1 shows the conduction of the electrical charges in the PEMFC. Protons H^+ freed from the HOR in the anode are sent to the cathode through the PEM and combine with the electrons and oxygen at the cathode.

Electrochemical Phenomena in the Cathode Impedance Spectrum of PEM Fuel Cells.
https://doi.org/10.1016/B978-0-323-90607-4.00003-3
Copyright © 2022 Elsevier Inc. All rights reserved.

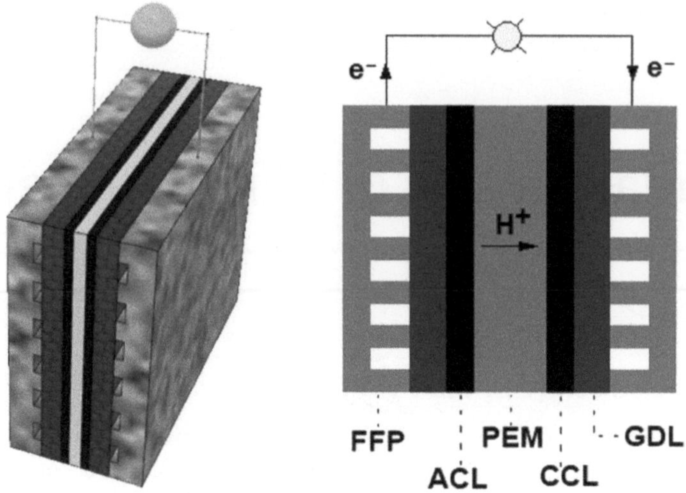

FIG. 2.1 Electrical charge flowing through the PEM fuel cell. Flow field plate (FFP), anode catalyst layer (ACL), polymer electrolyte membrane (PEM), cathode catalyst layer (CCL), and gas diffusion layer (GDL).

Water is the product resulting from the ORR taking part in the CCL. Water diffuses in vapour and liquid forms from cathode to anode through a difference in concentration between the electrodes and is transported from anode to cathode through an electroosmotic drag (EOD) effect. An optimal testing method is required to maintain water balance in the PEMFCs because an incorrect water balance could yield dehydration or flooding in the electrodes and a decrease in performance in the PEMFCs. The PEM is a quasiporous proton-conducting material that serves to simultaneously transport the protons lost in the ACL through to the CCL. The PEM is based on a PTFE backbone that is modified to hold sulfonic acid end groups and depends upon the presence of water for proton conductivity. The PEMs are classified as perfluorosulfonic acid (PFSA) membranes. Nafion is a common commercial example of a PFSA membrane. Structurally reinforced PEMs consist of an interconnected matrix that is impregnated with PFSA electrolyte. These types of PEMs are resilient to dimensional change on water uptake and are mechanically durable.

Modern fuel cells use multilayer assemblies such as the membrane electrode assembly (MEA), which is a PEM sandwiched by ACL and CCL. The MEA is then deployed with a GDL and an FFP on either side to make up a single cell as shown in Fig. 2.2. A gas diffusion electrode (GDE) is another type of assembly that consists of a GDL with a catalyst layer coating on one side. Two GDEs can be used to sandwich a PEM with an FFP on either side, thereby making up a single cell again. The book of Barbir [1] discusses in detail the structure of the PEMFC and the reader is able to gain an insight into the performance of the PEMFC at different operating conditions.

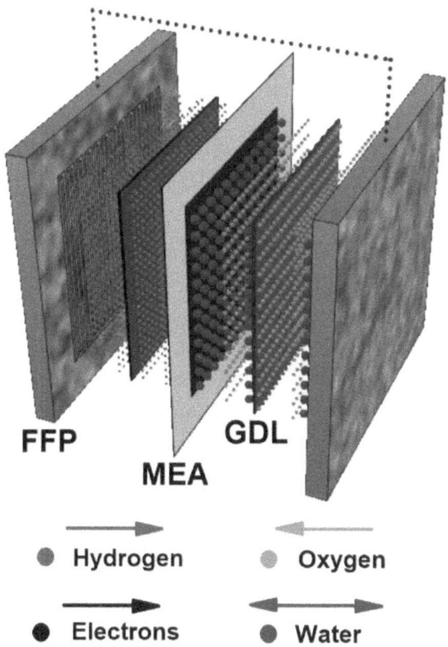

FIG. 2.2 Components of a PEM fuel cell. Flow field plate (FFP), membrane electrode assembly (MEA), and gas diffusion layer (GDL).

2.2 The voltage output in the PEM fuel cell

When an external circuit that allows the flow of electrons between the anode and cathode is not connected to the PEMFC, a reversible potential is present in the PEMFC. This reversible potential is derived from thermodynamic principles and is also known as the Open Circuit Potential (OCP).

2.2.1 Reversible potential

In an exothermal process such as the combustion of a fuel, the energy released is the heat of the chemical reaction, or enthalpy. Enthalpy of a chemical reaction is the difference between heat of formation of products and reactants. Although combustion is not present in the PEMFC, the enthalpy of the hydrogen combustion reaction is considered a way to measure the energy input in a PEMFC. Therefore the amount of reaction enthalpy that can be converted into electrical energy in a PEMFC is known as Gibbs free energy (G).

For each hydrogen molecule consumed and each water molecule produced, two electrons travel through an external circuit connected between the anode and the cathode. For 1 mol of hydrogen consumed in the HOR, $2N_A$ electrons travel between the anode and cathode, where N_A is Avogadro's number. The electrical charge that flows between the electrodes is represented as:

$$q = nN_{A}e = nF \quad (2.1)$$

where n is the number of electrons travelling between the electrodes, N_A is the number of molecules per mole 6.022×10^{23} mol^{-1}, e is the electron charge 1.602×10^{-19} coulombs and F is the Faraday constant 96485 C/mol. The electrical work in the PEMFC is defined as the product of the charge transferred in the fuel cell reaction and the potential:

$$-W_{el} = nFE \quad (2.2)$$

If the system is reversible, the electrical work is equal to the change in Gibbs free energy; therefore the reversible potential in the PEMFC is given by:

$$E_R = -\frac{\Delta G}{nF} \quad (2.3)$$

For standard temperature and pressure (STP, 298.15°K and 1 bar), the reversible potential is expressed as:

$$E^0 = -\frac{\Delta G^0}{nF} \quad (2.4)$$

The Gibbs free energy changes at different temperature and pressure. The reversible fuel cell potential at different temperature and pressure is obtained through a thermodynamic analysis [2]. For an isothermal process the change in Gibbs free energy is expressed as:

$$dG = V_m dp \quad (2.5)$$

where V_m is the molar volume of the reactant gas, and p is the pressure. Considering the reactant gas as a perfect gas $pV_m = RT$ and substituting into Eq. (2.5) yields:

$$dG = RT\frac{dp}{p} \quad (2.6)$$

where R is the gas constant and T is the operating temperature. Integrating Eq. (2.6) and considering a change in pressure from the standard pressure p^0 to a partial pressure p with the reactant gas yields:

$$G = G^0 + RT\ln\left(\frac{p}{p^0}\right) \quad (2.7)$$

where G^0 and p^0 are the Gibbs free energy and pressure at standard conditions (1 bar). For the general cell reaction:

$$H_2 + \frac{1}{2}O_2 \rightarrow H_2O \quad (2.8)$$

The change in Gibbs free energy of the electrochemical reaction (change between products and reactants) expressed in Eq. (2.8) is given by:

$$\Delta G = G_{H_2O} - G_{H_2} - \frac{1}{2} G_{O_2} \tag{2.9}$$

Substituting Eq. (2.7) into Eq. (2.9) yields:

$$\Delta G = \Delta G^0 + RT \ln \left(\frac{\left[\frac{p_{H_2O}}{p^0}\right]^1}{\left[\frac{p_{H_2}}{p^0}\right]^1 \left[\frac{p_{O_2}}{p^0}\right]^{\frac{1}{2}}} \right) \tag{2.10}$$

where p_{H_2} is the partial pressure of hydrogen, p_{O_2} is the partial pressure of oxygen and p_{H_2O} is the partial pressure of water. Substituting Eqs (2.3) and (2.4) into Eq. (2.10) yields:

$$E_R = E^0 + \frac{RT}{nF} \ln \left(\frac{\left[\frac{p_{H_2}}{p^0}\right]^1 \left[\frac{p_{O_2}}{p^0}\right]^{\frac{1}{2}}}{\left[\frac{p_{H_2O}}{p^0}\right]^1} \right) \tag{2.11}$$

When liquid water is produced in the fuel cell the partial pressure of the products is equal to $p_{H_2O} = 1$ bar. Also considering the reference pressure as $p^0 = 1$ bar, Eq. (2.11) can be represented as:

$$E_R = E^0 + \frac{RT}{nF} \left[\ln \left(\frac{p_{H_2}}{p^0} \right) + \frac{1}{2} \ln \left(\frac{p_{O_2}}{p^0} \right) \right] \tag{2.12}$$

Eq. (2.12) represents the reversible potential of the fuel cell for different partial pressures of the reactants. E^0 has a value of 1.229 and is the reversible potential of a fuel cell operated on pure hydrogen and oxygen at standard conditions (298.15°K and 1 bar). This reversible potential can also change for different temperatures; therefore Eq. (2.12) is modified as:

$$E_R = E^0 + E_T^0 + \frac{RT}{nF} \left[\ln \left(\frac{p_{H_2}}{p^0} \right) + \frac{1}{2} \ln \left(\frac{p_{O_2}}{p^0} \right) \right] \tag{2.13}$$

where E_T^0 is the reversible potential as a function of the temperature. From the thermodynamic theory, it is possible to represent the change in Gibbs free energy as:

$$\Delta G^0 = \Delta H^0 - T^0 \Delta S^0 \tag{2.14}$$

where ΔH^0 is the heat of formation of the electrochemical reaction at standard conditions, T^0 is the standard temperature at 298.15°K, and ΔS^0 is the entropy of the electrochemical reaction at standard conditions. If the temperature is different from the standard temperature, the change in Gibbs free energy can be expressed as:

$$\Delta G_T = \Delta H_T - T \Delta S_T \tag{2.15}$$

According to Brady and Holum [3], the magnitudes of ΔH_T and ΔS_T are insensitive for any change in temperature. This is because the heat of formation and entropy of both reactants and products increase equally with increasing temperature; therefore the difference between ΔH_T and ΔS_T at different temperatures is roughly the same. With an approximation given by:

$$\Delta G_T \approx \Delta H^0 - T\Delta S^0 \tag{2.16}$$

it is possible to consider that:

$$\Delta G_T^0 \approx \Delta G^0 - \Delta G_T \tag{2.17}$$

Substituting Eqs (2.14) and (2.16) into Eq. (2.17) yields:

$$\Delta G_T^0 \approx \Delta S^0 \left[T - T^0\right] \tag{2.18}$$

Substituting Eq. (2.18) into $-\Delta G_T^0 = nFE_T^0$ which considers a reversible process gives:

$$E_T^0 = \frac{-\Delta S^0 \left[T - T^0\right]}{nF} \tag{2.19}$$

Finally, substituting Eq. (2.19) into Eq. (2.13) gives the reversible potential of the fuel cell at different temperatures and pressures from standard conditions, as such:

$$E_R = E^0 - \frac{\Delta S^0 \left[T - T^0\right]}{nF} + \frac{RT}{nF}\left[\ln\left(\frac{p_{H_2}}{p^0}\right) + \frac{1}{2}\ln\left(\frac{p_{O_2}}{p^0}\right)\right] \tag{2.20}$$

Eq. (2.20) is known as the Nernst equation and can be expressed as:

$$\begin{aligned} E_{Nernst} &= 1.229 - 0.85 \times 10^{-3}\left[T - 298.15\right] \\ &\quad + 4.31 \times 10^{-5} T\left[\ln\left(\frac{p_{H_2}}{p^0}\right) + \frac{1}{2}\ln\left(\frac{p_{O_2}}{p^0}\right)\right] \end{aligned} \tag{2.21}$$

where E_{Nernst} is the reversible potential of the PEMFC. When an external circuit with a load is connected between the anode and cathode to allow electrical current to flow, losses in the reversible potential occur due to the following factors:

- The kinetics of the electrochemical reactions due to the energy required to overcome the barrier changes in the electrode potential.
- Electron and proton resistances.
- Hydrogen crossover through the membrane.
- Low reactant concentration in the electrodes due to mass transport limitations.

Commonly in the field of electrochemistry, the losses in the reversible potential are defined as overpotentials. This term has been coined because these are potentials superimposed over the reversible potential and in the PEMFC reduce the total output voltage. The relation between overpotential and reversible potential in the PEMFC can be expressed as:

$$V_{\mathrm{PEMFC}} = E_{\mathrm{Nernst}} - \eta_{\mathrm{act,ohm,con}} \tag{2.22}$$

where V_{PEMFC} is the PEMFC output voltage, E_{Nernst} is the reversible potential, and $\eta_{\mathrm{act,ohm,con}}$ is the total overpotential during PEMFC operation.

2.2.2 Activation overpotential

This overpotential is associated with the kinetics of the electrochemical reactions that take place on the electrode surface. The velocity of a chemical reaction is defined as the amount of substance in moles, which is either electro-oxidised or electro-reduced within a unit of time. This velocity is expressed as a function of the electrode area because the electrochemical reaction takes place between the electrode and the electrolyte interface. The rate of ORR being the slowest reaction and the most complicated reaction sequence determines the velocity of the electrochemical reaction in the PEMFC and therefore determines the activation overpotential. The activation overpotential can be expressed mathematically through the Tafel equation [2] as:

$$\eta_{\mathrm{act}} = \frac{RT}{\alpha z F} \ln \left(\frac{i}{i_0} \right) \tag{2.23}$$

where α is the charge transfer coefficient and determines the magnitude of the nonequilibrium between a redox pair process to transfer electric charge between the electrode-electrolyte interface, i is the rate of charge transfer (electrical current divided by the active area) at which the electrochemical reactions are taking place, z is the number of electrons released during the HOR or consumed during the ORR, and i_0 is the exchange current density and relates the rate of charge transfer at which a reversible reaction takes place. The exchange current density i_0 can change in orders of magnitude for different electrochemical reactions and different electrode composition. In this case, if the exchange current density is high, an increase in the activity of the electrode surface is expected. An empirical equation coined as the Tafel equation has been considered to estimate the exchange current density and charge transfer coefficient represented in the activation overpotential. The Tafel equation relates a slope with the curve representing the activation overpotential η_{act} vs a logarithm scale of the current density i as such:

$$\eta_{\mathrm{act}} = a + b \log_{10}(i) \tag{2.24}$$

where:

$$a = -\frac{2.3RT}{\alpha z F} \log_{10}(i_0) \tag{2.25}$$

and

$$b = \frac{2.3RT}{\alpha zF} \tag{2.26}$$

is known as the Tafel slope with units in V/dec.

The empirical Tafel equation represented in Eq. (2.24) is represented as a function of the logarithm of base 10. The general equation for activation overpotential represented in Eq. (2.23) is represented with the natural logarithm because it can be derived from the current-overpotential equation. This will be demonstrated later.

Fig. 2.3 shows a straight line generated through Eq. (2.24) to estimate the exchange current density i_0. It is also possible to estimate the charge transfer coefficient α from the slope of the straight line (Tafel slope) and is represented in Eq. (2.26). The curve shown in Fig. 2.3 represents the activation overpotential for an electrode (anode or cathode) and is obtained experimentally through a three-electrode configuration in the PEMFC. In the three-electrode configuration, one electrode is the working electrode at which species undergo a redox reaction to give the quantitative change in current or potential, a second electrode is called the counter electrode which carries the current (via a redox process). A third electrode called reference electrode is positioned close to the working electrode to minimise the effect of the PEM resistance on the activation overpotential curve. A more detailed discussion about the three-electrode configuration will be presented in the next chapter, Chapter 3.

To reduce the activation overpotential in the PEMFC, the following factors have to be considered: raising the operating temperature to improve the kinetics of the electrochemical reaction, using more effective catalysts (this will increase the exchange current density), increasing reactant concentration (using

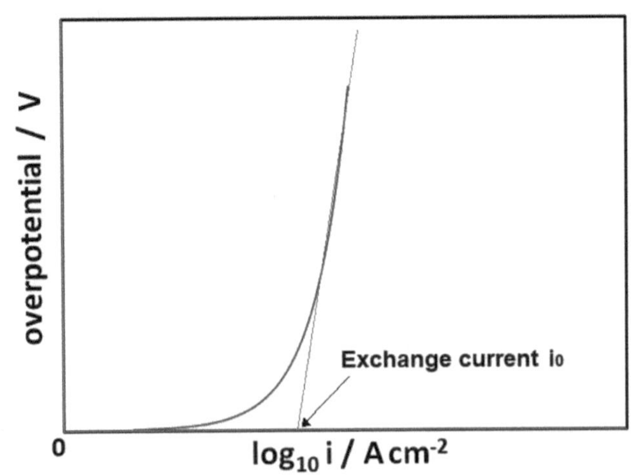

FIG. 2.3 Representation of Tafel slope.

pure oxygen instead of air), and increasing the operating pressure. This overpotential is dominant at low current densities and low temperature and occurs mainly at the cathode due to the ORR being a much more complicated reaction sequence.

2.2.3 Ohmic overpotential

Electronic resistance in the electrodes and FFPs and proton resistance in the electrolyte of the catalyst layer and PEM are related to this type of overpotential. The increase in overpotential due to the ohmic losses is proportional to the increase in current density, as such:

$$\eta_{ohm} = iR_{ohm} \tag{2.27}$$

where i is the total current density and R_{ohm} is the total ohmic resistance to the flow of protons and electrons in the PEMFC. To reduce the ohmic overpotential, electrodes and PEM with high electron and proton conductivities, and an appropriate selection of material for FFPs have to be considered.

2.2.4 Mass transport overpotential

The reactant gas after diffusing toward the electrode-electrolyte interface in the catalyst layer is consumed during the electrochemical reaction thereby a reactant concentration gradient is present. The gas concentration in the catalyst layer surface is less than the concentration present in the bulk of the GDL and FFP.

The reduction in gas concentration affects the potential to produce the required electrical current. The gas reactant concentration in the catalyst layer will drastically be reduced if the gas reactant is consumed faster than it can reach the electrode-electrolyte interface. This phenomenon will lead to a limited electrical current. Fick's First Law relates the diffusive flux to the concentration field assuming that the flux occurs from a region of high concentration to a region of low concentration. Fig. 2.4 shows a linear reduction in oxygen concentration from the channel of the FFP to the surface of the CCL.

The flux rate is a measure of the amount of substance that will flow through a defined area during a time interval.

$$v = \frac{D[c_b - c_s]}{\delta} \tag{2.28}$$

where v is the flux of the reactant gas through the electrode, D is the diffusion coefficient of the reactant gas, c_b is the bulk concentration, c_s is the reactant concentration at the surface of the catalyst layer, and δ is the diffusive distance. Substituting Faraday's law $v = i/zF$ into Eq. (2.28) yields a current density generated due to the difference in reactant concentration:

$$i = \frac{zFD[c_b - c_s]}{\delta} \tag{2.29}$$

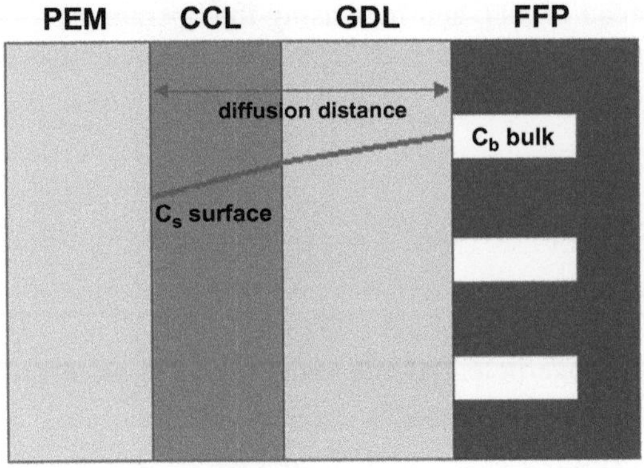

FIG. 2.4 Reduction in oxygen concentration across the cathode. Polymer electrolyte membrane (PEM), cathode catalyst layer (CCL), gas diffusion layer (GDL), and flow field plate (FFP).

A limiting current density is present when there is no reactant gas in the catalyst layer surface $c_s = 0$, as such:

$$i_L = \frac{zFDc_b}{\delta} \tag{2.30}$$

If the partial pressure of one of the reactant gases (hydrogen or oxygen) changes from p_1 to p_2 and the other reactant is unchanged, Eq. (2.12) can be expressed as:

$$\begin{aligned} E_{R1} &= E^0 + \frac{RT}{zF}\ln\left(\frac{p_{1H_2,O_2}}{p^0}\right) \\ E_{R2} &= E^0 + \frac{RT}{zF}\ln\left(\frac{p_{2H_2,O_2}}{p^0}\right) \end{aligned} \tag{2.31}$$

with $z = n$ if the hydrogen partial pressure changes and $z = 2n$ if the oxygen partial pressure changes.

Then the reversible potential will change as:

$$\Delta E_R = E_{R1} - E_{R2} = \frac{RT}{zF}\ln\left(\frac{p_{1H_2,O_2}}{p^0}\right) - \frac{RT}{zF}\ln\left(\frac{p_{2H_2,O_2}}{p^0}\right) \tag{2.32}$$

Eq. (2.32) can be expressed as a function of the reactant concentrations by considering the reactant gas a perfect gas, as such:

$$\Delta E_R = \frac{RT}{zF}\ln\left(\frac{c_{bH_2,O_2}}{c_{sH_2,O_2}}\right) \tag{2.33}$$

where c_{bH_2,O_2} is the bulk concentration and is related to the initial pressure supplied p_1, and c_{sH_2,O_2} is related to p_2 and represents the surface concentration of the reactants in the catalyst layer for anode or cathode. A deviation of the reversible potential due to the change in concentration between the catalyst layer surface and the bulk supplied is called concentration overpotential $\eta_{con} = \Delta E_R$. When the term inside the brackets of Eq. (2.33) is equal to 1 (low-current density), mass transport overpotential becomes negligible. At high current density the term inside the brackets becomes more than 1, and Eq. (2.33) results in an increase of the concentration overpotential. Substituting Eqs (2.29) and (2.30) into Eq. (2.33) yields the mass transport overpotential as a function of the current density:

$$\eta_{con} = \frac{RT}{zF} \ln \left(\frac{i_L}{i_L - i} \right) \tag{2.34}$$

2.2.5 Operating voltage in a PEM fuel cell

Substituting Eqs (2.23), (2.27) and (2.34) that define each overpotential into Eq. (2.22), it is possible to define a general equation for the operating voltage of a fuel cell at a current density i, as such:

$$V_{PEMFC} = E_{Nernst} - \eta_{act} - \eta_{ohm} - \eta_{con} \tag{2.35}$$

The PEM can conduct small amounts of electrons from the anode to the cathode. This effect is known as internal current in the PEMFC. In the same way, there is hydrogen that crosses through the PEM. This amount of hydrogen passing from anode to cathode reacts with oxygen molecules due to the Pt catalyst, producing no flow of electrons through the external circuit. Hydrogen that migrates through the PEM is known as fuel crossover. This effect of fuel crossover is similar to the internal current because the crossover of one hydrogen molecule from anode to cathode wastes two electrons, and is the same as two electrons crossing from anode to cathode internally, rather than through the external circuit. The magnitude of the internal current density is in the order of 10^{-3}A/cm^2 [1]. Barbir [1], Larmine and Dicks [2], and Balkin [4] have included this internal current density i_n in the irreversibilities or overpotentials defining the operating voltage of a fuel cell as:

$$V_{PEMFC} = E_{Nernst} - \frac{RT}{\alpha zF} \ln \left(\frac{i + i_n}{i_0} \right) - [i + i_n] R_{ohm} - \frac{RT}{zF} \ln \left(\frac{i_L}{i_L - i - i_n} \right) \tag{2.36}$$

Eq. (2.36) considers that only cathode losses reduce the voltage of the PEMFC and anode losses are negligible. The reversible potential calculated from Eq. (2.21), at standard temperature and pressure is equal to 1.23 V. However, in practical situations when the PEMFC is operated at open circuit, the reversible potential can drop to a value less than 1.23 V due to the internal current [1,2]. Zhang et al. [5] discussed the factors that yield the reduction of the open circuit voltage (OCV) with respect to the theoretical reversible voltage. The factors that affect the OCV are the PEMFC temperature, Pt catalyst surface in the electrodes, and hydrogen crossover.

The output voltage of a fuel cell can be represented through a voltage–current plot. In electrochemistry this plot is known as the polarisation curve because this curve represents any change in the equilibrium or reversible potential of the electrochemical reactions due to the irreversibilities or overpotentials during PEMFC operation. The voltage output of a PEMFC as a function of the current density calculated from Eq. (2.36) is shown in Fig. 2.5. The overpotentials that yield the reduction of the voltage output of a PEMFC are shown in Fig. 2.5.

Kim et al. [6] reported an empirical equation to fit experimental polarisation curves of PEMFCs at different operating conditions. The equation consisted of parameters representing kinetics of the electrode reactions, ohmic resistance and a term to predict the exponential voltage drop at high currents.

$$V_{\text{PEMFC}} = E_{\text{O}} - b\log_{10}(i) - ri - m\exp(di) \tag{2.37}$$

with

$$E_{\text{O}} = E_{\text{R}} + b\log_{10}(i_0) \tag{2.38}$$

where E_{R} is the reversible potential, i_0 is the exchange current density, b is the Tafel slope related to the activation overpotential during the ORR with units V/dec, r is the ohmic resistance of the PEMFC and can also comprise the charge transfer resistance during the HOR. The parameters m and d allow the prediction of the exponential voltage drop at high current.

MATLAB exercise. Apply the empirical Eq. (2.37) reported by Kim et al. [6] with a simulated polarisation curve reported in the study of Correa et al. [7].

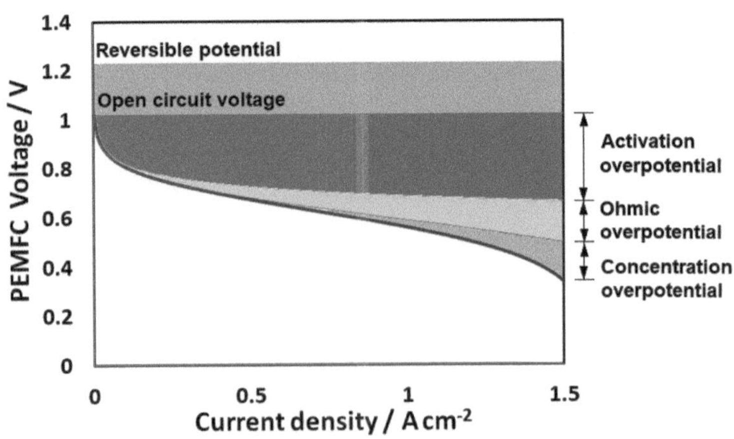

FIG. 2.5 Polarisation curve of a PEM fuel cell.

CHAPTER_2_polarisation_curve_fitting.m

```
% Parameters of the PEMFC model
I=0.09:0.1:75.7;      % Range of electrical current, A
T=343.15;             % Temperature, K
A=50.6;               % Active area, cm^2
L=17.8e-03;           % Membrane thickness, cm
co2=1e-04;            % Oxygen concentration, mol/cm^3
ch2=1e-04;            % Hydrogen concentration, mol /cm^3
po2=1;                % Oxygen pressure, atm
ph2=1;                % Hydrogen pressure, atm
g1=-0.948;            % Activation overpotential coefficients
g2=0.00286+0.0002*log(A)+4.3e-05*log(ch2);
g3=7.6e-05;
g4=-1.93e-04;
lambda=23;            % Parameter related to ohmic overpotential
B=0.016;              % Parameter related to concentration overpotential, V
Jmax=1.5;             % Limit current density, A/cm^2
J=I./A;               % Current density, A/cm^2
% Reversible potential
Nernst=1.229-(0.85e-03)*(T-298.15)+(4.31e-05)*T*(log(ph2)+0.5*log(po2));
% Activation overpotential
nact=-(g1+g2*T+g3*T*log(co2)+g4.*T.*log(I));
% Ohmic Overpotential
Pm=(181.6*(1+0.03.*(I./A)+0.062*(T/303)^2.*(I./A).^2.5))./((lambda-0.634-3*(I./A)).*exp(4.18*(T-303)/T));
Rm=(Pm.*L)./A;
nohmic=Rm.*I;
% Mass transport overpotential
ncon=-B.*log(1-(J./Jmax));
% Voltage output in PEMFC
V=Nernst-nact-nohmic-ncon;
% Plotting polarisation curve
plot(J,V,'k','LineWidth',2');
title('Polarisation curve');
xlabel('Current density / A cm^-^2','FontSize',12,'FontWeight','bold','Color','k');
ylabel('PEMFC voltage / V','FontSize',12,'FontWeight','bold','Color','k');
axis([0 1.5 0.4 1.4])
hold on
```

FIG. 2.6 Script in MATLAB software to simulate the polarisation curve considering parameters and model reported by Correa et al. [7].

Solution. Correa et al. [7] reported parameters and equations representing the overpotentials of the PEMFC to simulate a polarisation curve representing the performance of a single cell of a Ballard Mark V stack. Fig. 2.6 shows the MATLAB script to simulate the polarisation curve.

The empirical Eq. (2.37) reported by Kim et al. [6] is fitted to the simulated polarisation curve reported by Correa et al. [7]. The fitting process is carried out through the function *lsqcurvefit* defined in MATLAB software. This function solves nonlinear curve-fitting (data-fitting) problems in a least-squares sense and using the Levenberg–Marquardt algorithm. Fig. 2.7 shows the MATLAB script to fit a polarisation curve. The script in MATLAB shown in Fig. 2.7 is a continuation of the script shown in Fig. 2.6.

Fig. 2.8 shows a comparison between the simulated polarisation curve reported by Correa et al. [7] and the polarisation curve predicted by the empirical Eq. (2.37) reported by Kim et al. [6]. The estimated parameters of the empirical model Eq. (2.37) resulting from the fitting process are $E_O = 0.866$ V, $b = 0.147$ V/dec, $r = 0.1219$ Ωcm^2, $m = 3.955 \times 10^{-4}$ V, $d = 3.7289$ cm^2/A. The fitting process written in a script in MATLAB and shown in Fig. 2.7 can be applied to any experimental polarisation curve of the PEMFC to estimate kinetic and ohmic parameters. The scripts written in MATLAB shown in Figs 2.6 and 2.7 can be found in Appendix A.

```
%%% Fitting empirical equation, Eq. 2.37
xdata=J';                               % Current density
x0 =[0.5 0.5 0.5 0.5 0.5];        % Initial conditions
% Defining a function for empirical equation, Eq. 2.37
f_V=@(x,xdata) (x(1)-x(2).*log10(xdata)-x(3).*xdata-(x(4).*exp(x(5).*xdata)));
% Curve fitting
options = optimoptions('lsqcurvefit','Algorithm','levenberg-marquardt');
lb = [];
ub = [];
% V is the voltage of the polarisation curve to be predicted
x = lsqcurvefit(f_V,x0,xdata,V',lb,ub,options);
% Estimated parameters
E0=abs(x(1));
b=abs(x(2));
r=abs(x(3));
m=abs(x(4));
d=abs(x(5));
% Empirical equation, Eq. 2.37
Vs=E0-b.*log10(J)-r.*J-(m.*exp(d.*J));
% Plotting simulation of empirical equation
plot(J,Vs,'--r','LineWidth',2')
grid;
hlegl=legend(' Correa et al. [7]','Fitted, Kim et al. [6]');
hlegl=gca;
hlegl.FontSize = 16;
hlegl.FontWeight = 'bold';
clear all
```

FIG. 2.7 Script written in MATLAB software to fit the empirical equation reported by Kim et al. [6] with polarisation curves.

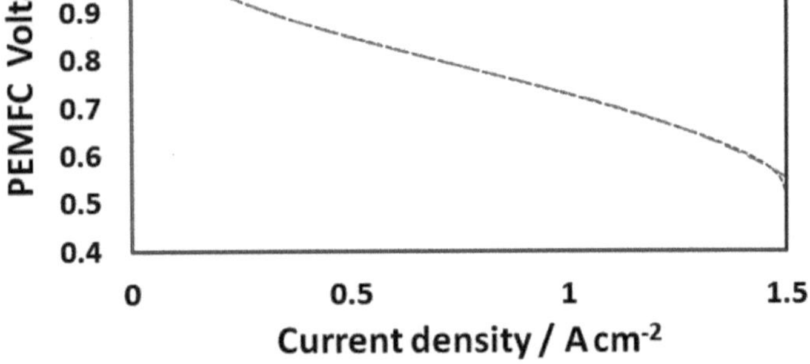

FIG. 2.8 Comparison between a polarisation curve simulated with the model and parameters reported by Correa et al. [7] and a simulated polarisation curve with an empirical model reported by Kim et al. [6]. The polarisation curve represents the steady-state performance of a single PEMFC in the Ballard Mark V stack.

2.3 Electrochemical phenomena in the catalyst layer

The catalyst layer is commonly formed by a composite structure of a matrix of carbon grains providing the electron conductivity, platinum Pt supported on carbon as the catalyst, PTFE as a binder stabilising the carbon matrix and as a hydrophobising agent, and an electrolyte network of perfluorosulfonate ionomer (PFSI) soaked with water. The carbon particles typically support noble metal or noble metal alloy deposits (i.e., Pt, platinum-ruthenium PtRu) with sizes in the range of 2–5 nm on its surface to drive the electrochemical reactions forward at the operating temperatures of the PEMFC. In addition, the catalyst layer contains a dispersion of polymer electrolyte to ensure continuity for proton conduction via the PEM. Fig. 2.9 shows the structure of the catalyst layer comprising the Pt supported on carbon and mixed with the electrolyte.

The resulting matrix structure is porous in nature and characterised by tortuous pathways for reactant transport. The matrix of carbon grains forms agglomerated structures where the agglomerated structure of the catalyst layer presents a bi-functional pore distribution. Primary pores are the pores existing inside the agglomerates between the Pt/carbon particles and secondary pores constitute the void spaces between agglomerates. It has been proposed that the molecules of the polymer electrolyte do not penetrate into the Pt/carbon particles [8] instead these molecules form a pathway of proton conduction attached to the surface of the agglomerated structure. Electrochemical reactions in the PEMFC rely upon the utilisation of Pt in the catalyst layer. The utilisation

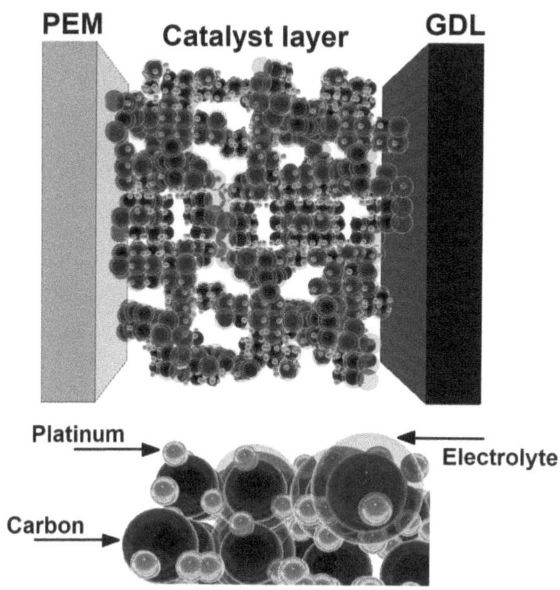

FIG. 2.9 Catalyst layer structure. Polymer electrolyte membrane (PEM), gas diffusion layer (GDL).

can be defined as the ratio of the electrocatalytic active surface area, which is accessible to the electrons and protons, to the total surface area of platinum. The literature suggests that it is possible to achieve a platinum utilisation of 86%–87% [9].

2.3.1 Faradaic process

The electrochemical reaction occurring at the electrode considering a simple reaction or a single-step electron transfer takes the following form:

$$\mathrm{Ox} + e^- \rightleftarrows \mathrm{Red} \tag{2.39}$$

where Ox is the oxidised form of the chemical species (reactants), and Red is the reduced form of the chemical species (products). The current-overpotential equation describes the cathodic and anodic currents at any potential as such:

$$i_F = i_0 \frac{c_{sOx}}{c_{bOx}} \exp\left(-\alpha f \eta\right) - i_0 \frac{c_{sRed}}{c_{bRed}} \exp\left([1-\alpha] f \eta\right) \tag{2.40}$$

with $f = zF/RT$ and $\eta = E - E_{eq}$ being the overpotential which relates the electrode potential E away from the equilibrium potential E_{eq}, i_0 is the exchange current density and relates the rate of charge transfer at which a reversible reaction takes place, c_{sOx} and c_{sRed} are the surface concentration of the reacting species and the bulk concentration of the reacting species are c_{bOx} and c_{bRed} respectively.

Dickinson and Wain [10] carried out a review on the use of the Butler–Volmer equation to describe electrode kinetics. Three versions of the Butler–Volmer equation were discussed: 'Engineer's' Butler–Volmer equation, 'Practical' Butler–Volmer equation, and 'Electroanalytical' Butler–Volmer equation. The concentration dependence is not defined in the 'Engineer's' Butler–Volmer equation and 'Practical' Butler–Volmer equation. The exchange current density and overpotential are measured at a chosen refence condition, which is an experimentally accessible composition of the reference solution. Eq. (2.40) is a form of the 'Electroanalytical' Butler–Volmer equation. The current density depends on the electrode polarisation with respect to a fixed reference electrode as well as the concentration of electroactive species. Dickinson and Hinds [11] recommended that a Nernst equation or linearised Butler–Volmer equation for the HOR at the anode and the Tafel equation for the ORR at the cathode should be considered as the kinetics of the HOR at the anode are fast and the kinetics during the ORR at the cathode are irreversible.

At large negative overpotential η, the second term on the right-hand side of Eq. (2.40) can be neglected considering that $\exp(-\alpha f \eta) \gg \exp([1-\alpha] f \eta)$ as such:

$$i_F = i_0 \frac{c_{sO_2}}{c_{bO_2}} \exp\left(-\alpha f \eta\right) \tag{2.41}$$

Eq. (2.41) is also known as the Tafel equation, it represents the faradaic current or cathodic current in the PEMFC during the ORR and takes place at the interface between dissimilar materials (electrolyte/Pt-carbon) in the CCL. c_{sO_2} is the concentration of oxygen on the surface of the CCL, and c_{bO_2} is the bulk concentration of oxygen in the FFP channel as depicted in Fig. 2.4.

Eq. (2.41) can be rearranged as:

$$\eta = -\frac{RT}{\alpha zF} \ln\left(\frac{i_F}{i_0}\right) - \frac{RT}{\alpha zF} \ln\left(\frac{c_{bO_2}}{c_{sO_2}}\right) \tag{2.42}$$

Eq. (2.42) can be expressed in terms of the limit current density when the oxygen concentration at the surface of the CCL is zero. The limit current density and current density derived from Faraday's law and Fick's first law have been previously described in Eqs (2.29) and (2.30). Substituting Eqs (2.29) and (2.30) into the second term of the right-hand side of Eq. (2.42) yields:

$$|\eta| = b \log_{10}\left(\frac{i_F}{i_0}\right) + b \log_{10}\left(\frac{i_L}{i_L - i_F}\right) \tag{2.43}$$

where b is the Tafel slope expressed in Eq. (2.26). Eq. (2.43) can be considered to estimate the Tafel slope b in current-overpotential curves of PEMFCs considering oxygen transport limitation. A Tafel slope commonly represents kinetics during the ORR [12], but also can be related to mass transport limitations [13]. A Tafel slope with values higher than 120mV/dec [14] up to 180mV/dec [15] is attributed to a potential loss related to electrode kinetics and mass transport. The Tafel slope b can change with different operating conditions. Wang et al. [16] reported a decrease in Tafel slope from 70 to 65mV/dec in the high potential region (low currents) in a PEMFC operated at 80°C with increasing oxygen partial pressure. Xu et al. [17] reported an increase of the Tafel slope when a PEMFC was operated at dry conditions.

Malevich et al. [18] reported a decrease in charge transfer coefficient α with decreasing operating voltage in a PEMFC as shown in Fig. 2.10a. The charge transfer coefficient α was calculated from electrochemical impedance spectroscopy (EIS) measurements carried out in a PEMFC. The Tafel slope b estimated from the polarisation curve is also shown in Fig. 2.10b. The curve shown in Fig. 2.10b is plotted on a log scale with the cell potential free of ohmic contributions. The charge transfer coefficient α is inversely proportional to the Tafel slope b as demonstrated in Eq. (2.26). Decreasing the charge transfer coefficient α and increasing the Tafel slope b can be attributed to mass transport losses. Commonly the Tafel slope has been calculated by relating only the first term on the right-hand side of Eq. (2.43) with current-overpotential curves. This has been demonstrated previously in Section 2.2.2. Nevertheless Malevich et al. [18] demonstrated through EIS measurements that oxygen transport limitations can also be present during PEMFC operation at low-medium current. Under this condition, a doubling of the Tafel slope is considered to account for oxygen transport limitations as shown in Fig. 2.10b.

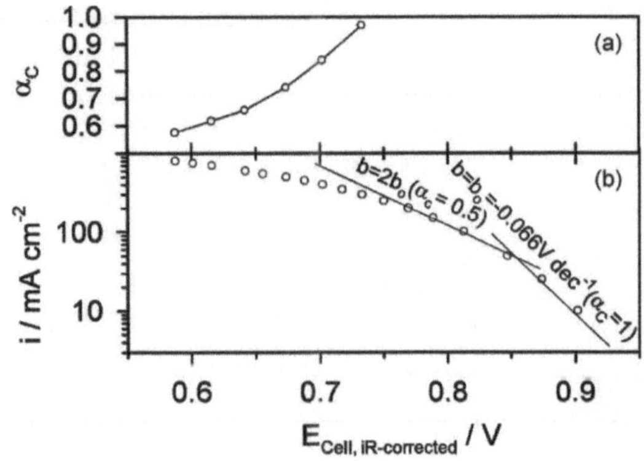

FIG. 2.10 Charge transfer coefficient α and Tafel slope b estimated from EIS measurements and polarisation curve measured from a PEMFC. *(Reproduced from D. Malevich, E. Halliop, B.A. Peppley, J.G. Pharoah, K. Karan, Investigation of charge-transfer and mass-transport resistances in PEMFCs with microporous layer using electrochemical impedance spectroscopy, J. Electrochem. Soc. 156 (2009) B222. © The Electrochemical Society, with permission of IOP Publishing. All rights reserved.)*

2.3.2 The charge double-layer in the catalyst layer

The catalyst layer is formed by a double-layer structure at the interface between the electrode (Pt supported carbon) and electrolyte (e.g. Nafion) materials. At this interface electrons will be collected at the surface of the electrode and ions will be attracted to the surface of the electrolyte. In the cathode, electrons and ions, together with the oxygen supplied will take part in the ORR. A potential difference is present because of the charge distribution between electrons and ions at the electrode-electrolyte interface. This potential difference has a determinant role in the charge distribution within the reactants as well as in the position and orientation of the reactant to form the desired product. Fig. 2.11 shows the electrode/electrolyte interface for the distribution of electric charge in the catalyst layer.

Fig. 2.11 shows that the electrolytic phase consists of two interfaces. The interface closer to the electrode, known as inner interface, contains water molecules and sometimes other species such as ions which are specifically adsorbed. This layer is also known as Helmholtz or Stern Layer. The locus of the electrical centres of the adsorbed ions is called the inner Helmholtz plane (IHP) which is at a distance x_1 from the electrode interface as shown in Fig. 2.11. The second interface presents solvated ions and can only approach the electrode at a distance x_2, the locus centres of these solvated ions is called the outer Helmholtz plane (OHP). When the OHP has the same amount of charge as in the electrode surface, there are two layers charged. The interaction of the solvated ions with the electrode is assumed to be electrostatic in nature, so that the interaction is independent of the chemical properties of the ions. Helmholtz considered that there is a potential drop confined just between the OHP in

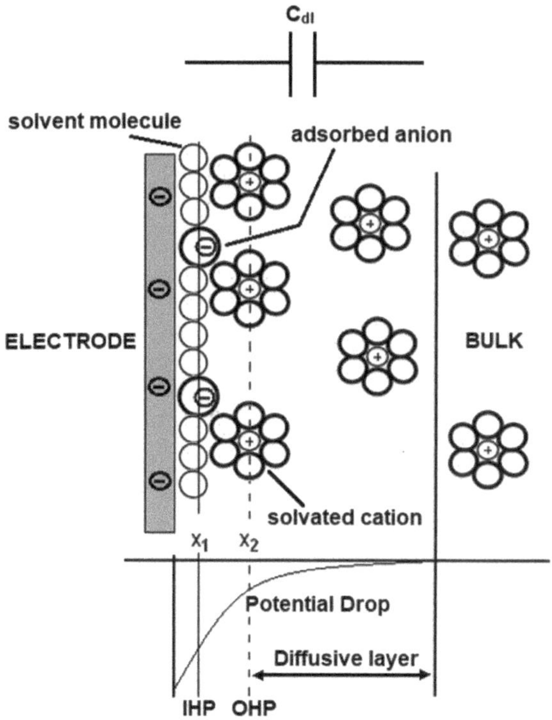

FIG. 2.11 Charge double-layer in the electrode-electrolyte interface.

the solution and the electrode surface. This phenomenon is analogous to an electrical capacitor that has two plates of charges separated by a distance. Hence this phenomenon is called the double-layer capacitance in the catalyst layer. However there are more factors to consider which can be in competition with the electrostatic interactions in this electrode-electrolyte interface such as diffusion/mixing in the solution. Under this factor, the ions are assumed to be able to move along the electrolyte. This region is called the diffusive layer which extends from the OHP into the bulk solution as shown in Fig. 2.11.

2.3.3 Faradaic and nonfaradaic processes

The catalyst layer is formed by a double-layer structure between the electrode-electrolyte interface. The electrode represents the electron-conducting network of carbon-supported platinum agglomerates while the electrolyte can be represented through the proton-conducting dispersion of Nafion. At this interface, an electric field is present and has a determinant role in the charge distribution within the reactants as well as in the position and orientation of the reactants to form the desired products. This double-layer can behave like a capacitor C_{dl} that is in parallel with the electrode reactions. The current passing from the electrode to the electrolyte can either take part in the charge transfer reaction or can contribute to the charge in the capacitive effect. The current due to the double-layer capacitance effect between electrode-electrolyte interface is expressed as:

$$i_{C_{dl}} = C_{dl}\frac{dE}{dt} \tag{2.44}$$

where $E = E_{\Phi} - E_{\phi}$ and E_{Φ} is the potential in the solution phase and E_{ϕ} is the potential in the solid phase of the CCL.

Both faradaic and nonfaradaic processes are present during the ORR. The total current density is defined as the sum between faradaic current (Eq. 2.41) and nonfaradaic current (Eq. 2.44) $i = i_F + i_{C_{dl}}$ as such:

$$i = i_0 \frac{c_{sO_2}}{c_{bO_2}} \exp(-\alpha f \eta) + C_{dl}\frac{dE}{dt} \tag{2.45}$$

Eq. (2.45) will be considered in Chapter 4 for the derivation of the equation representing the impedance response of the CCL in the frequency domain.

2.3.4 Multistep mechanisms during the oxygen reduction reaction

An electrochemical reaction can also be represented through a multistep electron transfer process:

$$\text{Ox} + \text{ne}^- \rightleftarrows \text{Red} \tag{2.46}$$

During a multistep electron transfer the overall electrochemical reaction can be coupled with reaction intermediates. The reaction of the intermediates can be influenced by the fractional surface coverage of the intermediate species γ_{X^+}. Kinetic models considering reactions of the intermediates in the overall ORR can assist the development of impedance models to interpret EIS measurements carried out in PEMFCs. Chapter 7 will discuss impedance models of the PEMFC considering the reaction of the intermediates such as hydrogen peroxide and platinum oxide formation during the ORR. The reaction for hydrogen peroxide and platinum oxide formation during the ORR are presented in the following sections.

2.3.4.1 Hydrogen peroxide formation

The ORR can be represented in a two-step process which yields the formation of hydrogen peroxide H_2O_2 and water H_2O [19,20]:

$$O_2 + 2H^+ + 2e^- \xrightarrow{k_1} H_2O_2 \tag{2.47}$$

$$H_2O_2 + 2H^+ + 2e^- \xrightarrow{k_2} 2H_2O \tag{2.48}$$

The faradaic current associated with reactions (2.47) and (2.48) have been reported as [21]:

$$i_{0,O_2} = zFk_1 c_{eq,O_2} c_{eq,H^+}^{3/2} \left[1 - \gamma_{eq,H_2O_2}\right] \exp\left(-E_{eq,O_2}/b_{O_2}\right) \tag{2.49}$$

$$i_{H_2O_2} = -zFk_2 c_{s,H_2O_2} c_{s,H^+}^{3/2} \gamma_{H_2O_2} \exp\left(-E/b_{H_2O_2}\right) \tag{2.50}$$

where k_i is the rate constant which depends on the global order of reaction, $\gamma_{H_2O_2}$ is the fractional surface coverage of H_2O_2 and b_{O_2} and $b_{H_2O_2}$ are the Tafel slopes associated with the reactions (2.47) and (2.48) with units in V. The total current density considering Eqs (2.49) and (2.50) can be represented as:

$$\begin{aligned} i_F = & -i_{0,O_2} \frac{1 - \gamma_{H_2O_2}}{1 - \gamma_{eq,H_2O_2}} \frac{c_{s,O_2}}{c_{eq,O_2}} \exp\left(-\left[E - E_{eq,O_2}\right]/b_{O_2}\right) \\ & - i_{0,H_2O_2} \frac{\gamma_{H_2O_2}}{\gamma_{eq,H_2O_2}} \exp\left(-\left[E - E_{eq,H_2O_2}\right]/b_{H_2O_2}\right) \end{aligned} \tag{2.51}$$

with the exchange current densities defined as:

$$i_{0,O_2} = zFk_1 c_{eq,O_2} c_{eq,H^+}^{3/2} \left[1 - \gamma_{eq,H_2O_2}\right] \exp\left(-E_{eq,O_2}/b_{O_2}\right) \tag{2.52}$$

$$i_{0,H_2O_2} = zFk_2 c_{eq,H_2O_2} c_{eq,H^+}^{3/2} \gamma_{eq,H_2O_2} \exp\left(-E_{eq,H_2O_2}/b_{H_2O_2}\right) \tag{2.53}$$

where the subscript eq represents equilibrium conditions and the subscript s represents surface concentration. The complete derivation of Eq. (2.51) can be found in the study from Cruz-Manzo et al. [22].

2.3.4.2 *Platinum oxide formation*

The CCL is formed by the interface between Pt-carbon/electrolyte where the ORR takes places. The formation of platinum oxide (PtO) during the ORR is also a multistep electron transfer process. The reaction for formation of platinum oxide (PtO) is represented [21] as:

$$Pt + H_2O \rightarrow PtO + 2H^+ + 2e^- \tag{2.54}$$

The reaction represented in Eq. (2.54) can be separated into a two-step process though with only one electron transfer [23]. The first step involves the formation of Pt-OH and secondly Pt-OH is oxidised to Pt-O.

The ORR can be represented as simple-step reaction:

$$O_2 + 4H^+ + 4e^- \rightarrow 2H_2O \tag{2.55}$$

The current associated with platinum oxide formation during the ORR can be represented as:

$$i_{O_2} = -i_{0,O_2} \frac{1 - \gamma_{PtO}}{1 - \gamma_{eq,PtO}} \frac{c_{s,O_2}}{c_{eq,O_2}} \exp\left(-\left[E - E_{eq,O_2}\right]/b_{O_2}\right) \tag{2.56}$$

where γ_{PtO} is the fractional surface coverage of PtO, b_{O_2} is the Tafel slope associated with the ORR represented in Eq. (2.55) with unit V and i_{0,O_2} is the exchange current density defined as:

$$i_{0,O_2} = zFk_{Pt}\left[1 - \gamma_{eq,PtO}\right] c_{eq,O_2} c_{eq,H^+}^{3/2} \exp\left(-E_{eq,O_2}/b_{O_2}\right) \tag{2.57}$$

The complete derivation of Eq. (2.56) can be found in the study from Cruz-Manzo et al. [24].

2.4 Electrochemical dynamic response

The electrochemical dynamic response of the PEMFC can be modelled through the time constant for the accumulation of charge in the electric double-layer of the catalyst layer and the diffusion of electroactive species. The dynamic output voltage of the PEMFC considers the dynamic change of overpotential with respect to the current demand. The electrochemical dynamic response of the PEMFC can be simulated by the electrical circuit shown in Fig. 2.12. The electrical circuit shown in Fig. 2.12 considers the reversible potential previously defined in Eq. (2.21), a resistor R_{ohm} to represent the ohmic resistance in the PEMFC connected in series with a parallel connection between a resistor R_F representing the faradaic resistance during the redox reaction and capacitor C representing an effective capacitance for the charge double-layer in the catalyst layer and diffusion of electroactive species, for example, capacitance calculated by dividing the diffusion time constant by the mass transport resistance. A load is connected between the terminals of the electrical circuit to simulate the current demand from the PEMFC. The circuit shown in Fig. 2.12 can be used to simulate the output voltage of the PEMFC during a step change in current. This will be demonstrated in the following section.

Where R_{ohm} is calculated from the ohmic overpotential η_{ohm} expressed in Eq. (2.27) and the resistor R_F is the faradaic resistance defined as:

$$R_F = \frac{\eta_{act} + \eta_{con}}{I_{cell}} \tag{2.58}$$

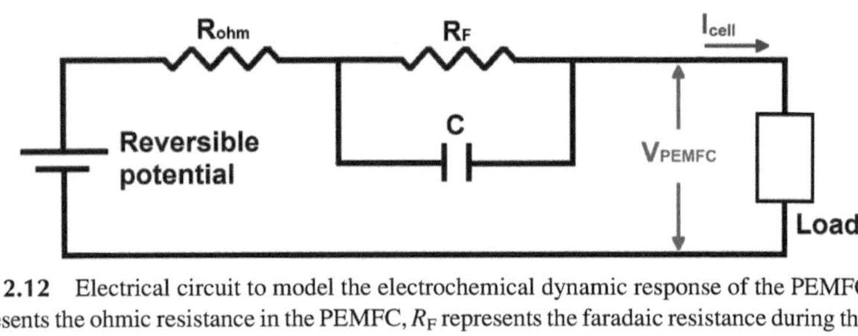

FIG. 2.12 Electrical circuit to model the electrochemical dynamic response of the PEMFC. R_{ohm} represents the ohmic resistance in the PEMFC, R_F represents the faradaic resistance during the redox reaction and C represents an effective capacitance for the charge double-layer in the catalyst layer and diffusion of electroactive species.

η_{act} is the activation overpotential defined in Eq. (2.23) and η_{con} is the concentration overpotential defined in Eq. (2.34) and I_{cell} is the current demand of the PEMFC.

The current and voltages associated with the electrical circuit shown in Fig. 2.12 can be represented from Kirchhoff's law as such:

$$\frac{d\eta_F}{dt} = \frac{1}{C} I_{cell} - \frac{\eta_F}{\tau_F} \tag{2.59}$$

where $\tau_F = R_F C$ is the electrochemical time constant, $\eta_F = \eta_{act} + \eta_{con}$ is the faradaic overpotential and represents the sum of activation and concentration overpotentials. Eq. (2.59) simulates the dynamic change of activation and concentration overpotentials with changing current demand of the PEMFC. The dynamic overpotential η_F expressed in Eq. (2.59) could only account for losses in the cathode and neglected losses in the anode. The dynamic output voltage of the PEMFC can be defined as:

$$V_{PEMFC} = E_{Nernst} - \eta_{ohm} - \eta_F \tag{2.60}$$

The solution of the differential equation (Eq. 2.59) to calculate the dynamic output voltage of the PEMFC (Eq. 2.60) can be carried out in commercial software MATLAB/Simulink.

2.4.1 Simulation of dynamic voltage of the PEMFC in MATLAB/Simulink environment

Simulink is a MATLAB-based graphical programming environment and a powerful programming environment for modelling and simulation of multidomain dynamical systems such as PEMFCs. Simulink allows the representation of differential equations through the use of graphical block diagramming tools and a customizable set of block libraries.

The equation that represents the dynamic output voltage of the PEMFC (Eq. 2.60) can be represented in a MATLAB/Simulink environment as shown in Fig. 2.13. The modelling architecture shown in Fig. 2.13 comprises an integrator block to represent the differential equation of the overpotential. In addition, blocks that perform arithmetic operations such as subtraction, division, and multiplication are considered in the modelling architecture shown in Fig. 2.13. Moreover, two scripts written in a MATLAB environment are interfaced with the modelling architecture from Simulink. The MATLAB scripts calculate the faradaic resistance represented in Eq. (2.58) and calculate the dynamic voltage output of the PEMFC represented in Eq. (2.60) considering the dynamic change of the faradaic overpotential (Eq. 2.59), ohmic overpotential (Eq. 2.27) and reversible potential (Eq. 2.21). The input of the modelling architecture is the current demand of the PEMFC. The dynamic modelling architecture shown in Fig. 2.13 considers the model of the polarisation curve of a single PEMFC in the Ballard Mark V stack reported by Correa et al. [7] and shown in Fig. 2.8.

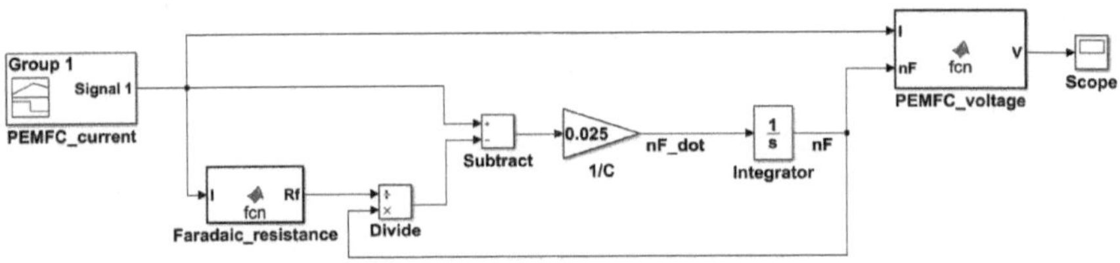

FIG. 2.13 Simulation of dynamic voltage of the PEMFC in MATLAB/Simulink environment.

Fig. 2.14 shows the script in MATLAB to calculate the faradaic resistance represented in Eq. (2.58) based on the activation and concentration overpotentials from the steady-state model of the PEMFC reported by Correa et al. [7]. Fig. 2.15 shows the script in MATLAB to calculate the dynamic output voltage of the PEMFC represented in Eq. (2.60) and considers the reversible potential and ohmic overpotential from the steady-state model of the PEMFC reported by Correa et al. [7]. The script shown in Fig. 2.6 to simulate the steady-state output voltage (polarisation curve) of the PEMFC has been considered in the script shown in Figs 2.14 and 2.15.

A capacitor of $C = 40$ F is considered in the dynamic modelling architecture shown in Fig. 2.13. The modelling architecture represented in Fig. 2.13 receives as an input the current demand of the PEMFC. Fig. 2.16 shows step current responses to simulate the dynamic output voltage of the PEMFC. A low value of the current $I_{cell} = 1 \times 10^{-4}$ A from $t = 0$ to $t = 5$ s is considered to avoid a division by zero in the equations comprising the Simulink model. A low value of current could simulate the effect of the internal current on the voltage output of the PEMFC. As previously discussed in Section 2.2.5, internal current can be a consequence of hydrogen crossover from anode to cathode. The pressure and

```
Faradaic_resistance × | PEMFC_voltage × | +
function Rf = fcn(I)
T=343.15;                  % Temperature, K
A=50.6;                    % Active area, cm^2
co2=1e-04;                 % Oxygen concentration, mol/cm^3
ch2=1e-04;                 % Hydrogen concentration, mol/cm^3
g1=-0.948;                 % Activation overpotential coefficients
g2=0.00286+0.0002*log(A)+4.3e-05*log(ch2);
g3=7.6e-05;
g4=-1.93e-04;
B=0.016;                   % Parameter related to concentration overpotential
Jmax=1.5;                  % Limit current density, A/cm^2
J=I./A;                    % Current density, A/cm^2
% Activation overpotential
nact=-(g1+g2*T+g3*T*log(co2)+g4.*T.*log(I));
% Mass transport overpotential
ncon=-B.*log(1-(J./Jmax));
% Faradaic resistance
Rf=(nact+ncon)./I;
```

FIG. 2.14 MATLAB script to calculate faradaic resistance.

Faradaic_resistance | PEMFC_voltage

```
function V = fcn(I,nF)
T=343.15;               % Temperature, K
A=50.6;                 % Active area, cm^2
L=17.8e-03;             % Membrane thickness, cm
po2=1;                  % Oxygen pressure, atm
ph2=1;                  % Hydrogen pressure, atm
lambda=23;              % Parameter related to ohmic overpotential
% Reversible potential
Nernst=1.229-(0.85e-03)*(T-298.15)+(4.31e-05)*T*(log(ph2)+0.5*log(po2));
% Ohmic Overpotential
Pm=(181.6*(1+0.03.*(I./A)+0.062*(T/303)^2.*(I./A).^2.5))./((lambda-0.634-3*(I./A)).*exp(4.18*(T-303)/T));
Rm=(Pm.*L)./A;
nohmic=Rm.*I;
% Dynamic PEMFC voltage
V=Nernst-nohmic-nF;
```

FIG. 2.15 MATLAB script to calculate voltage output of PEMFC.

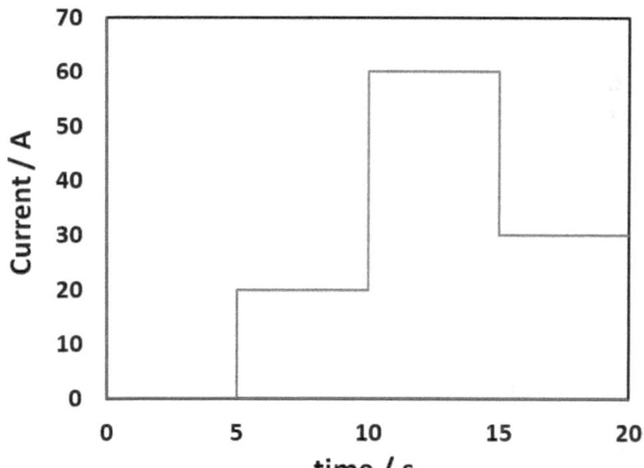

FIG. 2.16 PEMFC input current for simulation of dynamic voltage of a single PEMFC in the Ballard Mark V stack.

temperature are considered constant during the simulation of the dynamic voltage of the PEMFC. The step change in current is the only input considered in the dynamic modelling architecture as shown in Fig. 2.16.

Fig. 2.17 shows the simulation of the dynamic output voltage of the PEMFC. The voltage decreases with increasing current and vice versa. The simulation shown in Fig. 2.17 can be analogous to the voltage measured across the terminals/electrodes of the PEMFC during a current interrupt test. During a current interrupt test the transient voltage of the PEMFC is measured. The experimental equipment to perform the current interrupt tests are inexpensive and easily portable such as oscilloscopes and low-cost electronic equipment [25]. The transient response shown in Fig. 2.17 presents two features which can be associated with the overpotentials in the PEMFC [2]. During the sudden change in current, there is an immediate increase/decrease in voltage which is labelled 'Vo' in Fig. 2.17. Thereafter there is an exponential increase/decrease in voltage

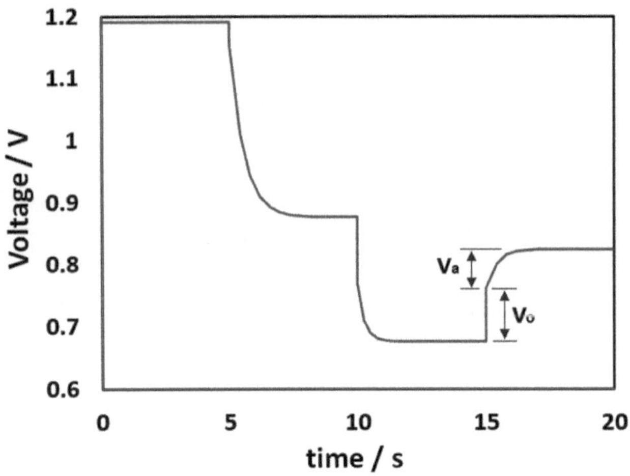

FIG. 2.17 Simulation of dynamic voltage response of a single PEMFC in the Ballard Mark V stack at different current steps.

which is labelled 'Va' in Fig. 2.17. The magnitude of the voltage labelled as 'Vo' is attributed to the ohmic overpotential and the magnitude of the voltage labelled as 'Va' is attributed to the activation and concentration overpotentials. In low temperature fuel cells such as PEMFCs the magnitude of the ohmic overpotential represented in the transient voltage during the current interrupt tests is considerable with respect to other types of fuel cells such as in direct methanol fuel cells [2].

The electrochemical model developed in MATLAB/Simulink and shown in Fig. 2.13 considers the parameters of the steady-state model of the PEMFC reported by Correa et al. [7]. The electrochemical model could be extended by incorporating the dependency of the steady-state parameters with different operating conditions such as pressure, temperature and flow rate [26]. Moreover, the electrochemical model developed in MATLAB/Simulink and shown in Fig. 2.13 can be implemented with a Simulink model representing the balance of plant of a PEMFC [27]. It would be possible to simulate the dynamic output voltage of the PEMFC during the dynamic change in reactant pressure, reactant flow rate and core temperature, which could be a valuable simulation tool for the development and optimisation of control strategies to achieve good PEMFC performance at different operating conditions [28].

References

[1] F. Barbir, PEM Fuel Cells, Academic Press, Burlington, MA, 2005.

[2] J. Larminie, A. Dicks, Fuel Cell System Explained, Wiley, England, 2003.

[3] J.E. Brady, J.R. Holum, Chemistry: The Study of Matter and Its Changes, John Wiley & Sons, Brisbane, 1996.

[4] A.R. Balkin, Modelling a 500W Polymer Electrolyte Membrane Fuel Cells (thesis), Faculty of Engineering, University of Technology, Sydney, 2002.

[5] J. Zhang, H. Zhang, J. Wu, J. Zhang, Fuel cell open circuit voltage, in: PEM Fuel Cell Testing and Diagnosis, Elsevier, United Kingdom, 2013 (Chapter 7).

[6] J. Kim, S.-M. Lee, S. Srinivasan, C.E. Chamberlin, Modeling of proton exchange membrane fuel cell performance with an empirical equation, J. Electrochem. Soc. 142 (1995) 2670–2674.

[7] J.M. Correa, F.A. Farret, L.N. Canha, An analysis of the dynamic performance of proton exchange membrane fuel cells using an electrochemical model, in: IECON'01, 27th Annual Conference of the IEEE Industrial Electronics Society (Cat. No.37243), Denver, CO, USA, vol. 1, 2001, pp. 141–146.

[8] K. Malek, M. Eikerling, Q. Wang, T. Navessin, Z. Liu, Self-organization in catalyst layers of polymer electrolyte fuel cells, J. Phys. Chem. C 111 (2007) 13627–13634.

[9] T.J. Schimidt, H.A. Gasteiger, G.D. Stab, P.M. Urban, D.M. Kolb, R.J. Behm, Characterization of high surface area electrocatalysts using a rotating disk electrode configuration, J. Electrochem. Soc. 145 (1998) 2354–2358.

[10] E.J.F. Dickinson, A.J. Wain, The Butler-Volmer equation in electrochemical theory: origins, value, and practical application, J. Electroanal. Chem. 872 (2020), 114145.

[11] E.J.F. Dickinson, G. Hinds, The Butler-Volmer equation for polymer electrolyte membrane fuel cell (PEMFC) electrode kinetics: a critical discussion, J. Electrochem. Soc. 166 (2019) F221–F231.

[12] H. Vrubel, T. Moehl, M. Gratzel, X. Hu, Revealing and accelerating slow electron transport in amorphous molybdenum sulphide particles for hydrogen evolution reaction, Chem. Commun. 49 (2013) 8985–8987.

[13] H. Chen, J.R. Elliott, H. Le, M. Yang, R.G. Compton, Super-Nernstian Tafel slopes: an origin in coupled homogeneous kinetics, J. Electroanal. Chem. 869 (2020), 114185.

[14] M. Chatenet, J. Benziger, M. Inaba, S. Kjelstrup, T. Zawodzinski, R. Raccichini, Good practice guide for papers on fuel cells and electrolysis cells for the journal of power sources, J. Power Sources 451 (2020), 227635.

[15] L. Qingfeng, X. Gang, H.A. Hjuler, R.W. Berg, N.J. Bjerrum, Oxygen reduction on gas-diffusion electrodes for phosphoric acid fuel cells by a potential decay method, J. Electrochem. Soc. 142 (1995) 3250–3256.

[16] J.X. Wang, F.A. Uribe, T.E. Springer, J. Zhang, R.R. Adzic, Intrinsic kinetic equation for oxygenreduction reaction in acidic media: the double Tafel slope and fuel cell applications, Faraday Discuss. 140 (2008) 347–362.

[17] H. Xu, Y. Song, H.R. Kunz, J.M. Fenton, Effect of elevated temperature and reduced relative humidity on ORR kinetics for PEM fuel cells, J. Electrochem. Soc. 152 (2005) A1828–A1836.

[18] D. Malevich, E. Halliop, B.A. Peppley, J.G. Pharoah, K. Karan, Investigation of charge-transfer and mass-transport resistances in PEMFCs with microporous layer using electrochemical impedance spectroscopy, J. Electrochem. Soc. 156 (2009) B216–B224.

[19] S.K. Roy, M.E. Orazem, B. Tribollet, Interpretation of low-frequency inductive loops in PEM fuel cells, J. Electrochem. Soc. 154 (2007) B1378–B1388.

[20] A. Damjanovic, V. Brusic, Electrode kinetics of oxygen reduction on oxide-free platinum electrodes, J. Electroanal. Chem. 12 (1967) 615–628.

[21] S.K. Roy, M.E. Orazem, Deterministic impedance models for interpretation of low frequency inductive loops in PEM fuel cells, ECS Trans. 3 (2006) 1031–1040.

[22] S. Cruz-Manzo, R. Chen, P. Greenwood, An impedance model for analysis of EIS of polymer electrolyte fuel cells under hydrogen peroxide formation in the cathode, J. Electroanal. Chem. 745 (2015) 28–36.

[23] H. Xu, R. Kunz, J.M. Fenton, Investigation of platinum oxidation in PEM fuel cells at various relative Humidities, Electrochem. Solid State Lett. 10 (2007) B1–B5.
[24] S. Cruz-Manzo, C. Perezmitre-Cruz, P. Greenwood, R. Chen, An impedance model for analysis of EIS of polymer electrolyte fuel cells under platinum oxidation and hydrogen peroxide formation in the cathode, J. Electroanal. Chem. 771 (2016) 94–105.
[25] M.A. Rubio, A. Urquia, S. Dormido, Diagnosis of PEM fuel cells through current interruption, J. Power Sources 171 (2007) 670–677.
[26] Z. Abdin, C.J. Webb, E. Gray, PEM fuel cell model and simulation in Matlab-Simulink based on physical parameters, Energy 116 (2016) 1131–1144.
[27] H.-S. Kim, D.-H. Lee, K. Min, M. Kim, Effects of key operating parameters on the efficiency of two types of pem fuel cell systems (high-pressure and low-pressure operating) for automotive applications, J. Mech. Sci. Technol. 19 (2005) 1018–1026.
[28] W.R.W. Daud, R.E. Rosli, E.H. Majlan, S.A.A. Hamid, R. Mohamed, T. Husaini, PEM fuel cell system control: a review, Renew. Energy 113 (2017) 620–638.

Chapter 3

Electrochemical impedance spectroscopy in PEM fuel cells

Electrochemical impedance spectroscopy (EIS) measurements are commonly applied in situ to characterise physical processes within an electrochemical system. Ex situ EIS measurements provide information about physical properties of electrically conductive materials. EIS in situ measurements allow the study of the electrochemical and diffusion mechanisms occurring during the polymer electrolyte membrane fuel cell (PEMFC) operation, whereas ex situ measurements allow the study of the electrical properties in the different layers constituting the PEMFC. The EIS technique measures the frequency dependence of the impedance or AC load opposition of a PEMFC. The objectives of EIS in PEMFCs are mainly:

1. To obtain information of the electrochemical and diffusion mechanisms within the PEMFC, which can help to optimise and select the most appropriate operating conditions during PEMFC operation.
2. To identify and differentiate the contribution of each component, such as the catalyst layer, polymer electrolyte membrane (PEM), gas diffusion layer (GDL), and flow field plate (FFP), in the overall impedance response of the PEMFC, which can assist in identifying failures in the mode of operation during the PEMFC performance.
3. To model the impedance response of the PEMFC with an appropriate mathematical model and consequently to interpret the electrochemical and diffusion mechanisms in the impedance response of the PEMFC.

In this chapter, in situ EIS measurements are only considered as the impedance response obtained during PEMFC operation is related to the different operating conditions and the electrochemical and diffusion phenomena within the PEMFC. Fig. 3.1 shows a conventional balance of plant required to carry out EIS measurements in a single PEMFC under different operating conditions. Other types of PEMFCs require fewer components in the balance of plant; for instance, open-cathode PEMFCs do not require a mass flow metre or humidifier for the supply of air as open-cathode PEMFCs are fed with air from ambient conditions through forced convection via the use of fans or blowers.

Electrochemical Phenomena in the Cathode Impedance Spectrum of PEM Fuel Cells.
https://doi.org/10.1016/B978-0-323-90607-4.00004-5
Copyright © 2022 Elsevier Inc. All rights reserved.

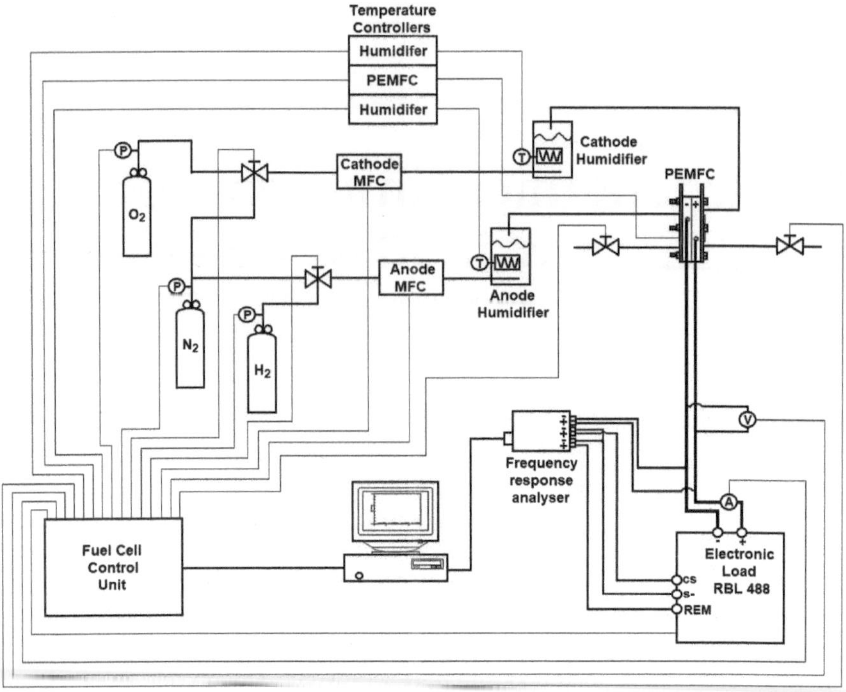

FIG. 3.1 Balance of plant of a single PEMFC to carry out EIS measurements under different operating conditions.

This chapter does not aim to describe laboratory experiments on PEMFC operation. Cooper et al. [1] reported a series of laboratory experiments on PEMFC operation to teach electrochemistry, thermodynamics, kinetics, and mass transport phenomena to engineering and research students new to the field of fuel cell technology. Cooper et al. [1] also reported a set of experiments to conduct EIS measurements. Further information about the application of EIS in PEMFCs can be found in the book published by Yuan et al. [2].

3.1 Representation of the impedance response of a PEM fuel cell

The mathematical relation to represent the impedance response of the PEMFC is through the transfer function describing the relation between the measured oscillating voltage $\widetilde{V}$ and oscillating current $\widetilde{I}_{\text{cell}}$ at different frequencies.

$$Z_{\text{PEMFC}} = \frac{\widetilde{V}}{\widetilde{I}_{\text{cell}}} \tag{3.1}$$

EIS measurements are represented with complex numbers in a Nyquist plot and can be represented in a Bode plot with modulus and phase angle as a function of the frequency. Chapter 1 has introduced the mathematical relation between oscillating signals in the time domain and frequency domain through Fourier transform. The frequency response of the PEMFC obtained from EIS tests is represented by a locus of points, where each datum corresponds to a different measurement frequency. Complex-impedance (Nyquist) plots are very popular because the loci of the points yield an insight into the possible mechanisms of governing phenomena. A disadvantage of the Nyquist format is that frequency dependence is obscured because multiple energy-controlled processes of an electrochemical system can be masked in the frequency plot. Characteristic frequencies should always be labelled to allow a better understanding of the time constant of the underlying phenomena. Fig. 3.2 shows an example of the impedance response of a 25 cm^2 H_2/air PEMFC represented in a Nyquist plot. The size of the impedance spectrum as shown in Fig. 3.2 will depend on different operating conditions (PEMFC temperature, stoichiometry of gas reactants, relative humidity (RH) of gas reactants, current density). The EIS measurements with positive imaginary components at high frequencies (e.g. 20–2 kHz in Fig. 3.2) are attributed to the inductance of electrical cables of the measurement system. The inductance effect from the measurement system on the impedance spectrum of the PEMFC will be discussed in more detail later in this chapter. The total ohmic resistance of the PEMFC is commonly represented, where the impedance spectrum crosses the real axis Z_r of the Nyquist plot at high frequencies (e.g. 2 kHz in Fig. 3.2). In other cases, if no positive imaginary components are present in the Nyquist plot at high frequencies, the ohmic resistance is represented with the real component of the EIS measurements at the highest frequency. Electrochemical mechanisms related to the charge transfer of the hydrogen-oxidation and oxygen-reduction (redox) processes can be represented at high to low frequencies (e.g. 2 kHz to 2 Hz in Fig. 3.2). Oxygen transport limitations are represented at the lowest frequencies (e.g. 2–0.1 Hz in Fig. 3.2). In some cases, the impedance response attributed to oxygen transport limitations is represented by a second semicircle at low frequencies. Fig. 3.2 shows that the frequency response of the PEMFC is

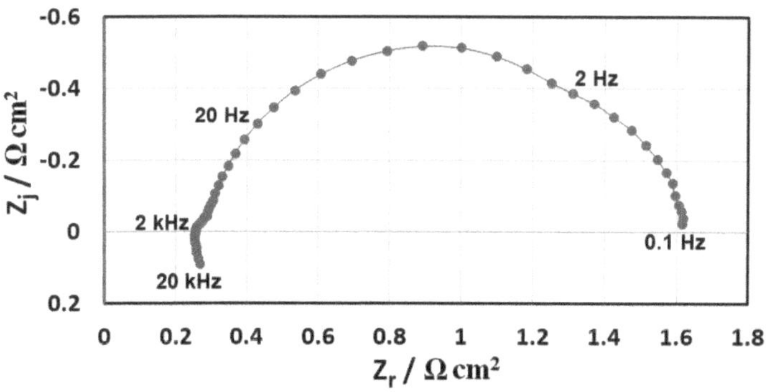

FIG. 3.2 Impedance response of a PEMFC represented in the complex-impedance plot.

resistive–capacitive as the measurements are mainly composed of negative imaginary components $-Z_j$. The impedance response of a resistor–capacitor circuit was discussed in Chapter 1 Section 1.4.4.

Fig. 3.3 shows the modulus of the complex quantity of the impedance response as a function of the frequency. The modulus shown in Fig. 3.3 corresponds to the impedance response presented in the Nyquist plot shown in Fig. 3.2. The asymptotic value of the modulus at high frequencies, for example, 2 kHz, provides a qualitative analysis of the total ohmic resistance of the PEMFC. The slope of the modulus, for example, between 100 and 1 Hz, is related to the distribution of charge (capacitive effect) between dissimilar materials, for example, electrolyte/carbon in the catalyst layers of PEMFCs. The difference between the asymptotic values of the modulus at a low frequency and a high frequency provides a qualitative value of the total electro-diffusion resistance in the PEMFC. Note that the difference between the moduli at the lowest frequency and at the highest frequency shown in Fig. 3.3 is equal to the diameter of the impedance spectrum shown in Fig. 3.2.

The phase angle shown in Fig. 3.4 is related to the complex quantity of the impedance response presented in the Nyquist plot shown in Fig. 3.2. Previously

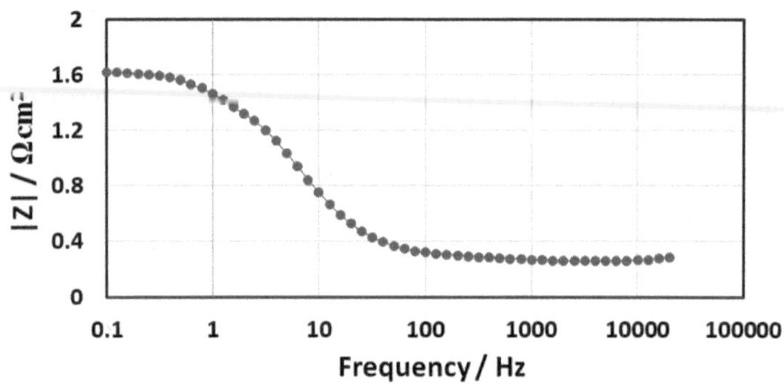

FIG. 3.3 Modulus of impedance response of a PEMFC as a function of frequency.

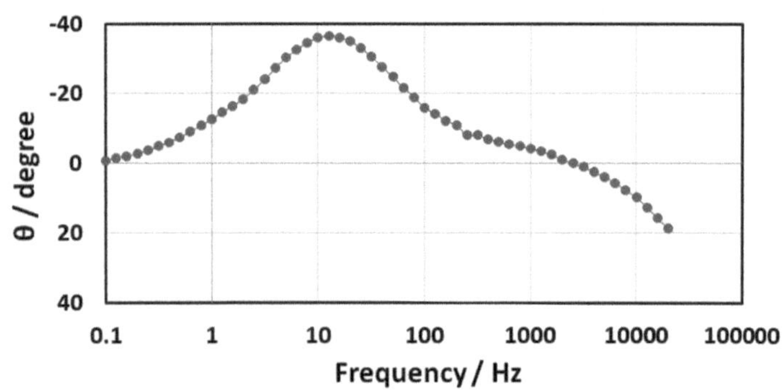

FIG. 3.4 Phase angle of impedance response of a PEMFC as a function of frequency.

in Chapter 1 Section 1.1.1, the relation between the phase angle and the real and imaginary components of the complex quantity of the impedance response was presented. The phase angle can be calculated with the inverse tangent of the ratio between the imaginary and real components of the impedance response as demonstrated in Eq. (1.6) from Chapter 1. The phase angle could provide information about the distribution of charge in the catalyst layers of PEMFCs. If the phase angle becomes more negative, there is a more homogeneous distribution (ideal-capacitive effect) of charge in the catalyst layers. This was demonstrated in Section 1.4.5 from Chapter 1 through the comparison between the phase angles of the impedance response of a resistor/capacitor circuit considering an ideal capacitor and a constant phase element (CPE).

The electrochemical phenomena represented in the impedance response of PEMFCs are studied through the use of physics-based impedance models and equivalent electrical circuits (EECs). Electrical components such as a resistor, Warburg impedance, an inductor, a capacitor, and CPE are considered in EECs. The impedance response of the components comprising EECs was introduced in Chapter 1 Section 1.4. The most common EEC configurations considered to characterise the impedance response of PEMFCs from EIS will be described later in this chapter. The development of physics-based impedance models to study the electrochemical phenomena represented in the impedance response of the PEMFC will be considered in the modelling section from this book.

3.2 Equipment for EIS measurements

One key advantage of the EIS technique is that it is noninvasive and can be applied in situ. Another advantage is that the frequency response tests are simple to carry out and can be easily tuned for greater accuracy by using readily available sinusoidal generators and precise measuring equipment. Impedance experiments involve the conversion of sinusoidal (time-domain) input and output signals into a complex quantity that is a function of frequency. A frequency response analyser (FRA) converts time-domain signals into frequency-domain signals through Fourier analysis and allows the measurement of the impedance in PEMFCs.

One-by-one frequency measurements are time-consuming because of the long AC wave period with decreasing frequency. If the PEMFC is not operated under steady-state conditions when applying one-by-one frequency measurements, the resulting frequency response measurements could be unreliable [3]. Instead of applying one-by-one single sinusoidal waves, a superposition signal containing different frequencies of sinusoidal waves can be applied. The method is called time-domain EIS, and the measurement time is significantly shorter than the one-by-one frequency measurements. The hardware required to carry out time-domain EIS measurements requires a signal generator to produce a broad band frequency-rich perturbation signal that is applied to the PEMFC. Also, a transient recorder, which is synchronised with the signal generator and measures the voltage $V(t)$ across the PEMFC and the current $I(t)$ that flows through the PEMFC in the time domain, is required. A fast Fourier transform (FFT) algorithm converts the resulting time-domain data into frequency-domain data. Time-domain EIS measurements based on fast Fourier transformation have been applied to electrochemical systems [4].

The performance of a PEMFC can be represented through a current–voltage diagram, which is commonly known as a polarisation curve, as previously discussed in Chapter 2 Section 2.2.5. When the electric current in a PEMFC is controlled and the potential is measured, the experimental test is described as galvanostatic, and a potentiostatic test is carried out when the potential is controlled and the current is measured. Commercial FRAs allow EIS measurements in either the potentiostatic or galvanostatic mode using FFT.

Fig. 3.5 shows a commercial FRA (Z#106 WonATech) to carry out EIS measurements in the galvanostatic mode. Other FRAs allow for carrying out EIS measurements in the potentiostatic mode, such as the Solartron 1280C, as shown in Fig. 3.6. During EIS measurements, a sinusoidal signal is superimposed onto the current or voltage of the PEMFC under steady-state conditions; therefore, the sinusoidal output voltage for galvanostatic tests and the sinusoidal output current for potentiostatic tests are measured. The PEMFC demonstrates nonlinear current–voltage behaviour as the current has an exponential dependence on potential. This was discussed in Chapter 2 through the polarisation curve shown in Fig. 2.5. For systems presenting a nonlinear current–voltage relationship, such as the PEMFC, a small amplitude of the sinusoidal signal is required for EIS measurements. For systems presenting a linear current–voltage relationship, a very large amplitude can be used. Commonly, a sinusoidal signal with a 10 mV amplitude is applied during EIS measurements

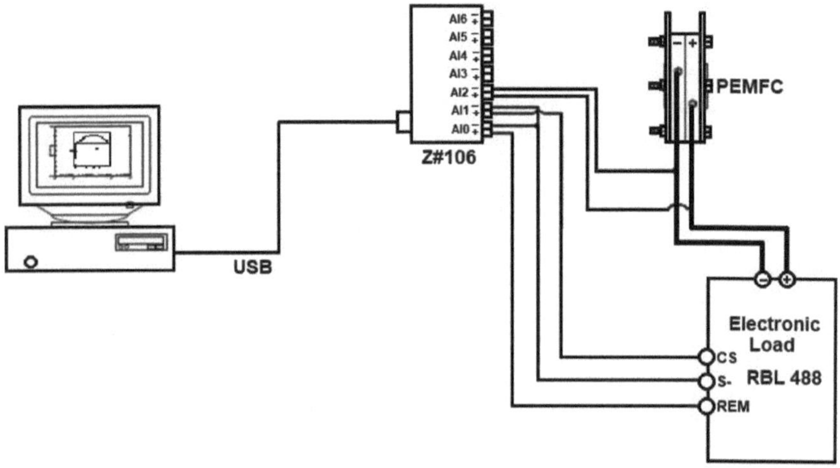

FIG. 3.5 Experimental set-up for EIS measurements in the galvanostatic mode using an FRA, Z#106 WonATech, connected with a RBL 488 Load.

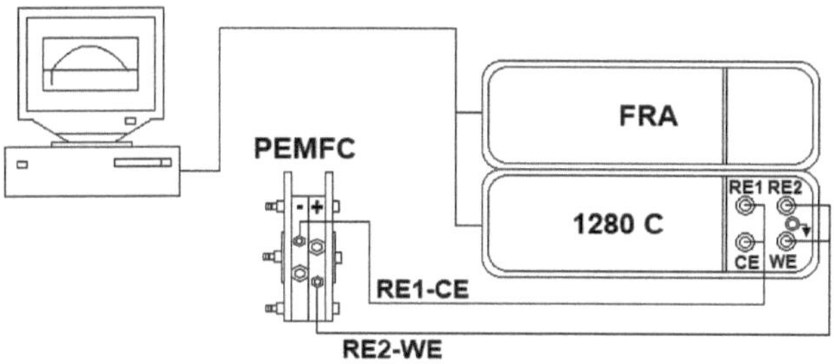

FIG. 3.6 Experimental set-up for EIS measurements in the potentiostatic mode using an FRA-potentiostat, Solartron 1280C. Reference electrode (RE), counter electrode (CE), and working electrode (WE).

of PEMFCs under potentiostatic tests. The amplitude of the sinusoidal signal in galvanostatic EIS tests of PEMFCs is commonly considered to be 5% of the DC current. A high amplitude of the sinusoidal signal superimposed onto the steady-state current or voltage during PEMFC-EIS measurements could violate the linearity condition for stable, linear, and time-invariant systems in the frequency domain. The conditions such as linearity, stability, and causality required for correct EIS data will be discussed in Section 3.12.

3.3 Reference electrodes

A reference electrode allows the measurement of potential related to the anode or cathode through a three-electrode configuration. Reference electrodes are also commonly used to isolate the impedance response of an anode or cathode from the total PEMFC impedance response. Some studies [5] have applied the three-electrode configuration in EIS measurements of PEMFCs and in EIS measurements of batteries [6]. Under such conditions, the signals are measured between the working electrode and the reference electrode, and the current induced is collected by the counter electrode.

As reported by Chan et al. [5], a reference electrode should be reversible, stable, and fast in response. Two types of reference electrodes can be used to measure individual contributions from the electrodes: internal type and external type. The internal type consists of connecting a small platinum wire in contact with the PEM of the PEMFC and using a potentiostat under a three-electrode configuration, as shown in Fig. 3.7. The experimental configuration shown in Fig. 3.7 also allows the estimation of individual electrode potential to calculate kinetic parameters related to the redox reaction, such as Tafel slope and exchange current density. The Tafel slope shown in Fig. 2.3 from Chapter 2 can be estimated through the three-electrode experimental configuration shown in Fig. 3.7.

In the external type, a liquid electrolyte is deposited in an acid bridge (commonly known as a salt bridge) and allows the connection between the membrane

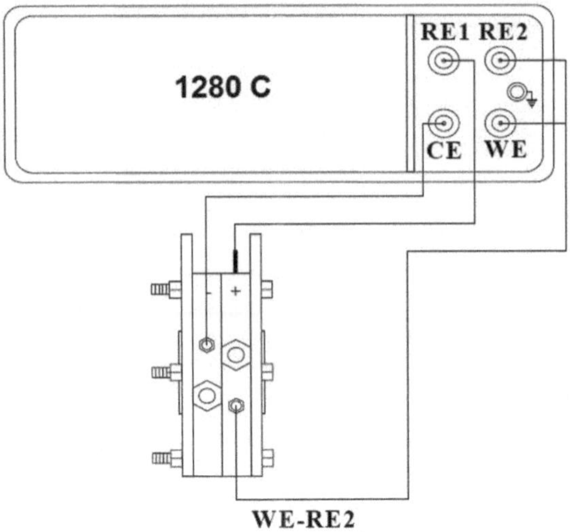

FIG. 3.7 Three-electrode configuration and reference electrode using a Solartron 1280C potentiostat. Reference electrode (RE), counter electrode (CE), and working electrode (WE).

and the reference electrode. In the external reference electrode, minor modifications in the hardware compared to the internal type are required. However, an external reference electrode can induce nonhomogeneous hydration and a proton concentration gradient in the PEM [7].

To overcome the issues related to the application of conventional (internal and external) reference electrodes, a dynamic hydrogen electrode (DHE) has been used in PEMFCs. It consists of a pair of platinum wires in contact with the PEM. A small current is applied between the two wires, and as a result, hydrogen is produced. Therefore, the coverage of hydrogen is constant and the potential between the terminals is constant, which in turn can be used as a reference electrode.

It is also possible to use a different experimental set-up, such as a multichannel FRA, to measure the simultaneous impedance responses from the total PEMFC impedance and from the individual electrodes (anode or cathode) using the reference electrode, as shown in Fig. 3.8.

Fig. 3.9 shows the impedance response of a PEMFC and the cathodic impedance response measured using a multichannel FRA (Fig. 3.8). One channel measures the impedance of the whole PEMFC, and another channel measures the impedance response of the cathode using the three-electrode configuration, as previously discussed. The EIS measurements shown in Fig. 3.9 were carried out in a 25 cm^2 commercially available fuel cell and test rig acquired from Baltic Fuel Cells as shown in Fig. 3.8. The membrane electrode assembly (MEA) consisted of a catalyst-coated membrane, DuPont Nafion-115 with a platinum loading of 0.4 mg/cm^2, and carbon black for the electrodes. The thickness of the

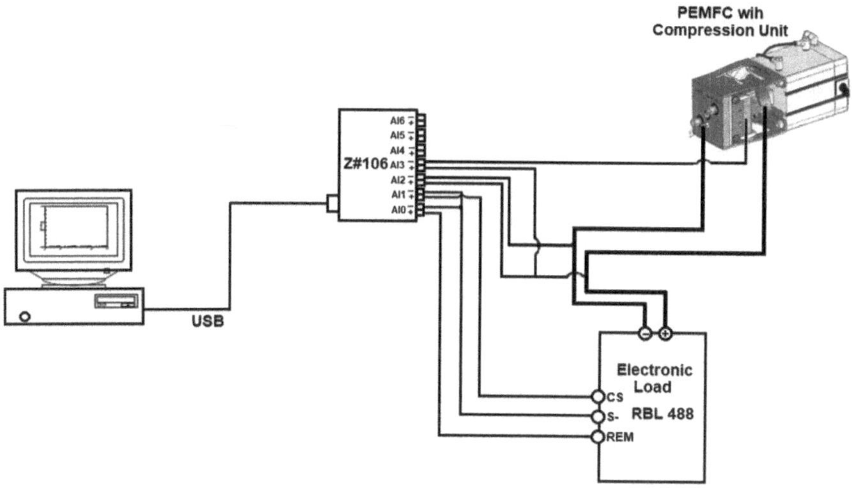

FIG. 3.8 Measurement of cathode impedance response and PEMFC impedance response using a multichannel FRA, Z#106 WonATech.

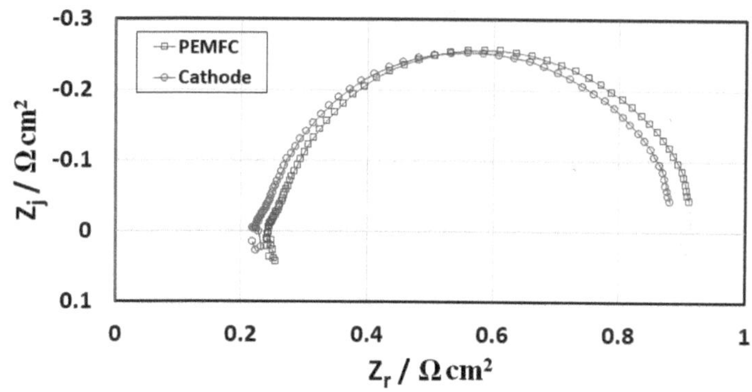

FIG. 3.9 Impedance spectra of PEMFC and the cathode measured with a multichannel FRA and a reference electrode.

catalyst layers are 12 μm. The contact pressure on the active cell area was adjusted through a pneumatic air cylinder from the Baltic Fuel Cell compression unit. The contact pressure on the active area was fixed to 1.4 N/mm^2. The PEMFC was operated with hydrogen at 100% RH in the anode; the air supplied in the cathode was dry. In terms of optimal performance of operational PEMFCs, the effect of RH in the reactant gases is not definite. The impedance measurements were carried out in the galvanostatic mode with 5% AC amplitude of the DC current at frequencies from 10 kHz to 0.2 Hz.

In Fig. 3.9, it can be observed that the impedance spectrum of the cathode is shifted towards the left or towards the lower end of the real axis Z_r with respect to

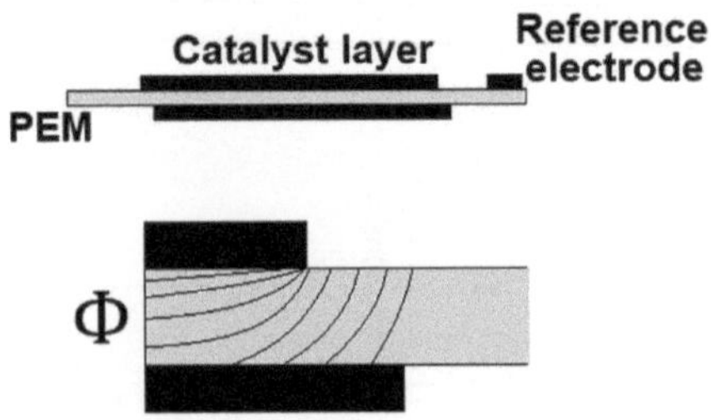

FIG. 3.10 Potential distribution Φ near the edges of the misaligned electrodes.

the PEMFC impedance spectrum. This difference between cathode and PEMFC spectra can be attributed to electrochemical mechanisms during the hydrogen oxidation reaction (HOR) in the anode. In addition, a higher ohmic resistance, which is represented in the real component value of the EIS measurements where the spectrum crosses the real axis Z_r of the Nyquist plot at high frequencies (transition from the negative to the positive imaginary components Z_j), is observed in the PEMFC spectrum. The ohmic resistance in the PEMFC impedance spectrum can be attributed to electro-osmotic drag of protons from the anode to the cathode, which could result in dehydration of the PEM near the anode catalyst layer interface.

It has been demonstrated that misalignment in the electrodes of the MEA can significantly shift the reference electrode signal [8]. The signal measured in a reference electrode can be shifted towards the electrode that protrudes whether the reference electrode is inserted in the anode or cathode side. Misalignment in the electrodes can lead to nonuniform potential distribution, which can affect the electrochemical measurements using the reference electrode. Fig. 3.10 shows that when the electrodes are misaligned, the potential distribution Φ through the PEM is affected; therefore, the potential measured by the reference electrode would be related to the potential of the electrode that protrudes. A study related to edge effects on reference electrodes in PEMFCs has been reported by He and Nguyen [8].

3.4 Inductive effect on the impedance spectrum

EIS measurements with positive imaginary components are related to inductive effects. In the high-frequency region, these measurements are a result of the inductance of the electrical cables of the measurement system. In some cases, EIS measurements with positive imaginary components at low frequencies are apparent in the impedance spectrum of the PEMFC. These EIS measurements in the low-frequency range are coined as inductive loops and have been related to electrochemical phenomena during PEMFC operation and not to artefacts during EIS measurements. Fig. 3.11 shows EIS measurements carried out in a H_2/air PEMFC featuring positive imaginary components at high and low frequencies.

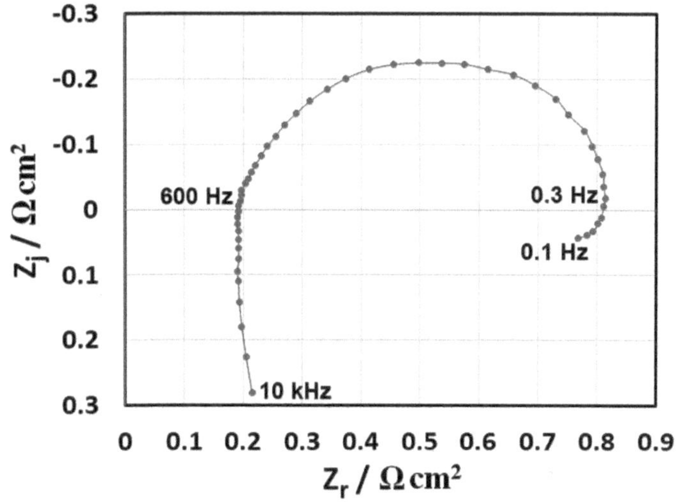

FIG. 3.11 EIS measurements carried out in a H_2/air PEMFC operated at 60°C and 0.45 V and featuring positive imaginary components at high and low frequencies.

The simulated effect of the inductive effect on the impedance spectrum of the PEMFC is shown in Fig. 3.12. The inductive effect is simulated through an impedance model of the PEMFC neglecting the anode contribution [9]. From Fig. 3.12, the electro-diffusion resistance of the PEMFC could be calculated by considering the end point of the impedance spectrum where the imaginary components are equal to zero $Z_j = 0$ at low frequencies. The total ohmic resistance of the PEMFC can be calculated by considering the end point of the impedance spectrum where the imaginary components are equal to zero

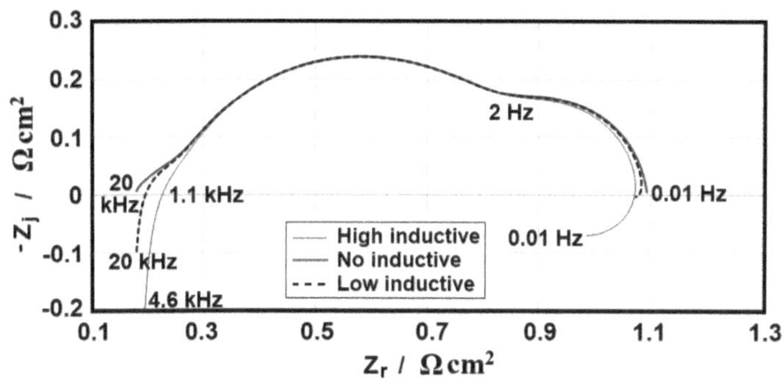

FIG. 3.12 Simulated inductive effect at high frequencies and the inductive loop at low frequencies on the impedance spectrum of a PEMFC. *(From S. Cruz-Manzo, Further understanding of uncertainties in the impedance spectrum of the polymer electrolyte fuel cell due to inductive effects and oxygen diffusion time constant, J. Energy Power Technol. 2 (2020) 9/21. Figure published under Creative Commons License (CC).)*

$Z_j = 0$ at high frequencies. However, Fig. 3.12 shows that the inductive effect at low and high frequencies on the impedance spectrum may yield an incorrect calculation of the electro-diffusion resistance and ohmic resistance considering the two end points of the impedance spectrum at $Z_j = 0$. The inductive effect at low and high frequencies deforms the end point of the impedance spectrum, as shown in Fig. 3.12.

3.4.1 Inductive effect of the measurement system

EIS measurements with positive imaginary components at the high-frequency end of the PEMFC spectrum have been attributed to the inductance of the electrical cables of the measurement system. Inductance is a property of an electric conductor generated by a magnetic field during the change in the current flowing. The effect of inductance is often seen at the highest frequencies of the impedance spectrum. In Chapter 1 Section 1.4.2, the impedance response of an inductor was discussed. It was demonstrated that the impedance of an inductor increases with increasing frequency. An erroneous interpretation on the impedance spectrum in the high-frequency region is attributed to a high inductive effect of the electrical cables during EIS measurements [10,11]. Fig. 3.13 shows the inductive effect of the electrical cables connected between the PEMFC and the load where the inductive effect is captured by the FRA during EIS measurements. Optimisation of the cables of the measurement system must be considered during EIS measurements to reduce the inductive effect, which can lead to an incorrect interpretation at high frequencies of the impedance complex plane [12–16].

Figs 3.14 and 3.15 show EIS measurements carried out in a single 25 cm^2 H_2/air PEMFC. The EIS measurements were carried out with an FRA-potentiostat, Solartron 1280C, as shown in Fig. 3.6. The range of

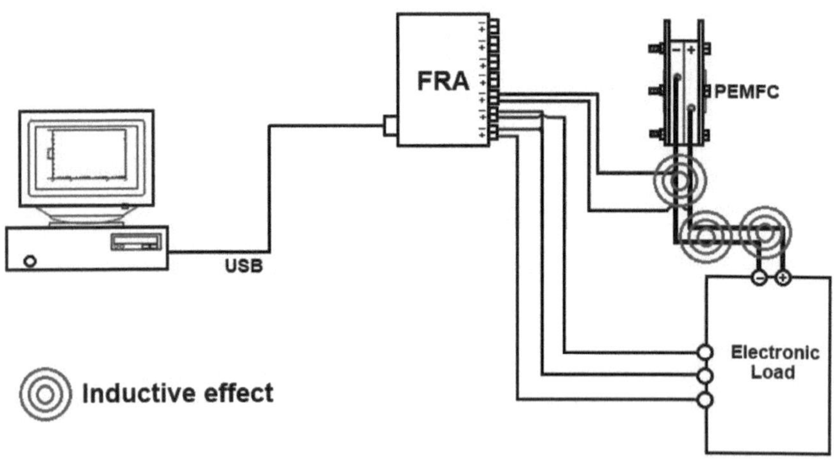

FIG. 3.13 Inductive effect from the EIS measurement system.

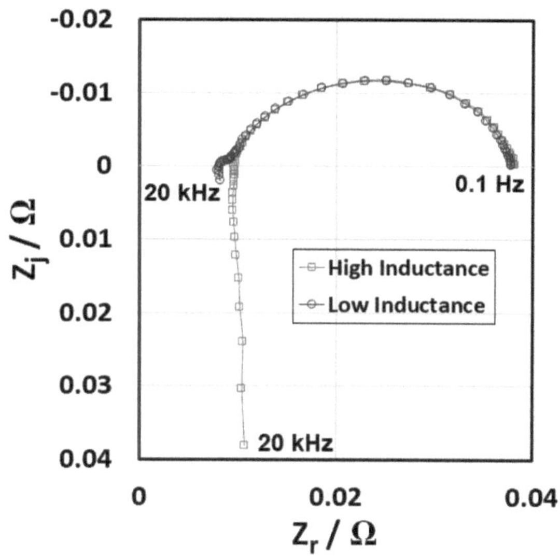

FIG. 3.14 Effect of inductance of the measurement system on the PEMFC impedance spectrum. The low inductance spectrum was measured by twisting the electrical cables of the measurement system.

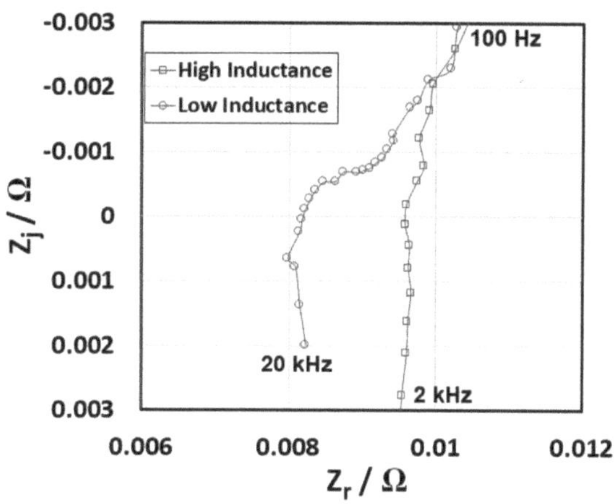

FIG. 3.15 Effect of inductance of the measurement system on the high-frequency region of the PEMFC impedance spectrum.

frequencies was from 20 kHz to 0.1 Hz. The amplitude of the sinusoidal signal was 10 mV under potentiostatic EIS measurements. The impedance spectrum labelled "Low Inductance" was measured by twisting the cables between the FRA-potentiostat and the PEMFC. Fig. 3.14 shows that by twisting the electrical cables of the measurement system, it is possible to reduce the inductance effect of the cables on the impedance spectrum.

The inductance effect of the cables of the measurement system on the high-frequency region of the impedance spectrum is observed in Fig. 3.15. The ohmic resistance of the PEMFC has been calculated from EIS measurements where the impedance spectrum crosses the real axis Z_r of the Nyquist plot or the imaginary component is equal to zero $Z_j=0$ at high frequencies [12,17,18]. Fig. 3.15 shows that this could lead to incorrect results in EIS measurements of PEMFCs under the influence of the high-frequency inductance. There is a clear discrepancy between the impedance spectra shown in Fig. 3.15, and this discrepancy is noticeable at frequencies higher than 100 Hz. The difference in impedance spectra at frequencies higher than 100 Hz demonstrates how the inductance effect of the cables can mask or hide electrochemical processes represented at high frequencies of the Nyquist plot.

A 45-degree straight line at high frequencies has also been revealed in cathodic EIS measurements through an impedance model that simulates a reduction of the inductive effect of the measurement system [19]. Fig. 3.16a shows cathodic EIS measurements carried out in a 25 cm^2 H_2/air PEMFC. The EIS measurements were carried out through a three-electrode configuration in the EIS measurement system. A low inductive cable connected between the load and the PEMFC resulted in the cathodic impedance spectrum with a low inductive effect. More information about the cathodic EIS measurements with high and low inductive effects shown in Fig. 3.16 can be found in the study of Cruz-Manzo et al. [19]. The high inductive effect shifts the value of the ohmic resistance (where the imaginary component is equal to zero $Z_j=0$) towards a higher value of the real component Z_r of the Nyquist plot as shown in Figs 3.15 and 3.16b. An impedance model representing the cathodic impedance response of the PEMFC allows the simulation of the impedance spectrum with a negligible inductive effect at high frequencies [9] as shown in Fig. 3.16b. The 45-degree straight line at high frequencies has been reported to represent the proton resistance in the cathode catalyst layer (CCL) with negligible electron resistance [14]. A further analysis about the features of the CCL impedance spectrum on the high-frequency region will be carried out in Chapter 4.

3.4.2 Inductive loops at low frequencies

EIS measurements with positive imaginary components at low frequencies of the Nyquist plot are coined inductive loops. Inductive loops in EIS measurements of PEMFCs have been attributed to side reactions with intermediate species [20,21], carbon monoxide poisoning [22], and water transport characteristics [23,24]. A literature review on inductive loops of EIS measurements in PEMFCs has been carried out by Pivac and Barbir [25]. Klots [26] demonstrated that electrical circuits considering an arrangement of resistors with an inductor or with a negative capacitor can reproduce low-frequency inductive loops in EIS measurements of electrochemical systems.

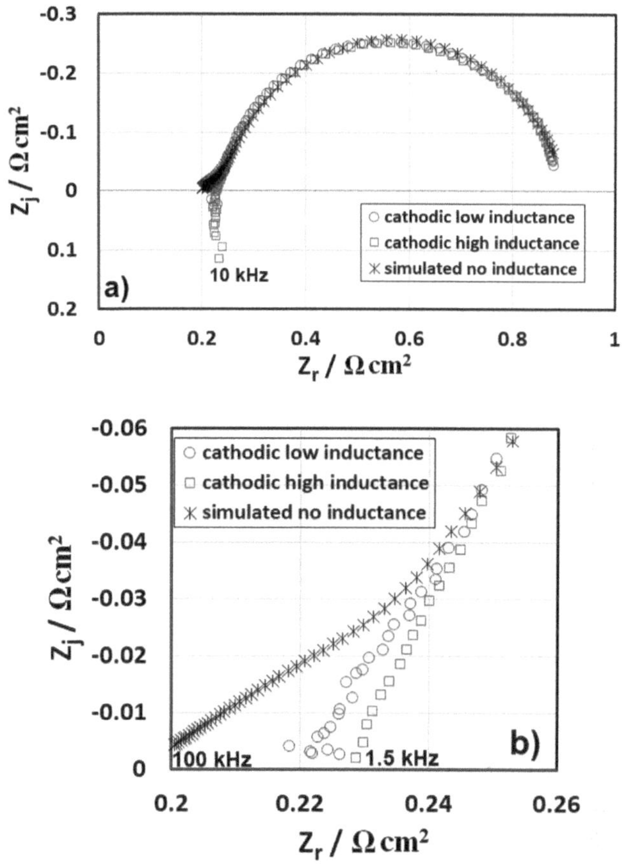

FIG. 3.16 Simulated reduction of the inductive effect on cathodic EIS measurements [9,19]: (a) full impedance spectrum and (b) high-frequency region.

Fig. 3.17 shows EIS measurements carried out in an open-cathode PEMFC stack operated at 0.47 A/cm^2. The open-cathode stack was operated in a through-flow mode on the hydrogen side. A fan attached to the stack provided air flow as the oxidant during the oxygen reduction reaction (ORR), cooling, and water removal. Simultaneous EIS measurements in the cells comprising the PEMFC stack were carried out using a multichannel FRA (Z#106 WonA-Tech Co). The difference in the EIS measurements can be attributed to different factors such as temperature gradient, hydrogen distribution, air distribution, and vapour–liquid water distribution in PEM, CCL, GDL, and FFP. No inductive effect of the measurement system is present in the impedance spectrum at high frequencies. The reduction of the inductive effect at high frequencies was achieved by connecting the sensing cables of the FRA directly to the FFPs to measure the impedance response of the cells. The value of the ohmic resistance

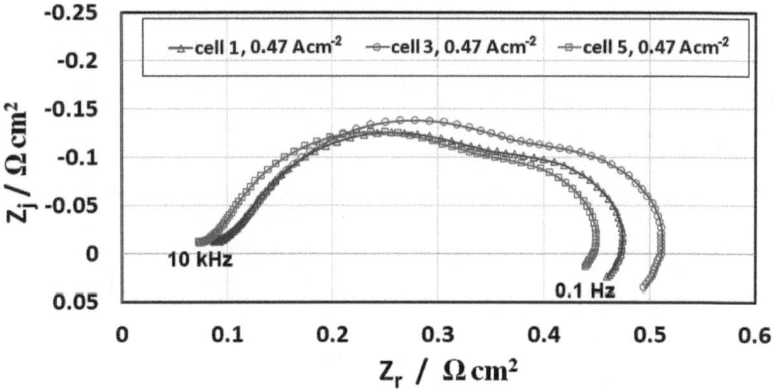

FIG. 3.17 EIS measurements featuring low-frequency inductive loops carried out in the individual cells of an open-cathode PEMFC stack at 0.47 A/cm^2 [27]. The stack is composed of 5 cells with an active area of 65 cm^2.

can be related to the value of the real component of the impedance spectrum at 10 kHz as shown in Fig. 3.17. More information about the EIS measurements of the open-cathode PEMFC stack shown in Fig. 3.17 can be found in the study of Cruz-Manzo and Greenwood [27].

The EIS measurements also shown in Fig. 3.17 featured inductive loops at low frequencies. The formation of the inductive loops could be attributed to mechanisms of water transport in the electrodes of the PEMFC, and the magnitude of the inductive loop could be related to the hydration state in the MEA. This can be demonstrated in the impedance spectrum of cell five which featured the lowest size of inductive loop and the lowest value of ohmic resistance compared to the rest of the cells of the open-cathode PEMFC stack. Wiezell et al. [28] developed an impedance model to demonstrate that low-frequency inductive loops in PEMFCs are attributed to rehydration of the membrane with water produced at the cathode. The inductive loop was apparent at a frequency proportional to $1/\ell_m^2$, where ℓ_m is the membrane thickness.

3.5 Effect of sinusoidal signal amplitude on EIS measurements

In the work reported by Yuan et al. [29], sinusoidal signals with amplitude values of 5%, 10%, and 15% of the DC current in a 500 W PEMFC stack were tested during galvanostatic EIS measurements. The results at a 50 A DC current showed a small expansion of the impedance spectra at low frequencies when the amplitude of the sinusidal signal was increased. Some works [30,31] have considered that a sinusoidal signal with an amplitude value of 5% of the DC current is a reasonable choice, but other works have considered an amplitude value of 10% of the DC current instead [32]. Fig. 3.18 shows EIS measurements carried out in a 25 cm^2 H_2/air PEMFC at 0.2 A/cm^2. The frequency range was from 5 kHz to 0.5 Hz. The EIS measurements were carried out in a galvanostatic

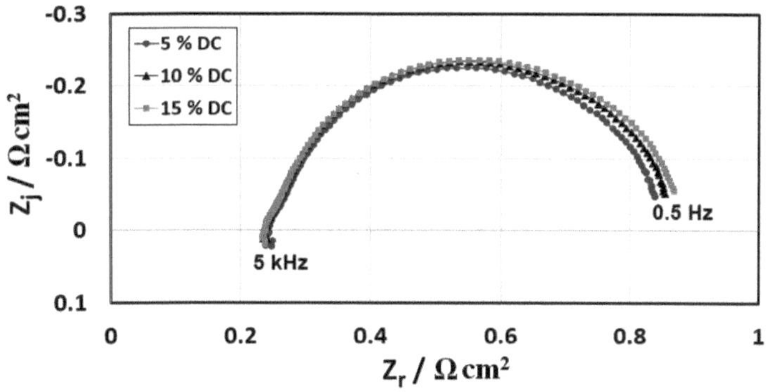

FIG. 3.18 Effect of the amplitude of the sinusoidal signal on the PEMFC impedance spectrum. EIS measurements carried out in a $25\,cm^2$ H_2/air PEMFC at $0.2\,A/cm^2$ with sinusoidal signals with amplitude values of 5%, 10%, and 15% of the DC current.

configuration with sinusoidal signals with amplitude values of 5%, 10%, and 15% of the DC current (5 A). Fig. 3.18 shows an expansion of the impedance spectrum with increasing amplitude of the sinusoidal signal during galvanostatic EIS measurements.

Fig. 3.19 shows EIS measurements carried out with two different sinusoidal perturbation amplitudes in a $16\,cm^2$ open-cathode four-cell PEMFC stack operated at 5 A. The frequency scan for the EIS measurements was carried out from 10 kHz to 0.1 Hz. The amplitudes selected of the sinusoidal signals were 5% and 15% of the DC current (5 A).

At high frequencies, there is no inductive effect from the electrical cables (EIS measurements with positive imaginary components), as shown in Fig. 3.19. The sensing cables from the FRA to carry out EIS measurements were connected directly to the FFPs of the PEMFC stack. The straight line at high frequencies is noticeable and relates to the proton resistance in the catalyst layer

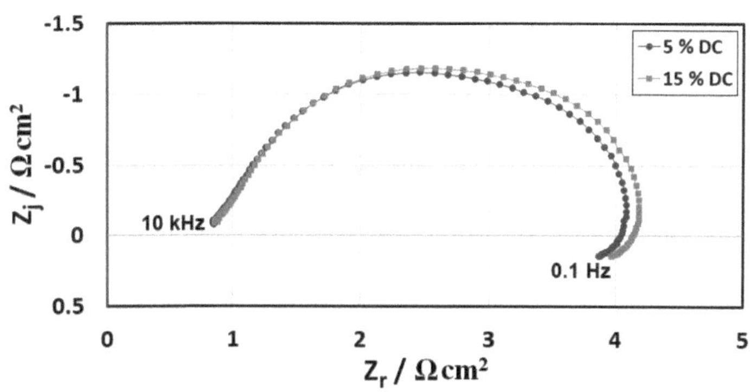

FIG. 3.19 EIS measurements carried out in a $16\,cm^2$ open-cathode four-cell PEMFC stack operated at $0.3125\,A/cm^2$ with sinusoidal signals with amplitude values of 5% and 15% of the DC current.

[14,33]. Fig. 3.19 shows an increase in the spectrum when the amplitude of the sinusoidal signal is increased. From Figs 3.18 and 3.19, the increase of the size of the impedance spectra with increasing amplitude of the sinusoidal perturbation during EIS measurements is consistent with the work reported by Yuan et al. [29]. The increase of the impedance spectrum with increasing sinusoidal signal amplitude can be attributed to a nonlinear effect in the EIS response of the PEMFC. A nonlinear EIS response measured with a high amplitude of the sinusoidal perturbation is not consistent with Kramers–Kronig (K–K) relations for linear systems [34].

3.6 Relation between polarisation resistance from EIS and the polarisation curve

A property from EIS measurements is that as the frequency is reduced and reaches a value of zero, the performance of the PEMFC must converge to steady-state conditions [35]. This means that a resistive value (no frequency dependence) for the impedance response of the PEMFC is expected as the frequency is reduced and reaches a value of zero. This resistive value is named polarisation resistance and is related to the slope of the polarisation curve [36] at a defined current density and voltage from the polarisation curve, as shown in Fig. 3.20. Nevertheless, it has been demonstrated that there is a discrepancy between the polarisation resistances from EIS measurements and from the polarisation curve when a PEMFC is operated at low oxygen stoichiometry [35]. The discrepancy is attributed to an increase in the amplitude of the

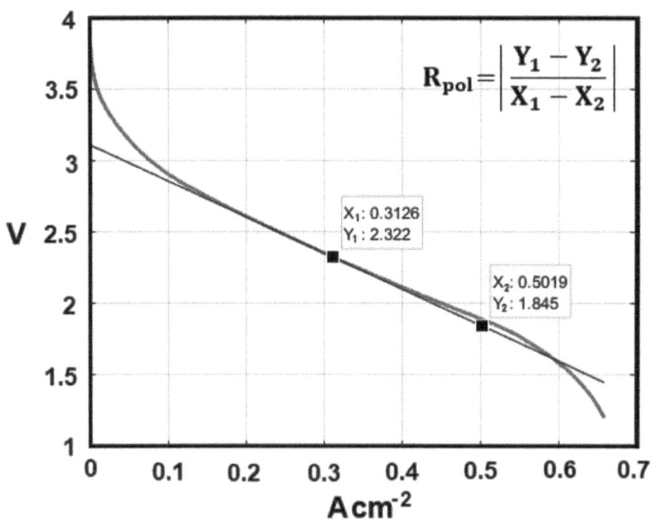

FIG. 3.20 Estimation of polarisation resistance R_{pol} from the polarisation curve of an open-cathode four-cell PEMFC stack at 0.3125 A/cm^2.

oscillating oxygen concentration along the gas channel during EIS measurements. The effect of oxygen transport on the gas channels and represented in EIS measurements will be analysed in Chapter 5.

EIS measurements are commonly carried out at the minimum frequency of 0.01 Hz. However, to validate EIS measurements with the DC polarisation resistance, which is calculated from the slope of the polarisation curve, it would be necessary to carry out EIS measurements at a lower frequency value, for example, 1×10^{-6} Hz, until the imaginary components of the EIS data become negligible with reducing frequency. It is not practical to carry out EIS measurements under low frequency values $<1 \times 10^{-4}$ Hz as it would take a very long time to complete the EIS test. The validation of the DC polarisation resistance becomes more complicated if EIS measurements with positive imaginary components Z_j are present at the lowest frequencies (inductive loops), as shown in Fig. 3.21. In this case, a physics-based impedance model derived from electrode theory would allow the simulation and extrapolation of the experimental EIS measurements at the lowest frequency to validate the DC polarisation resistance resulting from EIS with the one calculated from the polarisation curve. The DC polarisation resistance calculated from the polarisation curve shown in Fig. 3.20 corresponds to the DC polarisation resistance from the EIS measurements shown in Fig. 3.21. In addition, an impedance model would allow the characterisation of the electrochemical mechanisms represented in the EIS measurements with positive imaginary components Z_j (inductive loop) at low frequencies. Fig. 3.21 shows that two different impedance models [37,38] can simulate the low-frequency inductive loop until this converges with the DC polarisation resistance R_{pol}. Fig. 3.21 shows that there may be an ambiguity within the physical interpretation of low-frequency inductive loops in EIS

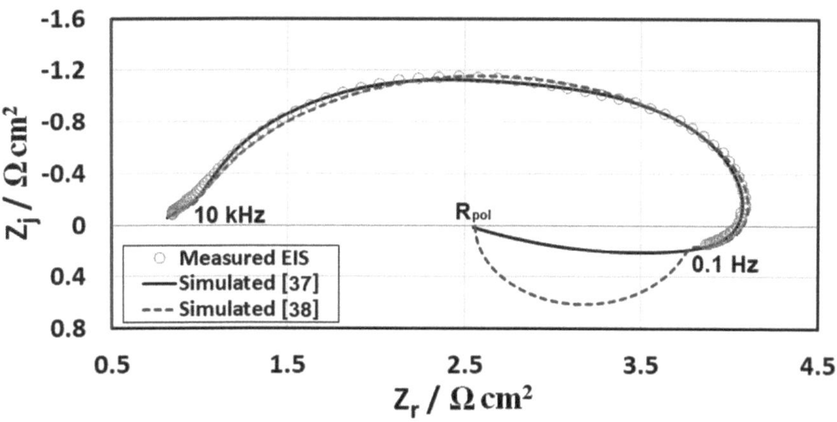

FIG. 3.21 EIS measurements of an open-cathode four-cell PEMFC stack featuring an inductive loop. Impedance models [37,38] to simulate the impedance spectrum and converge the inductive loop with the DC polarisation resistance R_{pol} at the lowest frequency.

measurements of PEMFCs as the relation between inductive loops and electrochemical processes in the PEMFC is still open to different interpretation. Chapter 7 will discuss the different electrochemical phenomena of the PEMFC that yield the formation of inductive loops in EIS measurements at low frequencies.

3.7 Local EIS measurements

Localised EIS measurements can provide information about heterogeneous physical properties in the electrodes. Localised EIS can also assist in the design of gas channels as the local conditions in a PEMFC can change because of manufacturing inconsistencies during cell design. However, the main challenge of the localised EIS technique remains in the redesign of a fuel cell to carry out spatially resolved measurements. The mathematical expression to represent the total impedance of a segmented cathode along the gas channel will be discussed in Chapter 5.

Localised EIS measurements along a single gas channel of a PEMFC were carried out for the first time by Brett et al. [39]. A single gas channel machined out of a printed circuit board that has 10 segmented current collectors running adjacent to and along the gas channel was used. The EIS results showed that the diameter of the impedance spectra increased with distance along the gas channel from the gas inlet. The increase in diameter was attributed to the oxygen diffusion limitation in the electrode diffuser. Then, reactant starvation occurs along the gas channel. The results reported by Brett et al. [39] show how conventional EIS measurements on the bulk fuel cell are not representative and not sufficient to properly characterise the operation of a PEMFC.

A spatially resolved in situ diagnostic method for PEMFCs using EIS measurements was carried out in the study reported by Schneider et al. [40]. This method pinpointed inhomogeneities in the current distribution. The combination of locally resolved current density measurements and local EIS allows the identification of local inhomogeneities in the performance of a PEMFC. A segmented PEMFC of 29.2 cm^2 was used for the tests in which the anode and cathode FFPs were segmented into 11 electrically isolated segments. A multichannel FRA for both data acquisition and frequency response analysis was built up. The analyser consists of ten voltage and ten current channels and can carry out impedance measurements for up to nine segments in the FFPs and the entire cell in parallel. The results show differences in the local cell impedance spectra. For low humidity and H_2/O_2 PEMFC operation, the locally resolved impedance spectra show that the performance loss along the gas channels can be attributed to drying effects. For H_2/air operation, an increasing size of the low-frequency loop on the local impedance spectra along the gas channels from the air inlet to the outlet is observed. This effect can be related to the cumulative flooding on the GDL and a decreasing molar fraction of oxygen in the cathode gas stream along the gas channels.

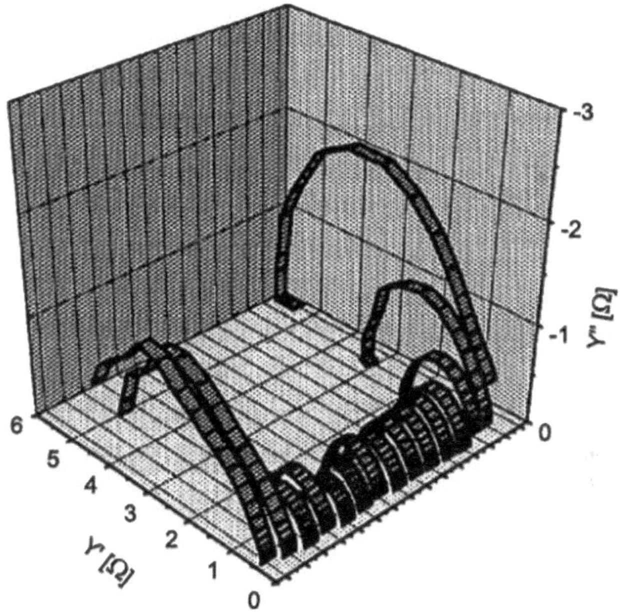

FIG. 3.22 Local impedance spectra measured in the middle row of the segmented straight-line channel PEMFC. The gas channel design consisted of 45 segments arranged in three rows of 15. *(Reproduced from A. Hakenjos, C. Hebling, Spatially resolved measurement of PEM fuel cells, J. Power Sources 145 (2005) 309, with permission of Elsevier. Copyright Elsevier.)*

Local EIS measurements in a PEMFC were also carried out in the study reported by Hakenjos and Hebling [41] as shown in Fig. 3.22. The geometry of the flow field channels of the PEMFC has an area of 29 cm^2. The air in the PEMFC is supplied with 80% RH and a flow rate of 300 sccm. The cathodic flow field temperature was maintained at 55°C. The EIS measurements were carried out at a cell voltage of 0.5 V. The segmented cell consisted of 45 graphite segments arranged in 3 rows of 15. Each segment is connected with a separate current line and voltage sensor. Fig. 3.22 shows EIS measurements carried out in the middle row of the segmented cell. The high-frequency resistance of the local EIS measurements decreases along the gas channel because of an increase in the humidity in the membrane.

3.8 EIS measurements in PEMFCs under different operating conditions

Different operating conditions have been reported in the literature to study different electrochemical phenomena during PEMFC operation using EIS [42]. In this section, the change in the size of the impedance spectrum under different PEMFC operating conditions is discussed. In Section 3.1, the impedance response of a PEMFC was represented at different frequencies in the Nyquist and Bode plots. The impedance response of the PEMFC represented in the complex-impedance (Nyquist) plot is mainly composed of two overlapped semicircles. The first semicircle at high-low frequencies represents the charge transfer during the hydrogen-oxidation and oxygen-reduction (redox) processes. The second semicircle at low frequencies is mainly attributed to oxygen transport limitations in the different layers comprising the cathodic side of the PEMFC. The size of the overlapped semicircles comprising the overall impedance spectrum of the PEMFC will change at different PEMFC temperatures, current density, stoichiometry of gas reactants, and RH of gas reactants.

3.8.1 Anodic impedance response

The kinetics of HOR at the anode is faster than the kinetics of the ORR at the cathode. In addition, the ORR is a more complicated reaction sequence than the HOR. The contribution of the impedance response of the anode on the PEMFC impedance spectrum is considered to be minimal and has been commonly neglected [13,43,44]. Andreaus et al. [45] reported that the anode impedance cannot be neglected in PEMFCs with thicker membranes and operated at high currents as the water content in the anode side of the membrane is reduced and an increase of losses in the anode is expected because of low proton transport in the active zone of the anode catalyst layer.

In Fig. 3.9, the impedance response of the cathodic side was separated from the overall impedance response of the PEMFC through a multichannel FRA and a reference electrode. Fig. 3.23 shows the difference between the cathode impedance spectrum and the PEMFC impedance spectrum depicted in Fig. 3.9. Fig. 3.23 shows that the contribution of the anodic impedance spectrum compared to that of the impedance spectra of the PEMFC and cathode shown in Fig. 3.9 is minimal. Springer et al. [46] reported a negligible contribution of the anode in EIS measurements of a PEMFC when measuring the cathode impedance relative to the anode or relative to a reference electrode placed on the side of the anode.

The impedance response of the anode can contribute to the impedance spectrum of the PEMFC under certain operating conditions, for example, impurities

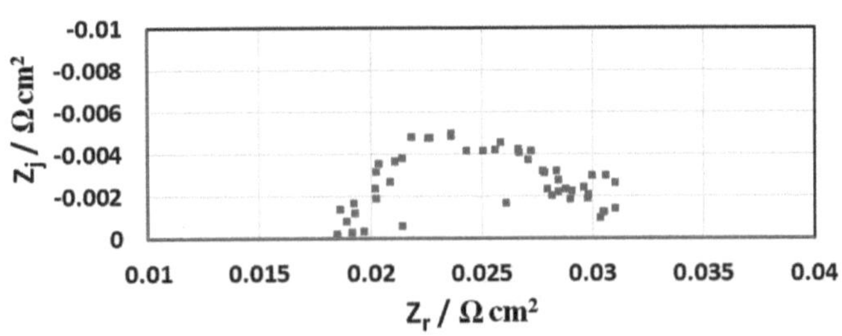

FIG. 3.23 Anode impedance response estimated by subtracting cathodic impedance response from PEMFC impedance response shown in Fig. 3.9.

in hydrogen. The poisoning effect of carbon monoxide, CO, at the anode has been studied through EIS measurements [47–49]. The anode electrode process can be studied by using a symmetrical gas-feeding mode, for example, H_2+CO/H_2, or by applying a reference electrode to the cell [7]. The effect of CO poisoning on the anode has been related to inductive loops at low frequencies in EIS measurements carried out in PEMFCs. The formation of the inductive loop at low frequencies has been attributed to the competitive oxidation between CO and hydrogen in the anode catalyst layer [22]. Fig. 3.24 shows EIS measurements carried out in a PEMFC operated with H_2/100 ppm CO in the anode and oxygen in the cathode and reported by Wagner and Gülzow [47]. The poisoning effect in the anode increases with time and becomes more dominant than the cathodic impedance response. Fig. 3.24 shows that the CO poisoning effect in the anode increases the overall impedance spectrum of the PEMFC and the inductive loop at low frequencies.

Wiezell et al. [28] reported and demonstrated through an impedance model that changes in water content in the PEMFC can yield a change in anode kinetics and the formation of inductive loops in EIS measurements at low frequencies.

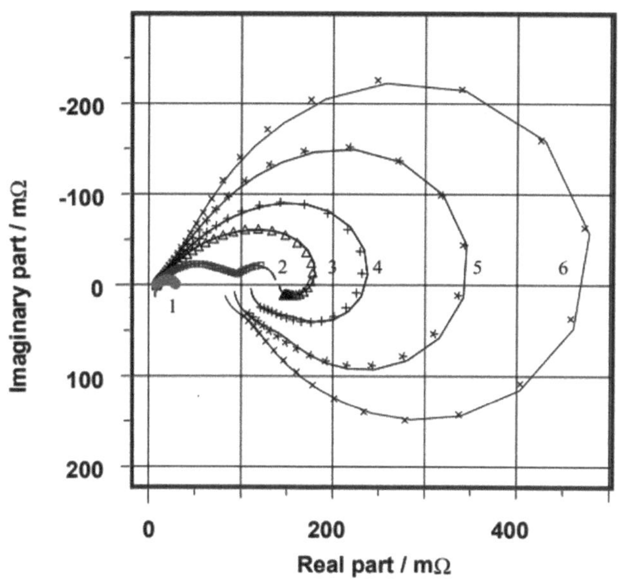

FIG. 3.24 Nyquist plot of the fuel cell impedance at different times; the influence of the carbon monoxide poisoning leads to an increase of both the real and imaginary parts of the impedance during the experiment; at the beginning (left part of the diagram), the anodic and cathodic impedances are separable (two half cycles in the Nyquist plots); with increasing time, the anodic cell impedance increases and becomes more and more dominant, 'hides' the cathodic cell impedance, and exhibits an enlarged pseudo-inductive behaviour. *(Reproduced from N. Wagner, E. Gülzow, Change of electrochemical impedance spectra (EIS) with time during CO-poisoning of the Pt-anode in a membrane fuel cell, J. Power Sources 127 (2004) 344, with permission of Elsevier. Copyright Elsevier.)*

Kim et al. [50] reported that losses in the anode are expected during long-term PEMFC operation (e.g. after 250h). Loss in the electrochemical surface area in the anode has been reported during start-up and shut-down events during PEMFC operation [51].

3.8.2 Current density

The semicircle at high-low frequencies in the impedance spectrum of the PEMFC and represented in the complex-impedance plot can mainly be related to the charge-transfer resistance during the ORR. The charge-transfer resistance during the HOR can have a major contribution on the impedance spectrum of the PEMFC under certain conditions (e.g. during start-up and shut-down events and long-term PEMFC operation, a degradation in the active surface area of the anode can result). The size of the semicircle at high-low frequencies attributed to the charge-transfer resistance during the ORR is reduced with increasing current density during PEMFC operation. The size of the semicircle related to oxygen transport limitations at low frequencies increases with increasing current density. Yan et al. [42] reported EIS measurements carried out in a 20 cell H_2/air PEMFC stack with an active area of 270cm^2. Fig. 3.25 shows the EIS measurements of the PEMFC stack at different currents. The impedance spectra shown in Fig. 3.25 are mainly composed of two overlapped semicircles that represent the charge-transfer resistance during the ORR and oxygen transport resistance in the GDL and CCL. The size of the impedance spectra initially decreases

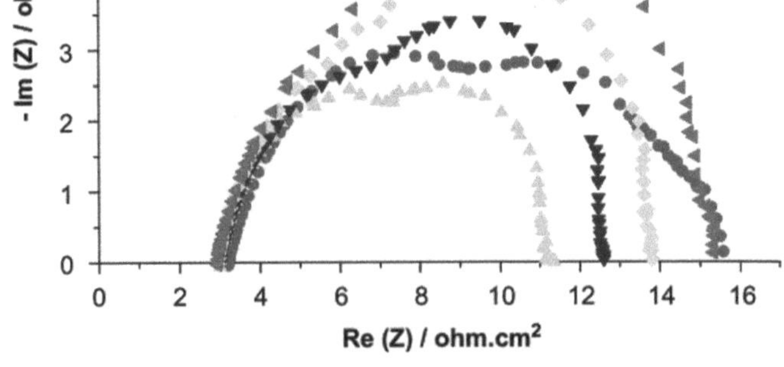

FIG. 3.25 EIS measurements carried out in a 2kW PEMFC stack at different currents. *(Reproduced from X. Yan, M. Hou, L. Sun, D. Liang, Q. Shen, H. Xu, P. Ming, B. Yi, AC impedance characteristics of a 2kW PEM fuel cell stack under different operating conditions and load changes, Int. J. Hydrogen Energy 32 (2007) 4360, with permission of Elsevier. Copyright Elsevier.)*

with increasing current from 54 to 108 A. The decrease of the spectrum at 108 A is attributed to an increase in the driving force for the ORR which reduces the charge-transfer resistance. The increase of the size of the spectra at higher currents is attributed to the increase in water generated during the ORR which hinders the transport of oxygen in the GDL towards the CCL.

Malevich et al. [52] reported EIS measurements in a H_2/air PEMFC with microporous layers (MPLs) attached to GDLs. The semicircle related to oxygen transport limitations at low frequencies increased with increasing current density whether or not a MPL is attached to the GDL. The size of the low-frequency semicircle decreased with the use of MPLs compared to GDLs without MPLs. The presence of the MPL reduced water saturation in the GDL and enhanced oxygen transport towards the CCL.

A single semicircle represented in the complex-impedance plot and representing the charge-transfer resistance during the ORR and oxygen transport limitations has been reported in EIS measurements of PEMFCs. Yuan et al. [53] reported EIS measurements carried out in a H_2/air PEMFC stack featuring a single semicircle with increasing current. Under these conditions, it is difficult to visualise in the Nyquist plot the processes related to the kinetics of the ORR and oxygen diffusion. Fig. 3.26 shows experimental cathodic EIS measurements carried out in a single PEMFC and featuring a single semicircle with increasing current density. The EIS measurements shown in Fig. 3.26 were carried out in a 25 cm^2 H_2/air PEMFC operated at 0.2 and 0.32 A/cm^2 and using a three-electrode configuration in the measurement system to separate cathodic EIS measurements from the total PEMFC-EIS measurements. Fig. 3.26 also shows that an impedance model allowed the separation of the semicircles related to the charge-transfer resistance during the ORR and oxygen transport limitations by increasing a parameter related to the oxygen diffusion time constant. Further

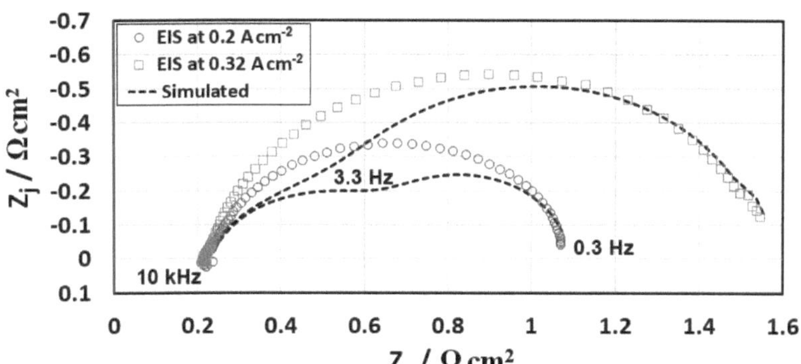

FIG. 3.26 EIS measurements in a H_2/air PEMFC featuring a single semicircle. An impedance model allowed the separation of charge transfer and diffusion processes during the oxygen-reduction reaction. Further information about the EIS measurements and impedance model can be found in the study of Cruz-Manzo et al. [54].

information about the EIS measurements and simulation of the impedance responses shown in Fig. 3.26 can be found in the study of Cruz-Manzo et al. [54]. The semicircle associated to the charge-transfer effect from 10 kHz to 3.3 Hz decreased with increasing current density. The semicircle associated with oxygen transport limitations from 3.3 to 0.3 Hz increased with increasing current density.

3.8.3 PEMFC temperature

The effect of PEMFC temperature on the impedance response of the PEMFC is shown in Fig. 3.27. The EIS measurements shown in Fig. 3.27 were carried out in a commercial (Electrochem, Inc.) 5 cm^2 PEMFC operated with H_2/O_2 and a dead-ended configuration at the anode and cathode (a stoichiometric ratio of 1). The temperature of the PEMFC was controlled through the station ElectroChem MTS-150. The EIS measurements were carried out using a Solartron 1287 electrochemical interface and a Solartron 1260 FRA with a potentiostatic configuration at a PEMFC voltage of 0.8 V. Strictly speaking, the performance of the PEMFC under a dead-ended configuration in the gas reactants will not be steady and will not be consistent with the frequency response of steady systems. The PEMFC-EIS measurements shown in Fig. 3.27 were carried out at a low current density. The steady-state condition in a PEMFC under a dead-ended configuration could only endure a limited frame of time during low current operation [55]. The diameter of the semicircle of the impedance response of the PEMFC shown in Fig. 3.27 is mainly attributed to the charge-transfer resistance during the ORR. The semicircle decreases with increasing temperature of the PEMFC because increasing the temperature enhances the kinetics of the ORR and reduces the charge-transfer resistance. With increasing temperature, the ohmic resistance of the PEMFC (EIS measurements with the imaginary component

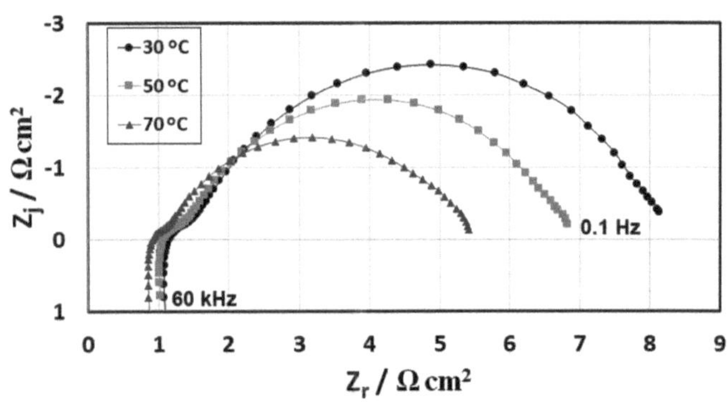

FIG. 3.27 Effect of temperature on the impedance response of the PEMFC. EIS measurements carried out in a 5 cm^2 H_2/O_2 PEMFC under the dead-ended configuration at the anode and cathode. EIS measurements provided by The National Institute of Electricity and Clean Energy (Instituto Nacional de Electricidad y Energías Limpias), Cuernavaca, Mexico.

equal to zero $Z_j = 0$) decreased as shown in Fig. 3.27. The temperature improved the proton conduction through the CCL and PEM and reduced the ohmic resistance of the PEMFC [56].

3.8.4 Oxygen stoichiometry

Rubio et al. [57] reported EIS measurements carried out in a 100 cm^2 H_2/air PEMFC. The PEMFC temperature was not controlled, and the atmospheric pressure of hydrogen and air was considered. The air was supplied at three different stoichiometries of 2.78, 2.3, and 1.46 at a current of 12.5 A. Fig. 3.28 shows the EIS measurements of the PEMFC reported by Rubio et al. [57] at three different oxygen stoichiometries. The semicircle at high-low frequencies related to the charge-transfer resistance increases with the decrease of the oxygen stoichiometry. In addition, the semicircle at low frequencies attributed to oxygen transport limitations in the GDL and CCL increases with decreasing oxygen stoichiometry. Similar results have been reported by Yan et al. [42]; the increase of the low-frequency semicircle with decreasing oxygen stoichiometry was attributed to an increase in water accumulation which hinders the transport of oxygen in the GDL and towards the active sites in the CCL. However, no reason was provided by Yan et al. [42] about the change of the first semicircle attributed to the charge-transfer resistance with changing oxygen stoichiometry. According to Weiß et al. [58], a change in oxygen stoichiometry causes a change in partial pressure of oxygen and hence a change in exchange current density and a change in the Butler–Volmer curve. The change in exchange

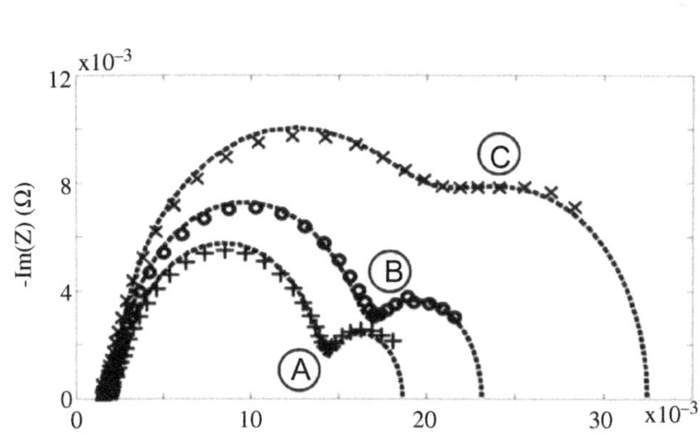

FIG. 3.28 Effect of oxygen stoichiometry on the impedance response of the PEMFC. The oxygen stoichiometry decreases from A to C. A is the highest stoichiometry (2.78), and C is the lowest stoichiometry (1.46). *(Reproduced from M.A. Rubio, A. Urquia, R. Kuhn, S. Dormido, Electrochemical parameter estimation in operating proton exchange membrane fuel cells, J. Power Sources 183 (2008) 123, with permission of Elsevier. Copyright Elsevier.)*

current density with changing oxygen stoichiometry during EIS could yield a change in the semicircle at high-low frequencies attributed to the charge-transfer resistance during the ORR.

3.8.5 Relative humidity

External gas humidification is desirable to ensure apposite humidification of the ionomer in the catalyst layer and the PEM. A high RH in the reactant gases enhances the proton conductivity in the PEM and improves catalyst utilisation and overall performance. Nevertheless, a high RH can result in pore saturation in the GDL and catalyst layer, which will impede reactant transport. In terms of optimal performance of operational PEMFCs, the effect of RH in the reactant gases is not definite. For instance, Yan et al. [59] reported that optimal performance in a PEMFC can occur at low air relative humidity supplied at the cathode and high hydrogen relative humidity supplied at the anode. Fig. 3.29 shows EIS measurements carried out in a 6 cm^2 PEMFC operated at a current density of 500 mA/cm^2, a temperature of 80°C, 20 psig anode and cathode inlet pressures, and 20% O_2/80% N_2 (nitrox) as a gas reactant on the cathode side. The RH for hydrogen at the anode was 100%. The RHs for nitrox in the cathode were 100% and 30%. The impedance spectrum considering nitrox at 30% RH is shifted towards the right or towards the higher end of the real component Z_r of the complex-impedance plot, as shown in Fig. 3.29. Therefore, a higher ohmic resistance is present in the impedance spectrum considering nitrox at 30% RH than the impedance spectrum considering nitrox at 100% RH. High RH increases the water concentration in the PEMFC and enhances the conductivity of protons through the PEM and catalyst layers. An increase in ohmic resistance represented in EIS measurements when the PEMFC is operated at high RH in the anode and low RH in the cathode has been reported in some studies [60–62]. This effect can be related to a higher effect of the water back diffusion through

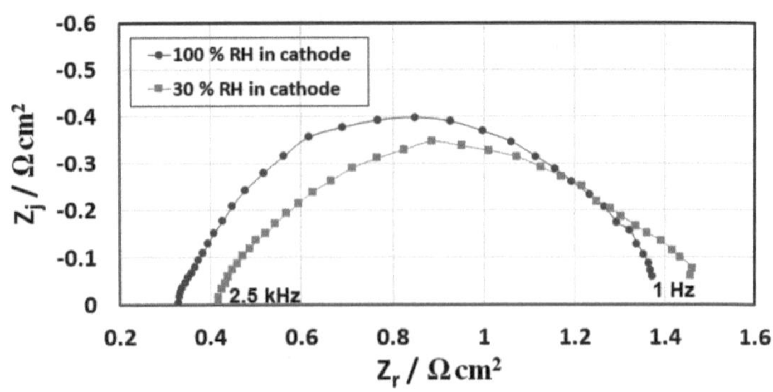

FIG. 3.29 EIS measurements carried out in a H_2/nitrox PEMFC with a relative humidity (RH) of 100% at the anode and 100/30% RH at the cathode.

the electrolyte compared to the electro-osmotic drag effect of protons [60]. The impedance spectrum considering nitrox at 30% RH also presents a depressed semicircle compared to the impedance spectrum considering nitrox at 100% RH. A depressed semicircle can be attributed to a nonhomogeneous charge distribution in the charge double-layer of the CCL. Increasing RH results in a more homogeneous charge distribution at the interface between dissimilar materials in the CCL, for example, Nafion/carbon [63]. Mocoteguy et al. [60] demonstrated through EIS measurements carried out in a PEMFC and represented in the Bode plot that the frequency at which the phase angle reaches its minimum value decreases with increasing RH in the reactant gases.

3.8.6 Nitrox or heliox as a gas reactant

For practical applications, air (21% O_2, 79% N_2) is used as a reactant gas on the cathode side during PEMFC operation. EIS measurements carried out in PEMFCs operated with oxygen diluted with helium (heliox) can be compared with EIS measurements carried out in PEMFCs operated with air. This experimental comparison allows the study of the mass transport effect represented in the low-frequency region of the impedance spectrum of the PEMFC. Fig. 3.30 shows EIS measurements carried out in a 6 cm^2 PEMFC operated at a current density of 500 mA/cm^2, a temperature of 80°C, 20 psig anode and cathode inlet pressures, 100% RH for hydrogen in the anode, and 30% RH for the cathode. The PEMFC was operated with oxygen diluted with helium (heliox with 40% O_2/60% He) at the cathode, held for a few minutes to achieve a steady-state condition, and then switched to oxygen diluted with nitrogen (nitrox with 40% O_2/60% N_2). EIS measurements with a range of frequencies from 65 kHz to 1 Hz in a two-electrode configuration were measured for each experimental condition, as shown in Fig. 3.30.

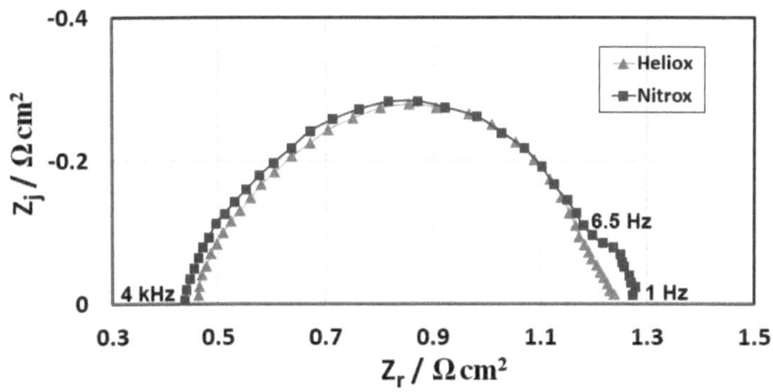

FIG. 3.30 EIS measurements in a PEMFC with O_2 diluted in He (heliox 40% O_2/60% He) and N_2 (nitrox 40% O_2/60% N_2). The PEMFC was operated with helium, held for a few minutes to achieve a steady-state condition, and then switched to nitrox.

The oxygen concentration in the CCL may be considered similar for the two cases (heliox and nitrox). Thus, concentration-related effects, for example, the variation in charge-transfer resistance during the ORR because of different concentrations, are expected to be minimal. This can be demonstrated in Fig. 3.30, in which both semicircles present similar characteristics from frequencies of 4 kHz to 6.5 Hz. A difference in the spectra for frequencies between 6.5 and 1 Hz is noticeable. This difference can be attributed to diffusion mechanisms of O_2 diluted in N_2 and He.

The diffusivity of O_2 diluted in N_2 at 80°C has been reported [64] to be 2.8×10^{-5} m^2/s, and the diffusion coefficient of water vapour in N_2 is 3.9×10^{-5} m^2/s under the same conditions; the oxygen diffusivity in He is 1.13×10^{-4} m^2/s, and the diffusion coefficient of water vapour in He is 3.14×10^{-4} m^2/s. Gas-phase diffusion of oxygen in heliox (O_2/He) is considerably faster than in nitrox (O_2/N_2) because of the difference in diffusion coefficients of oxygen diluted in nitrogen and in helium. Therefore, when the background gas in the cathode is switched to nitrogen from helium, a difference in impedance measurements is expected, as shown in Fig. 3.30. The small overlapped semicircle presented from 6.5 to 1 Hz for the nitrox case may be attributed to processes related primarily to low diffusivity of oxygen through the GDL and CCL. EIS measurements carried out in H_2/air PEMFCs and featuring a single semicircle in the negative complex half-plane of the Nyquist plot have been reported [29]. Under these conditions, it is difficult to visualise at which frequency the contribution of oxygen diffusion processes on EIS measurements is apparent. In addition, it is not straightforward to determine the role of the effective diffusion coefficient of oxygen in the CCL and GDL of the EIS measurements shown in Fig. 3.30 because this parameter is related to the different modes of transport through the multiphase material in the CCL and diffusion through the porous media in the GDL. An impedance model developed from electrode diffusion theory could help to estimate the diffusion mechanisms in the CCL and GDL that are represented in the Nyquist plot at low frequencies.

3.9 Errors during the interpretation of EIS measurements

Boillat et al. [65] reported that two errors during the interpretation of PEMFC-EIS measurements are commonly found in the literature. The first error is attributed to the relation between the charge-transfer resistance and activation losses during the ORR in a PEMFC. The charge-transfer resistance is commonly calculated from the linearisation of the Butler–Volmer equation around the operation point where EIS measurements are carried out. Fig. 3.31a shows that there is a relation between the charge-transfer resistance and the charge-transfer overpotential near equilibrium conditions (a low charge-transfer overpotential). Boillat et al. [65] concluded that these two parameters become uncorrelated in the region where the Tafel approximation of the Butler–Volmer equation is valid. Therefore, the charge-transfer resistance estimated from EIS

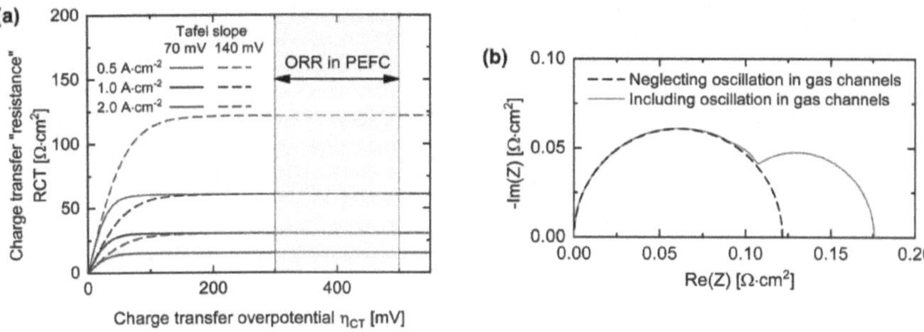

FIG. 3.31 (a) charge-transfer resistance calculated from the Butler–Volmer equation for different current densities and Tafel slopes, (b) impact of oxygen concentration oscillation in the gas channels on the cell impedance in the absence of diffusion transport losses. *(Reproduced from P. Boillat, F.N. Büchi, L. Gubler, T.J. Schmidt, Use and misuse of electrochemical impedance spectroscopy (EIS) in fuel cell research, in: ECS Meeting Abstracts, MA2019-02, 2019, p. 1435, with permission of IOP Publishing. © The Electrochemical Society. All rights reserved.)*

and considering an operating point away from equilibrium conditions cannot be related in terms of ORR activity.

The second error during the interpretation of PEMFC-EIS measurements is the omission of the effect of the oscillation of oxygen concentration in the gas channels on the impedance spectrum. Fig. 3.31b shows the effect of the oscillating oxygen concentration in the gas channels on the impedance spectrum. The oscillating oxygen concentration in the gas channels can be represented as a low-frequency loop in the impedance spectrum of PEMFCs. However, the low-frequency loop in the impedance spectrum of PEMFCs is commonly related to oxygen diffusion in the GDL, and the effect of oscillating oxygen concentration in the gas channels on the impedance spectrum is commonly neglected.

3.10 Electrochemical pressure impedance spectroscopy

The conventional use of the EIS technique considers the superimposition of a sinusoidal perturbation on the steady-state current (galvanostatic test) and the measurement of the sinusoidal voltage or superimposition of a sinusoidal perturbation on the steady-state voltage (potentiostatic test) and the measurement of the sinusoidal current at different frequencies. The characterisation of mass transport processes in the impedance response of a PEMFC through the use of the conventional EIS test can result in a challenging task because different electrochemical mechanisms during PEMFC operation can be overlapped in the low-frequency region of the impedance response of the PEMFC.

A new impedance test approach coined as electrochemical pressure impedance spectroscopy (EPIS) has been reported to investigate transport processes of the PEMFC represented in the low-frequency regimen [66–68]. During an EPIS test, a sinusoidal perturbation is superimposed on the back pressure of the cathode outlet and the fluctuations of the voltage output of the cell are measured. Increasing the back pressure can enhance the performance of the PEMFC, but it can have a negative effect on the PEMFC system, such as increasing parasitic power loss and sealing problems [69]. The transfer function describing the relation between the voltage variation and cathodic outlet pressure fluctuation of the PEMFC at different frequencies defines the EPIS as such:

$$Z_{\mathrm{EPIS}} = \frac{\widetilde{V}}{\widetilde{p}_{\mathrm{out}}} \tag{3.2}$$

EPIS can study mass transport processes at low frequencies in which it is difficult with the use of the conventional EIS. Fig. 3.32 shows a comparison between measurements carried out in a PEMFC using the conventional EIS and EPIS. Fig. 3.32d shows that conventional EIS measurements in the frequency range from 1 to 0.1 Hz can provide limited information related to diffusion processes of the PEMFC. EPIS measurements as shown in Fig. 3.32b demonstrate that relevant information related to diffusion processes of the PEMFC can be obtained in the low-frequency range of 1–0.001 Hz.

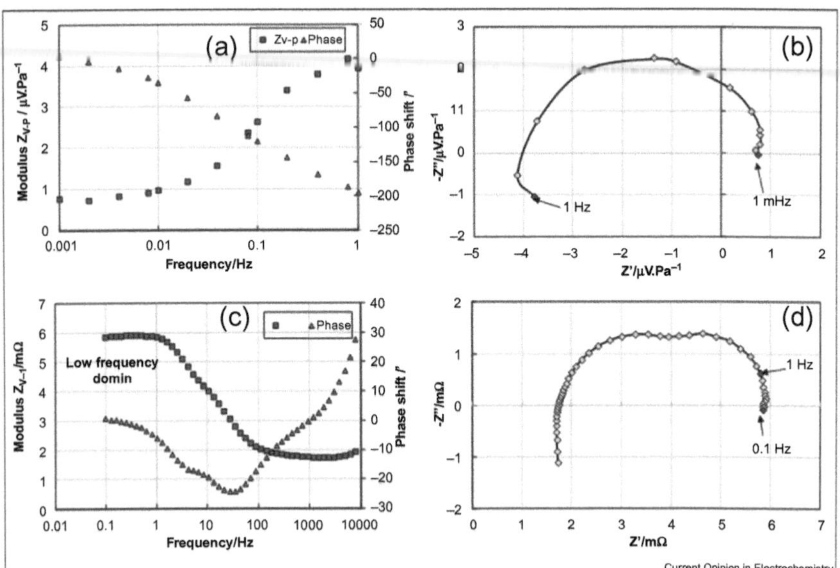

FIG. 3.32 Comparison between EIS and EPIS measurements in a PEMFC operated at 0.4 A/cm^2, with air humidified at 55%; T (fuel cell) = 55°C. (a) Bode plot; (b) Nyquist plot, with EPIS measurements; (c) Bode plot; (d) Nyquist plot with EIS measurements. The *solid lines* in Nyquist plots (b) and (d) are only for illustrative purposes and not related to modelling. *(Reproduced from Shirsath, et al., Electrochemical pressure impedance spectroscopy for investigation of mass transfer in polymer electrolyte membrane fuel cells, Curr. Opin. Electrochem. 20 (2020) 84, with permission of Elsevier. Copyright Elsevier.)*

Shirsath et al. [67] reported that there are operating constrains during the application of EPIS. In particular, some criteria related to the pressure modulation amplitude have to be complied with in order to find the optimal back-pressure amplitude for the application of EPIS on the PEMFC system. Physics-based modelling can assist the interpretation of the EPIS results and provide an insight into the transport mechanisms of the PEMFC. A physics-based model using the in-house multiphysics code DENIS (Detailed Electrochemistry and Numerical Impedance Simulation) has been reported by Grübl et al. [70] for the interpretation of EPIS. Schiffer et al. [71] reported a 1D + 1D continuum model of a fuel/air channel pair with GDL and MEA to demonstrate that features of the EPIS spectrum can be attributed to different transport mechanisms in the layers comprising the PEMFC. A recent study from Kulikovsky [72] demonstrated that an analytical model for concentration/pressure impedance can simulate the experimental EPIS spectrum reported by Engebretsen et al. [68].

Sorrentino et al. [73] studied mass transport dynamics in PEMFCs using a technique coined as concentration-alternating frequency response analysis (cFRA). The experimental technique measures the response of the PEMFC to partial pressure perturbations of oxygen and water vapour. The partial pressure perturbations in the cFRA reported by Sorrentino et al. [73] are carried out in the main stream of the cathode unlike the EPIS technique previously mentioned in which the pressure fluctuations are carried out in the cathode outlet stream. A recent study from Sorrentino et al. [74] reported that analysis of the experimental spectra from the cFRA technique using a mathematical model allows the study and decoupling of the dynamics of oxygen transport in the cathode channel and water sorption in Nafion.

Other methodologies such as electrothermal impedance spectroscopy (ETIS) [75] and hydro-electrochemical impedance imaging (HECII) [76] based on the fluctuation of nonelectrical variables have been considered to decouple different processes during PEMFC operation. For instance, mass transport, heat transport, and liquid water dynamics can be decoupled from other PEMFC processes using cFRA, ETIS, and HECII [77].

3.11 Electrical circuits for EIS analysis

The EIS measurements shown previously in Section 3.8 demonstrate that it is possible to conduct a qualitative assessment of the PEMFC operated under different conditions through a visual inspection of the impedance spectrum represented in the complex-impedance plot. However, this approach does not give a clear understanding of the electrochemical mechanisms represented in the frequency domain. For instance, it is not possible to obtain an insight into the limitations of the oxygen transport across the different layers comprising the PEMFC which are present at low frequencies of the EIS spectrum.

The measured results from EIS are widely interpreted through various EECs, but the information provided by the EECs on the electrochemical processes of the PEMFC studied can be limited. The problem when choosing an EEC is that it must suit the bulk EIS results of the PEMFC, but the impedance response of the PEMFC may need to be analysed with an EEC with very different parameters or may require a different EEC.

One of the challenges in the EIS technique is to separate the oxygen transport limitations between the CCL and the GDL. EECs with Warburg components have been broadly used in PEMFC-EIS measurements to estimate the oxygen transport limitations. However, it has not been clear how to relate the effect of oxygen transport limitations in the CCL and GDL from EIS measurements and using EECs. For a higher water content in the CCL which could subsequently reduce the water content in the GDL with MPL, oxygen transport limitations in the GDL would be reduced and vice versa, where a high water content in the GDL with MPL could reduce the water content and oxygen transport limitations in the CCL. Low-frequency PEMFC-EIS measurements with negative imaginary components have been commonly attributed to oxygen transport limitations in the GDL, neglecting any oxygen transport effect in the CCL and have been modelled with the Warburg component.

Different arrangements of EECs from simple components [12,13,36,47,52] to more complicated transmission-line representations [14,18,78–81] have been reported in the literature to model and characterise different phenomena in a PEMFC. Each electrical component in the EEC can represent a specific physical process occurring in a PEMFC. EECs have been applied within EIS measurements to estimate the charge-transfer resistance and oxygen transport resistance under different PEMFC operating conditions. However, the circuits are limited to a certain range of operating conditions, and a simple circuit cannot characterise all the electrochemical phenomena of the PEMFC represented in EIS measurements. The EECs are commonly fitted to the impedance response of the PEMFC through a nonlinear regression strategy, for example, Levenberg–Marquardt algorithm. The fitting process is commonly carried out through commercial software, for example, ZView (Scribner Associates Inc) and MATLAB. The resulting values of the electrical components comprising the EECs that are associated with the electrochemical phenomena of the PEMFC are represented at different frequencies of the EIS measurements.

3.11.1 Randles circuit

The most common electrical circuit configuration to characterise the PEMFC is based on the Randles circuit [82] as shown in Fig. 3.33. This electrical circuit consists of an ohmic resistance accounting for the opposition to the flow of protons and electrons in the catalyst layer, PEM, GDL, and FFP. The charge-transfer resistances during the HOR and ORR are connected in parallel with a capacitor for the double-layer capacitance for the interface between dissimilar

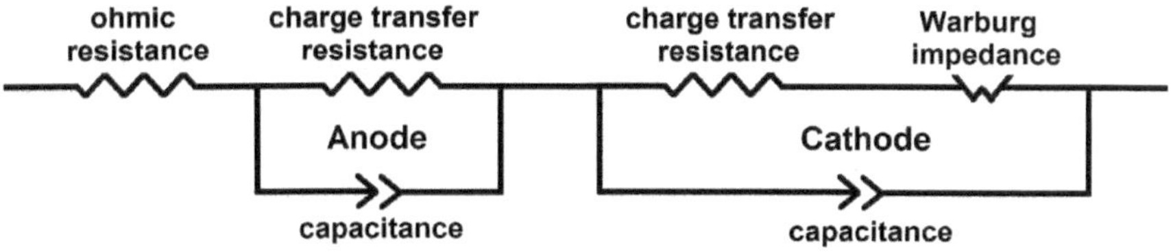

FIG. 3.33 Randles electrical circuit with the Warburg element.

materials in the electrode/electrolyte interface. A CPE is commonly considered instead of a capacitor to account for the nonhomogeneous charge distribution between the electrode/electrolyte interface. The impedance response of a capacitor and CPE has already been demonstrated in Chapter 1 Section 1.4.5. To include the oxygen transport effect in the Randles circuit, a Warburg impedance which models the finite diffusion of oxygen in the cathode is connected in series with the charge-transfer resistance of the ORR. The EEC shown in Fig. 3.33 was also considered in the study reported by Malevich et al. [52] to study the effect of MPLs in GDLs on the impedance response of a PEMFC. The authors reported EIS measurements with three semicircles, and the semicircle at the highest frequencies in the complex-impedance plot was modelled through the resistor/CPE configuration representing the anode impedance as shown in Fig. 3.33. The Warburg impedance was considered to quantify the oxygen transport resistance in the GDL with and without MPLs. The results reported by Malevich et al. [52] show that the charge-transfer resistance in the cathode decreases with increasing current density, whereas the mass transport resistance increases with increasing current density for cells with and without MPL. The charge-transfer resistance in the cathode for cells with an MPL is lower than that for cells without MPL. Cells with MPL demonstrate lower mass transport resistance compared with cells without MPL. In the study of Malevich et al. [52], the mass transport resistance is most likely because of the gas transport in the GDL. The presence of an MPL helps to reduce the water saturation in the GDL, thereby improving the oxygen transport to the CCL.

The electrical components representing the anode impedance have been neglected in the Randles circuit to characterise the impedance response of the PEMFC and diagnosis of PEMFC operation [13,57]. As discussed in Section 3.8.1, a contribution of the anode impedance on the overall PEMFC impedance response under certain operating conditions such as long-term PEMFC operation and start-up and shut-down events which increase the degradation of the anode catalyst layer is expected. The impedance response of the Randles circuit neglecting the anodic contribution shown in Fig. 3.33 was presented in Chapter 1 Section 1.4.6. Fouquet et al. [13] monitored the flooding and drying out of a PEMFC stack using the Randles circuit with EIS measurements, but neglected the anode contribution in the PEMFC and neglected the electrical components representing the anode impedance in the Randles circuit shown in

Fig. 3.33. Parameters of the Randles circuit were fitted to the EIS data and the EIS measurements at nominal conditions showed a full depressed semicircle at high-low frequencies, followed by a smaller semicircle at low frequencies. In the case of a flooded PEMFC, both real and imaginary components of the impedance response increase, and the two semicircles are overlapped in the complex-impedance plot. With a dry PEMFC, the whole spectrum is shifted towards the higher end of the real component of the complex-impedance plot. The process of flooding is characterised by the modification of the low-frequency part of the PEMFC stack's impedance which, in turn, is associated with the diffusion process through the GDL. In the case of a drying PEM, the impedance spectrum is progressively shifted towards the higher end of the real component of the complex-impedance plot, whereas the low-frequency semicircle increases to a comparable size as the high-low frequency semicircle. The variations of the Randles parameters while the PEM is drying are less dramatic than in the case of flooding. Robust and reliable PEMFC's state of hydration monitoring was demonstrated in the study of Fouquet et al. [13] using the parameters defining the Randles circuit.

EIS measurements have limitations and present disadvantages as the low impedance values are obscured for low frequencies and some effects are not visible because of a masking effect in the complex-impedance plot. Yuan et al. [53] attempted to separate both processes, charge transfer and diffusional effects during the ORR, in the impedance spectra of a H_2/air PEMFC stack operated at high current densities and featuring a single semicircle through an electrical circuit. The results showed that the EEC model cannot account for the exact electrode processes at high current densities because the charge-transfer resistance of the ORR increased with increasing current density.

EECs combined with EIS measurements can deliver information about the opposition for the charge transfer during the ORR and the opposition for the proton and electron to conduct through the different layers comprising the PEMFC. However, the interpretation of frequency-dependent phenomena related to oxygen transport limitations in PEMFCs and using the Randles circuit shown in Fig. 3.33 is still a subject to debate as this has been principally attributed to the gaseous oxygen transport limitations in the GDL neglecting any oxygen transport limitations in the CCL [83–85]. However, it is suspected that the CCL becomes flooded before the GDL because water is generated in the CCL and transported into the GDL, where the CCL has lower porosity and a smaller pore size and tends to have higher flooding levels than the GDL [86].

A modification of the Randles circuit to study the effect of carbon monoxide, CO, on the anode when carrying out EIS measurements in a PEMFC operated with H_2+CO/H_2 in the anode and cathode, respectively, is presented in the study of Jiang et al. [87]. This EEC includes a resistance and CPE $R_{ad}-CPE_{ad}$ accounting for H_2 adsorption (ad) connected in series with an $R_{ox}-CPE_{ox}$ to account for lowering of the H_2 adsorption because of CO oxidation (ox) and also connected in series with the charge-transfer resistance R_{CT} of the

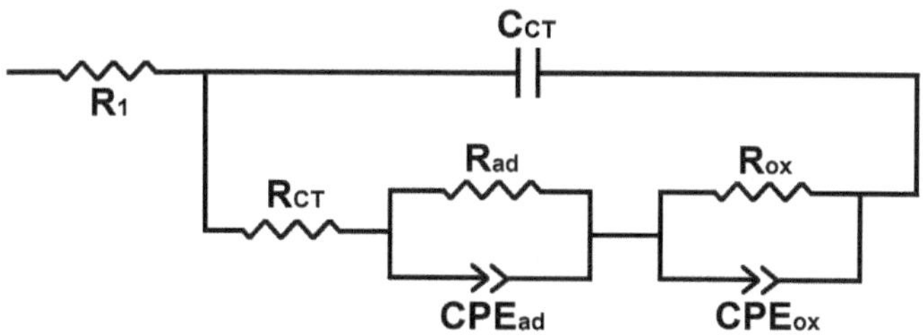

FIG. 3.34 Electrical circuit to model the anode impedance during carbon monoxide poisoning.

electrode/electrolyte interface, as shown in Fig. 3.34. The EEC shown in Fig. 3.34 has also been considered by Ciureanu et al. [88–90] and Venkataraman et al. [91] to study the influence of CO poisoning of the platinum gas diffusion electrodes. The EEC shown in Fig. 3.34 has been applied to EIS measurements presenting an inductive loop (EIS measurements with positive imaginary components) at low frequencies, which is attributed to CO electro-oxidation and removal by oxygenated species on the electrode surface. The EIS measurements reported by Wagner and Gülzow [47] and shown in Fig. 3.24 demonstrated that the CO poisoning effect in the anode increases the overall impedance spectrum of the PEMFC and the inductive loop at low frequencies. Bai and Conway [92] have demonstrated that an inductive loop in EIS measurements at low frequencies of electrochemical systems with adsorbed species is related to a change in the sign of the coverage dependence on potential.

3.11.2 Transmission-line model

Porous electrodes are used in PEMFCs because they have the advantage of an increased effective active area. A porous electrode may have the same structure as a porous layer; however, the pore walls are electroactive for a porous electrode, whereas the pore walls are inert for a porous layer. The porous electrode is usually represented by a simplified single-pore model. The impedance of the pore can be represented by the transmission-line model (TLM) presented in Fig. 3.35.

The general solution of the impedance of a pore can be obtained by a numerical calculation of the corresponding TLM. de Levie [93] calculated analytically the impedance of one pore considering the TLM shown in Fig. 3.35 as such:

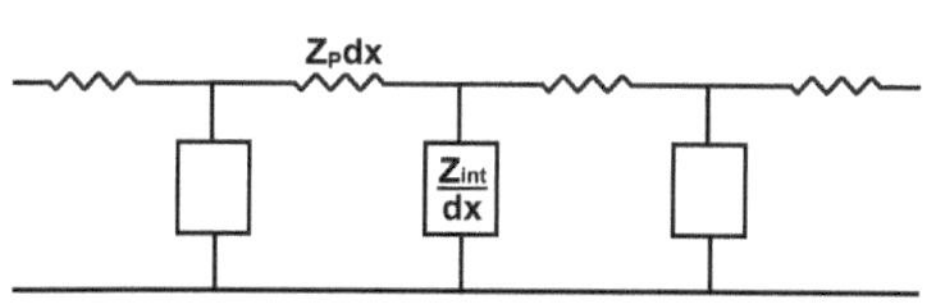

FIG. 3.35 A small section dx of the equivalent circuit of a pore; x is the distance along the pore axis, Z_P is the electrolyte resistance, and Z_{int} is the interfacial impedance.

$$Z_{pore} = \sqrt{Z_P Z_{int}} \coth\left(\ell_P \sqrt{\frac{Z_P}{Z_{int}}}\right) \tag{3.3}$$

where Z_P is the electrolyte resistance with units of Ω/cm, Z_{int} is the interfacial impedance with units of Ωcm, and ℓ_P is the length of the pore with units of cm.

Eikerling and Kornyshev [78] proposed a TLM to study the impedance response of the porous catalyst layer at low overpotentials. The authors modelled the catalyst layer as an agglomerate structure of carbon particles bound together by polytetrafluoroethylene (PTFE), which serves, in addition, as a hydrophobising agent in order to ensure good gas porosity of the layer. The voids between the agglomerates are penetrated by the electrolyte, which forms the proton conducting pathways. The TLM shown in Fig. 3.36 consists of an upper rail to account for the proton conduction and a lower rail to represent the electron access in the catalyst layer. In the upper rail, distributed resistances to account for the proton resistance in the electrolyte of the catalyst layer are considered. The lower rail considers the distributed electron resistances for the electron transfer in the carbon of the catalyst layer. Between the upper rail and lower rail, parallel distributed resistances and capacitances to account for the charge-transfer process and the capacitance effect between the electrode/electrolyte interface are considered. These four elements are then repeated a finite number of times as shown in Fig. 3.36.

The TLM shown in Fig. 3.36 has also been considered to study the impedance response of the CCL of PEMFCs using EIS. The distributed electron resistances in the lower rail of the TLM have been considered negligible for CCLs with high electron conductivity [14,81,94,95]. Under this configuration, the TLM is available in the commercial software ZView (Transmission-Line Open Circuit Terminus, DX-Type 6, Scribner Associates, Inc., version 2.3) for analysis of EIS. The TLM shown in Fig. 3.36 to study the impedance response of the CCL at high currents cannot be used because it does not account for oxygen transport limitations in the CCL. At low currents, this phenomenon does not occur because the rate of water production is low and so does not appear as an equivalent electrical component in the TLM.

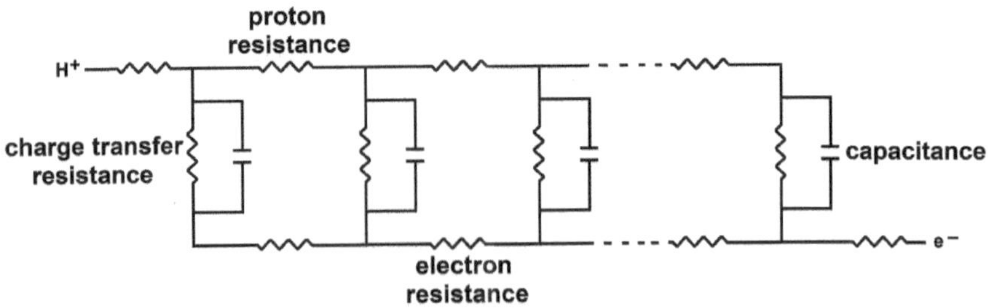

FIG. 3.36 Transmission-line model.

Other studies [18,79,80,96,97] have removed the charge-transfer resistance connected between the upper and lower rails of the TLM to estimate the proton resistance of the catalyst layer in EIS measurements carried out in H_2/N_2 PEMFCs. Under such H_2/N_2 PEMFC operating conditions, there is no faradaic reaction during the ORR. Therefore, the impedance response of the H_2/N_2 PEMFC is represented through a 45-degree straight line at high frequencies in the complex-impedance plot and represents the proton resistance in the catalyst layer. A vertical line at low frequencies is also represented in the complex-impedance plot and represents the double-layer capacitance in the catalyst layer.

The TLM shown in Fig. 3.36 can be applied to characterise the proton resistance in the CCL electrolyte for PEMFCs operated at low currents. At high currents, the product water that is formed in the CCL can begin to saturate the tortuous pathways of the porous network, which then acts as a resistance to the mass transport of oxygen from the interface with the GDL to the catalyst sites. At low currents, this phenomenon does not occur because the rate of water production is low. The TLM with Warburg components has been considered in EIS measurements [33,94,98] to analyse oxygen transport limitations in the CCL. In Chapter 4, the mathematical equation to represent a TLM comprising distributed Warburg components to characterise the impedance response of the CCL with oxygen transport limitations will be presented.

3.11.3 Voigt circuit

A measurement model based on the Voigt circuit to interpret EIS data in PEMFCs is reported in the study of Agarwal and Orazem [99]. The Voigt circuit has been introduced previously in Chapter 1, and its configuration and mathematical representation have been described in Section 1.5.1. A Voigt circuit consists of a resistor connected in series with a resistor/capacitor parallel configuration and then repeated a finite number of times, as shown in Fig. 1.22 from Chapter 1. By including a sufficient number of elements in the Voigt circuit, it is possible to fit impedance data for typical stationary electrochemical systems such as PEMFCs. The lack of complete agreement between the data and the impedance response of the Voigt circuit could be attributed to the stochastic error of the experimental data or nonstationary processes. The Voigt circuit can be used to identify the frequency-dependent error structure of impedance spectra; the error structure can be used to weight the data during regression and to provide a means of deciding whether a given regression provided a good fit [100,101]. The Voigt circuit can also be used to check the consistency of electrochemical impedance data to the K–K relations because the Voigt circuit itself satisfies the K–K relations and provides a statistically adequate fit to electrochemical impedance data. The application of the Voigt circuit to assess K–K consistency was demonstrated in Chapter 1 Section 1.5.1. Montella [102] demonstrated that the Voigt circuit with an infinite number of resistors/capacitors in a parallel configuration is an exact model of the finite-length Warburg impedance. The impedance response of the finite-length Warburg impedance has been discussed previously in Chapter 1 Section 1.4.3.

3.12 Evaluation of validity of EIS measurements

In Chapter 1 Section 1.3.1, the mathematical equations of K–K transformation to relate the real and imaginary components of the impedance response resulting from the electrochemical system studied were presented. K–K relations can be applied to the impedance response of a PEMFC to evaluate correctness of the EIS data. During K–K transformation, the measured real component of the impedance response of the PEMFC is transformed into the imaginary component and vice versa. If the impedance data of a PEMFC can be transformed (real to imaginary $Z_r \rightarrow Z_j$ and imaginary to real $Z_j \rightarrow Z_r$) by K–K, the EIS response of the PEMFC complies with linearity, causality, and stability properties.

3.12.1 Kramers–Kronig analysis

The application of the K–K transformations requires integration of the frequency from zero to infinity, but a finite and reduced range of frequencies is commonly considered during EIS measurements in PEMFCs. Therefore, the use of EECs that comply with K–K transformations is an accepted approach to evaluate validity of EIS measurements from PEMFCs. Agarwal and Orazem [99] and later Boukamp [103] reported that it is possible to evaluate K–K relations by fitting the Voigt circuit previously introduced in Chapter 1 and shown in Fig. 1.22 into EIS measurements. Boukamp also reported that if the EIS data cannot be well approximated by the Voigt model containing a reasonable number of R-C pairs, it means that data are not transformable and do not represent the physical processes of the electrochemical system studied. ZView software (Scribner Associates Inc.) is a powerful tool to fit any electrical circuit with EIS measurements of PEMFCs. However, the electrical components defined in ZView software such as a resistor, capacitor, and inductor do not provide sufficient information about the electrochemical phenomena of PEMFCs. For instance, the Randles circuit constructed in ZView software can represent the impedance response of an electron-transfer reaction between the electrode/electrolyte interface, but the resistors do not provide enough information about electrochemical mechanisms of the electron-transfer reaction (e.g. Tafel slope, exchange current). Nevertheless, the ZView software allows the evaluation of K–K consistency with measured EIS. The construction of the Voigt circuit in ZView software for K–K evaluation was discussed in Chapter 1 Section 1.5.3. More detailed information about the application of the K–K transform in EIS data using ZView software can be found in the technical report from Cruz-Manzo and Greenwood [104]. Previously in Chapter 1 Section 1.5.1, the application of the Voigt circuit with the impedance response of the Randles circuit using a script written in MATLAB software was demonstrated.

3.12.1.1 Stability analysis

The analysis described in this section has also been considered in the study of Cruz-Manzo et al. [105]. A PEMFC was operated under two different conditions to evaluate K–K transformation in the EIS measurements using the Voigt circuit and ZView software. First, EIS measurements were carried out in a 25 cm^2 H_2/air PEMFC operated at a current density of 0.15 A/cm^2 with 100% RH in the cathode, and the hydrogen supplied in the anode was dry. Second, EIS measurements were carried out in the 25 cm^2 H_2/air PEMFC operated at 0.23 A/cm^2 with 100% RH in the anode, and the air supplied in the cathode was dry. The operating temperature was controlled at 50°C using a test rig from Baltic Fuel Cells. EIS measurements were carried out from 10 kHz to 0.1 Hz for both RH cases, as shown in Fig. 3.37.

The ohmic resistance in a PEMFC attributed to the opposition of protons and electrons in the PEM, catalyst layer, GDL, and FFP is estimated, where the EIS measurements cross the real component Z_r of the Nyquist plot at high frequencies or where the imaginary components of the measurements are equal to zero $Z_j = 0$. Fig. 3.37 shows that there is a difference in ohmic resistance between both experimental cases with different RHs in the anode and cathode. When the PEMFC is operated with dry hydrogen in the anode and 100% RH in air supplied to the cathode, the ohmic resistance is higher than the case for 100% RH in the anode and dry cathode. This increase in ohmic resistance could be attributed to dehydration in the PEM in the anode side and attributed to

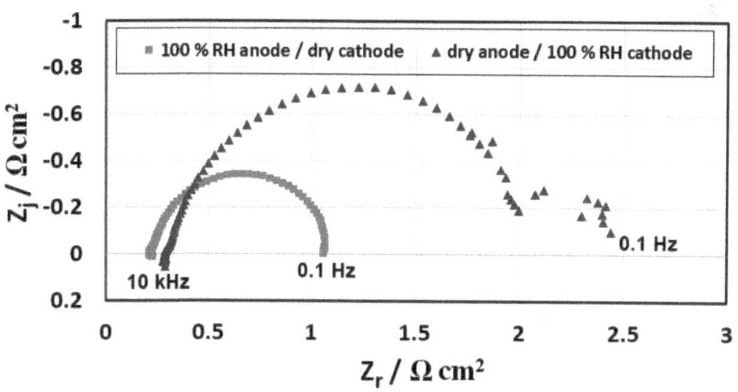

FIG. 3.37 EIS measurements carried out in a 25 cm^2 H_2/air PEMFC operated under two different relative humidity (RH) conditions in the hydrogen and air supplied. 100% RH in the anode and the dry cathode at 0.23 A/cm^2. The dry anode and 100% RH in the cathode at 0.15 A/cm^2. *(From S. Cruz-Manzo, R. Chen, P. Greenwood, Evaluate the validity of electrochemical impedance measurements of polymer electrolyte fuel cells using a computational algorithm based on fast Fourier transform, Insights Anal. Electrochem. 1 (2015) 9. Figure published under license of Creative Commons Attribution 4.0 International License.)*

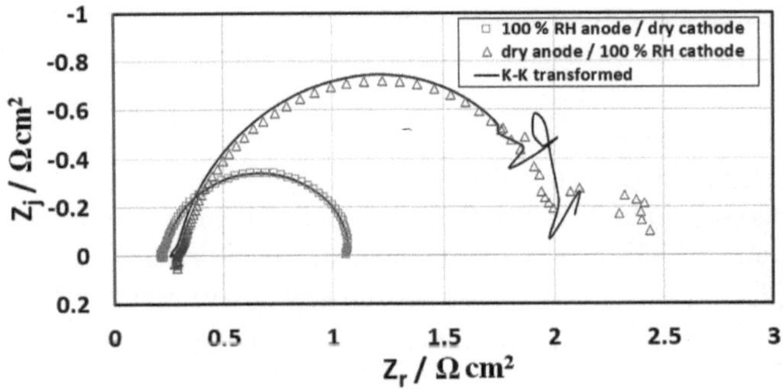

FIG. 3.38 Comparison between EIS measurements carried out in a 25 cm^2 H_2/air PEMFC under two different relative humidity (RH) conditions in the hydrogen and air supplied and Kramers–Kronig (K–K) transformed data. 100% RH in the anode and the dry cathode at 0.23 A/cm^2. The dry anode and 100% RH in the cathode at 0.15 A/cm^2.

electro-osmotic drag from the anode to the cathode. The effect of oxygen transport limitations in the GDL and CCL is commonly represented at low frequencies in the complex-impedance plot. The decrease in diameter under the 100% RH anode/dry cathode conditions is attributed to an increase in the driving force for the interfacial oxygen reduction process with increasing current density.

EIS measurements shown in Fig. 3.37 for a PEMFC during two different RH conditions were considered to evaluate the measurements through K–K in ZView software. The application of K–K in EIS measurements using ZView software has been discussed in Chapter 1 Section 1.5.3. The K–K analysis for the EIS measurements for dry anode and 100% RH cathode conditions was carried out, and this resulted in the K–K transformation shown in Fig. 3.38. Fig. 3.38 also shows the K–K transformation for the EIS measurements for 100% RH under the anode and dry cathode conditions. The results show that the K–K transformation fails at low frequencies for EIS measurements carried out when fully saturated air and dry hydrogen are supplied in the cathode and the anode, respectively. This failure in data transformation at low frequencies which leads to invalid experimental data can be a result of instability in the PEMFC during EIS measurements. Therefore, the impedance response of the PEMFC at low frequencies when fully saturated air and dry hydrogen are supplied in the cathode and the anode is not time-invariant. Water in a PEMFC is produced on the cathode side because of the ORR, so when air is supplied with a high RH, an increase in water concentration in the cathode is present. Within the PEMFC, there is competition mainly between back transport of water from the cathode to the anode and electro-osmotic drag of water by protons from the anode to the cathode, which could lead to instability during PEMFC operation. Yan et al. [59] reported that that the best performance during PEMFC operation (e.g. current limit) occurs at low RH in air and high RH in hydrogen.

3.12.1.2 Linearity analysis

In Section 3.5, the effect of the amplitude of the sinusoidal perturbation during EIS on the impedance response of the PEMFC was demonstrated. Increasing the amplitude of the sinusoidal input signal during EIS measurements yields an expansion of the impedance spectrum of the PEMFC as previously shown in Figs 3.18 and 3.19. The PEMFC demonstrates nonlinear current–voltage behaviour as the current has an exponential dependence on potential. For systems presenting a nonlinear current–voltage relationship, a small amplitude of the sinusoidal signal is needed for EIS measurements. For systems presenting a linear current–voltage relationship, a very large amplitude of the sinusoidal signal can be used. The PEMFC represents a nonlinear voltage–current response; therefore, a small amplitude of the sinusoidal input signal has to be considered during EIS in PEMFCs to measure the impedance response as a pseudo-linear system. In this section, a K–K analysis is carried out to evaluate the linearity property of the resulting impedance response of the PEMFC with increasing amplitude of the sinusoidal input signal during EIS. The impedance spectra of the PEMFC were previously shown in Fig. 3.19. The K–K analysis is carried out using ZView software as discussed in Chapter 1 Section 1.5.3.

Fig. 3.39 shows that it is possible to achieve K–K transformation in spite of the increase of the sinusoidal signal amplitude. However, some degree of discrepancy can be observed during the K–K transformation of the EIS measurements with a high sinusoidal signal amplitude (15% of DC current). In the work reported by Macdonald et al. [106], different amplitudes of the sinusoidal perturbation in EIS potentiostatic tests were changed from 5 mV to 150 mV to test the linearity condition of data for a rapidly corroding aluminium alloy through the K–K relations. The results showed a successful K–K transformation (real to imaginary component and vice versa) of the impedance data, in spite of a clear violation of linearity when the amplitude was increased. This success in the K–K transformation was an artefact of the experimental equipment used for

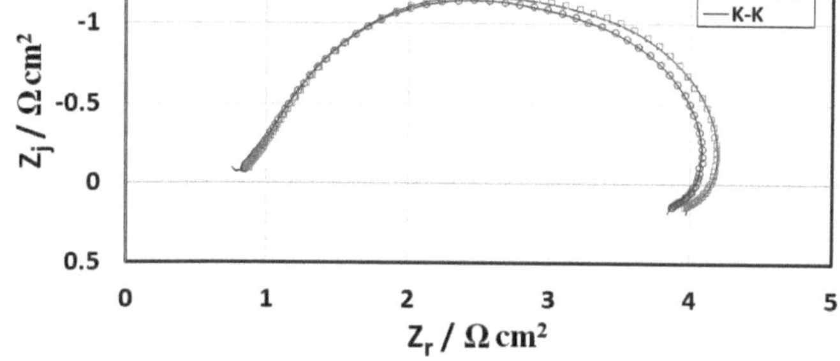

FIG. 3.39 Comparison between PEMFC-EIS measurements carried out in a galvanostatic configuration at two different sinusoidal signal amplitudes (5% and 15% of the DC current) and Kramers–Kronig (K–K) transformed data. The impedance spectra of the PEMFC were previously shown in Fig. 3.19.

the impedance measurements. The authors concluded that some commercial FRAs are narrow band devices that reject inputs at frequencies other than that designated; in other words, some frequency response analysers can reject harmonic components of the current response that manifest nonlinearity. The K–K relations cannot be used to determine if the impedance measurements were corrupted by instrument artefacts.

Fig. 3.39 shows that it is possible to achieve K–K transformation in the impedance measurements with increasing sinusoidal signal amplitude during EIS. However, the high sinusoidal signal amplitude could bias the agreement between the DC polarisation resistances estimated from the slope of the polarisation curve and the DC polarisation resistance estimated from EIS measurements as the frequency is reduced to a value near zero [35]. The effect of sinusoidal signal amplitude on the low-frequency limit of the impedance response of the PEMFC will be analysed in Chapter 5.

It is possible to obtain a correct data transformation at the lowest frequencies where the EIS data resulted in positive imaginary components Z_j as shown in Fig. 3.39. Roy and Orazem [20] concluded that there is a K–K consistency (linearity, causality, and stability) for low-frequency inductive loops in the entire range (current density) of operation of the PEMFC. The electrochemical phenomena that yield the formation of inductive loops at low frequencies in EIS measurements of PEMFCs will be discussed in Chapter 7.

3.12.2 Hilbert transformation

It is possible to calculate the imaginary component from the measured real component and vice versa and evaluate the validity of EIS measurements for steady, causal, and linear systems using the Hilbert transformation of impedance data [107]. In Chapter 1 Section 1.5.2, the application of the Hilbert transformation method reported by VanderNoot [107] was demonstrated. The computational algorithm based on the Hilbert transformation of impedance data for evaluating data transformation (real to imaginary $Z_r \rightarrow Z_j$ and imaginary to real $Z_j \rightarrow Z_r$) and hence validity of impedance measurements is shown in Appendix A and has been written as a script in MATLAB software.

The MATLAB script requires a vector for the range of frequencies applied during the EIS measurements as an input. A vector for the measured real component and a vector for the measured imaginary component of the EIS data are required in the MATLAB script as inputs as well. The EIS measurements described in the previous Section 3.12.1.1 and shown in Fig. 3.37 have been considered to carry out data transformation $Z_r \rightarrow Z_j$, $Z_j \rightarrow Z_r$ through the Hilbert transformation method.

Fig. 3.40 shows a comparison between the experimental data when the PEMFC was operated at 0.23 A/cm^2 with 100% RH in the anode and that the air supplied in the cathode was dry and the transformed data $Z_r \rightarrow Z_j$, $Z_j \rightarrow Z_r$ using the Hilbert transformation method. The transformed imaginary

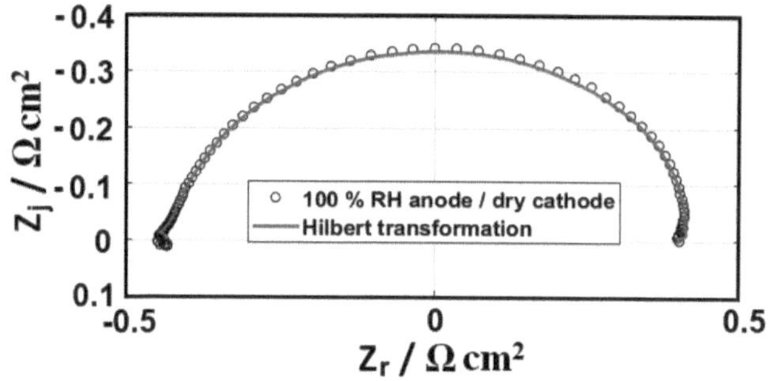

FIG. 3.40 Comparison between EIS measurements carried out in a 25 cm^2 H_2/air PEMFC with 100% relative humidity (RH) in the anode and the dry cathode at 0.23 A/cm^2 and transformed data using Hilbert transformation.

and transformed real data obtained from the Hilbert transformation method are calculated from a centre frequency and correspond to the frequency where the imaginary component of EIS measurements reaches its minimum value or becomes more negative $-Z_j$. The real component Z_r calculated from the Hilbert transformation is zero at the centre frequency. The real component Z_r of the measured impedance spectrum where the imaginary component is most negative $-Z_j$ was shifted to zero. This modification in the measured data allowed the comparison with the transformed data from the Hilbert transformation, as shown in Fig. 3.40. This data modification (real component shifted to zero where the imaginary component reaches its minimum value $-Z_j$) has been considered in the script shown in Appendix A to compare experimental EIS measurements with the transformed data from the Hilbert transformation. The results shown in Fig. 3.40 demonstrate that the experimental EIS measurements satisfy real to imaginary $Z_r \rightarrow Z_j$ and imaginary to real $Z_j \rightarrow Z_r$ data transformations and therefore comply with linearity, stability, and causality properties and truly represent the frequency response of the electrochemical mechanisms of the PEMFC to the sinusoidal perturbation.

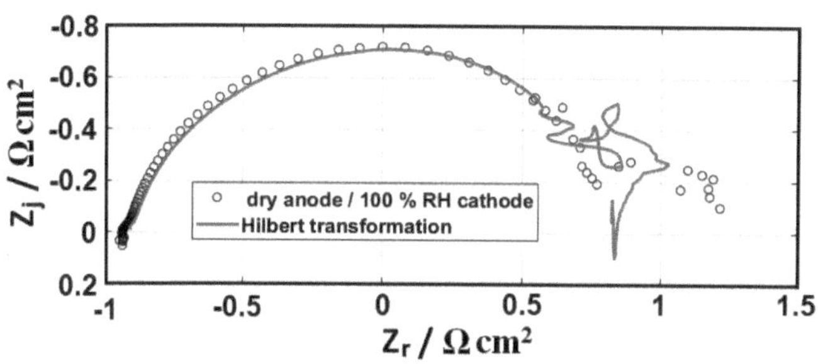

FIG. 3.41 Comparison between EIS measurements carried out in a 25 cm^2 H_2/air PEMFC with the dry anode and 100% relative humidity (RH) in the cathode at 0.15 A/cm^2 and transformed data using Hilbert transformation.

The Hilbert transformation algorithm was also applied to EIS measurements carried out when the hydrogen supplied to the anode was dry and with 100% RH of air supplied to the cathode, as shown in Fig. 3.41. The results shown in Figs 3.40 and 3.41 using the Hilbert transformation algorithm are in agreement with the results shown in Fig. 3.38 using the K–K transformation in ZView software. More information about the results shown in Figs 3.38, 3.40, and 3.41 can be found in the study of Cruz-Manzo et al. [105].

References

[1] K.R. Cooper, V. Ramani, J.M. Fenton, H.R. Kunz, Experimental Methods and Data Analyses for Polymer Electrolyte Fuel Cells, Scribner Associates, 2005.

[2] X.-Z. Yuan, C. Song, H. Wang, J. Zhang, Electrochemical Impedance Spectroscopy in PEM Fuel Cells: Fundamentals and Applications, Springer, London, 2010.

[3] G.S. Popkirov, R.N. Schindler, A new impedance spectrometer for the investigation of electrochemical systems, Rev. Sci. Instrum. 63 (1992) 5366–5372.

[4] G.S. Popkirov, R.N. Schindler, Effect of sample nonlinearity on the performance of time domain electrochemical impedance spectroscopy, Electrochim. Acta 40 (1995) 2511–2517.

[5] S.H. Chan, X.J. Chen, K.A. Khor, Reliability and accuracy of measured overpotential in a three-electrode fuel cell system, J. Appl. Electrochem. 31 (2001) 1163–1170.

[6] M. Dolle, F. Orsini, A.S. Gozdz, J. Tarascon, Development of reliable three-electrode impedance measurements in plastic Li-ion batteries, J. Electrochem. Soc. 148 (2001) A851–A857.

[7] X. Yuan, H. Wang, J.C. Sun, J. Zhang, AC impedance technique in PEM fuel cell diagnosis—a review, Int. J. Hydrogen Energy 32 (2007) 4365–4380.

[8] W. He, V.T. Nguyen, Edge effects on reference electrode measurements in PEM fuel cells, J. Electrochem. Soc. 151 (2004) A185–A195.

[9] S. Cruz-Manzo, Further understanding of uncertainties in the impedance spectrum of the polymer electrolyte fuel cell due to inductive effects and oxygen diffusion time constant, J. Energy Power Technol. 2 (2020) 21.

[10] B. Savova-Stoynov, Z.B. Stoynov, Analysis of the inductance influence on the measured electrochemical impedance, J. Appl. Electrochem. 17 (1987) 1150–1158.

[11] H. Göhr, M. Mirnik, C.A. Schiller, Distortions of high frequency electrode impedance their causes and how to avoid them, J. Electroanal. Chem. Interfacial Electrochem. 180 (1984) 273–285.

[12] M. Ciureanu, R. Roberge, Electrochemical impedance study of PEM fuel cells. Experimental diagnostics and modeling of air cathodes, J. Phys. Chem. B 105 (2001) 3531–3539.

[13] N. Fouquet, C. Doulet, C. Nouillant, D. Dauphin-Tanguy, B. Ould-Bouamama, Model based PEM fuel cell state-of-health monitoring via AC impedance measurements, J. Power Sources 159 (2006) 905–913.

[14] R. Makharia, M.F. Mathias, D.R. Baker, Measurement of catalyst layer electrolyte resistance in PEMFCs using electrochemical impedance spectroscopy, J. Electrochem. Soc. 152 (2005) A970–A977.

[15] W. Mérida, D.A. Harrington, J.M. Le Canut, G. McLean, Characterisation of proton exchange membrane fuel cell (PEMFC) failures via electrochemical impedance spectroscopy, J. Power Sources 161 (2006) 264–274.

[16] S. Asghari, A. Mokmeli, M. Samavati, Study of PEM fuel cell performance by electrochemical impedance spectroscopy, Int. J. Hydrogen Energy 35 (2010) 9283–9290.

[17] F. Liu, B. Yi, D. Xing, J. Yu, Z. Hou, Y. Fu, Development of novel self-humidifying composite membranes for fuel cells, J. Power Sources 124 (2003) 81–89.

[18] G. Li, P.G. Pickup, Ionic conductivity of PEMFC electrodes, J. Electrochem. Soc. 150 (2003) C745–C752.

[19] S. Cruz-Manzo, R. Chen, P. Rama, Inductive effect on the fuel cell cathode impedance spectrum at high frequencies, J. Fuel Cell Sci. Technol. 9 (2012), 051002.

[20] S.K. Roy, M.E. Orazem, B. Tribollet, Interpretation of low-frequency inductive loops in PEM fuel cells, J. Electrochem. Soc. 154 (2007) B1378–B1388.

[21] S. Cruz-Manzo, R. Chen, P. Greenwood, An impedance model for analysis of EIS of polymer electrolyte fuel cells under hydrogen peroxide formation in the cathode, J. Electroanal. Chem. 745 (2015) 28–36.

[22] N. Wagner, M. Schulze, Change of electrochemical impedance spectra during CO poisoning of the Pt and Pt–Ru anodes in a membrane fuel cell (PEFC), Electrochim. Acta 48 (2003) 3899–3907.

[23] K. Wiezell, P. Gode, G. Lindbergh, Steady-state and EIS investigations of hydrogen electrodes and membranes in polymer electrolyte fuel cells I. Modeling, J. Electrochem. Soc. 153 (2006) A749–A758.

[24] I.A. Schneider, M.H. Bayer, A. Wokaun, G.G. Scherer, Impedance response of the proton exchange membrane in polymer electrolyte fuel cells, J. Electrochem. Soc. 155 (2008) B783–B792.

[25] I. Pivac, F. Barbir, Inductive phenomena at low frequencies in impedance spectra of proton exchange membrane fuel cells—a review, J. Power Sources 326 (2016) 112–119.

[26] D. Klotz, Negative capacitance or inductive loop?—a general assessment of a common low frequency impedance feature, Electrochem. Commun. 98 (2019) 58–62.

[27] S. Cruz-Manzo, P. Greenwood, Low frequency inductive loop in EIS measurements of an open-cathode polymer electrolyte fuel cell stack. Impedance of water vapour diffusion in the cathode catalyst layer, J. Electroanal. Chem. 900 (2021), 115733.

[28] K. Wiezell, N. Holmstrom, G. Lindbergh, Studying low-humidity effects in PEFCs using EIS II. Modeling, J. Electrochem. Soc. 159 (2012) F379–F392.

[29] X. Yuan, J.C. Sun, H. Wang, J. Zhang, AC impedance diagnosis of a 500 W PEM fuel cell stack: part II: individual cell impedance, J. Power Sources 161 (2006) 929–937.

[30] P. Gode, F. Jaouen, G. Lindbergh, A. Lundblad, G. Sundholm, Influence of the composition on the structure and electrochemical charasteristics of the PEFC cathode, Electrochim. Acta 48 (2003) 4175–4187.

[31] F. Jaouen, G. Lindbergh, K. Wiezell, Transient techniques for investigating mass-transport limitations in gas diffusion electrodes, J. Electrochem. Soc. 150 (2003) A1711–A1717.

[32] Q. Meyer, I. Pivac, F. Barbir, C. Zhao, Detection of oxygen starvation during carbon corrosion in proton exchange membrane fuel cells using low-frequency electrochemical impedance spectroscopy, J. Power Sources 470 (2020), 228285.

[33] S. Cruz-Manzo, R. Chen, A generic electrical circuit for performance analysis of the fuel cell cathode catalyst layer through electrochemical impedance spectroscopy, J. Electroanal. Chem. 694 (2013) 45–55.

[34] F. Fasmin, R. Srinivasan, Detection of nonlinearities in electrochemical impedance spectra by Kramers–Kronig transforms, J. Solid State Electrochem. 19 (2015) 1833–1847.

[35] G. Maranzana, J. Mainka, O. Lottin, J. Dillet, A. Lamibrac, A. Thomas, S. Didierjean, A proton exchange membrane fuel cell impedance model taking into account convection along the Air Channel: on the bias between the low frequency limit of the impedance and the slope of the polarization curve, Electrochim. Acta 83 (2012) 13–27.

[36] N. Wagner, Characterization of membrane electrode assemblies in polymer electrolyte fuel cells using A.C. impedance spectroscopy, J. Appl. Electrochem. 32 (2002) 859–863.

[37] S. Cruz-Manzo, R. Chen, An electrical circuit for performance analysis of polymer electrolyte fuel cell stacks using electrochemical impedance spectroscopy, J. Electrochem. Soc. 160 (2013) F1109–F1115.

[38] S. Cruz-Manzo, C. Perezmitre-Cruz, P. Greenwood, R. Chen, An impedance model for analysis of EIS of polymer electrolyte fuel cells under platinum oxidation and hydrogen peroxide formation in the cathode, J. Electroanal. Chem. 771 (2016) 94–105.

[39] D.J.L. Brett, S. Atkins, N.P. Brandon, V. Vesovic, N. Vasileiadis, A. Kucernak, Localized impedance measurements along a single channel of a solid polymer fuel cell, Electrochem. Solid St. 6 (2003) A63–A66.

[40] I.A. Schneider, H. Kuhn, A. Wokaun, G.G. Scherer, Fast locally resolved electrochemical impedance spectroscopy in polymer electrolyte fuel cells, J. Electrochem. Soc. 152 (2005) A2092–A2103.

[41] A. Hakenjos, C. Hebling, Spatially resolved measurement of PEM fuel cells, J. Power Sources 145 (2005) 307–311.

[42] X. Yan, M. Hou, L. Sun, D. Liang, Q. Shen, H. Xu, P. Ming, B. Yi, AC impedance characteristics of a 2kW PEM fuel cell stack under different operating conditions and load changes, Int. J. Hydrogen Energy 32 (2007) 4358–4364.

[43] J. Mainka, G. Maranzana, J. Dillet, S. Didierjean, O. Lottin, On the estimation of high frequency parameters of proton exchange membrane fuel cells via electrochemical impedance spectroscopy, J. Power Sources 253 (2014) 381–391.

[44] L. Pisani, G. Murgia, M. Valentini, B. D'Aguanno, A new semi-empirical approach to performance curves of polymer electrolyte fuel cells, J. Power Sources 108 (2002) 192–203.

[45] B. Andreaus, A.J. McEvoy, G.G. Scherer, Analysis of performance losses in polymer electrolyte fuel cells at high current densities by impedance spectroscopy, Electrochim. Acta 47 (2002) 2223–2229.

[46] T.E. Springer, T.A. Zawodzinski, M.S. Wilson, S. Gottesfeld, Characterization of polymer electrolyte fuel cells using AC impedance spectroscopy, J. Electrochem. Soc. 143 (1996) 587–599.

[47] N. Wagner, E. Gülzow, Change of electrochemical impedance spectra (EIS) with time during CO-poisoning of the Pt-anode in a membrane fuel cell, J. Power Sources 127 (2004) 341–347.

[48] H. Nakajima, T. Konomi, T. Kitahara, H. Tachibana, Electrochemical impedance parameters for the diagnosis of a polymer electrolyte fuel cell poisoned by carbon monoxide in reformed hydrogen fuel, J. Fuel Cell Sci. Technol. 5 (2008), 041013.

[49] M. Chandesris, L. Guetaz, P. Schott, M. Scohy, S. Escribano, Investigation of degradation heterogeneities in PEMFC stack aged under reformate coupling in situ diagnosis, post-mortem ex situ analyses and multi-physic simulations, J. Electrochem. Soc. 165 (2018) F3290–F3306.

[50] M. Kim, N. Jung, K. SupEom, S.J. Yoo, J.Y. Kim, J.H. Jang, H.-J. Kim, B.K. Hong, E. Cho, Effects of anode flooding on the performance degradation of polymer electrolyte membrane fuel cells, J. Power Sources 266 (2014) 332–340.

[51] J.N. Schwämmlein, P.J. Rheinländer, Y. Chen, K.T. Freyer, H.A. Gasteiger, Anode aging during PEMFC start-up and shut-down: H_2-air fronts vs voltage cycles, J. Electrochem. Soc. 165 (2018) F1312–F1322.

[52] D. Malevich, E. Halliop, B.A. Peppley, J.G. Pharoah, K. Karan, Investigation of charge-transfer and mass-transport resistances in PEMFCs with microporous layer using electrochemical impedance spectroscopy, J. Electrochem. Soc. 156 (2009) B216–B224.

[53] X. Yuan, J.C. Sun, M. Blanco, H. Wang, J. Zhang, D.P. Wilkinson, AC impedance diagnosis of a 500 W PEM fuel cell stack: part I: stack impedance, J. Power Sources 161 (2006) 920–928.
[54] S. Cruz-Manzo, R. Chen, P. Rama, Study of current distribution and oxygen diffusion in the fuel cell cathode catalyst layer through electrochemical impedance spectroscopy, Int. J. Hydrogen Energy 38 (2013) 1702–1713.
[55] S. Cruz-Manzo, U. Cano-Castillo, P. Greenwood, Impedance study on estimating electrochemical mechanisms in a polymer electrolyte fuel cell during gradual water accumulation, Fuel Cells 19 (2019) 71–83.
[56] L. Xing, X. Liu, T. Alaje, R. Kumar, M. Mamlouk, K. Scott, A two-phase flow and non-isothermal agglomerate model for a proton exchange membrane (PEM) fuel cell, Energy 73 (2014) 618–634.
[57] M.A. Rubio, A. Urquia, R. Kuhn, S. Dormido, Electrochemical parameter estimation in operating proton exchange membrane fuel cells, J. Power Sources 183 (2008) 118–125.
[58] A. Weiß, S. Schindler, S. Galbiati, M.A. Danzer, R. Zeis, Distribution of relaxation times analysis of high-temperature PEM fuel cell impedance spectra, Electrochim. Acta 230 (2017) 391–398.
[59] Q. Yan, H. Toghiani, H. Causey, Steady state and dynamic performance of proton exchange membrane fuel cells (PEMFCs) under various operating conditions and load changes, J. Power Sources 161 (2006) 492–502.
[60] P. Mocoteguy, B. Ludwig, D. Beretta, T. Pedersen, Study of the impact of water management on the performance of PEMFC commercial stacks by impedance spectroscopy, Int. J. Hydrogen Energy 45 (2020) 16724–16737.
[61] N. Holmstrom, K. Wiezell, G. Lindbergh, Studying low-humidity effects in PEFCs using EIS: I. Experimental, J. Electrochem. Soc. 159 (2012) F369–F378.
[62] S.S. Araya, F. Zhou, S.L. Sahlin, S. Thomas, C. Jeppesen, S.K. Kær, Fault characterization of a proton exchange membrane fuel cell stack, Energies 12 (2019) 152.
[63] T. Reshetenko, A. Kulikovsky, On the origin of high frequency impedance feature in a PEM fuel cell, J. Electrochem. Soc. 166 (2019) F1253–F1257.
[64] R.S. Fu, X. Zhang, U. Pasaogullari, Heat and mass transfer in polymer electrolyte fuel cells in ultra-low humidity operation, ECS Trans. 25 (2009) 323–332.
[65] P. Boillat, F.N. Büchi, L. Gubler, T.J. Schmidt, Use and misuse of electrochemical impedance spectroscopy (EIS) in fuel cell research, in: ECS Meeting Abstracts, MA2019-02, 2019, p. 1435.
[66] A.M. Niroumand, W. Mérida, M. Eikerling, M. Saif, Pressure–voltage oscillations as a diagnostic tool for PEFC cathodes, Electrochem. Commun. 12 (2010) 122–124.
[67] A.V. Shirsath, S. Raël, C. Bonnet, F. Lapicque, Electrochemical pressure impedance spectroscopy applied to polymer electrolyte membrane fuel cells for investigation of transport phenomena, Electrochim. Acta 363 (2020), 137157.
[68] E. Engebretsen, T.J. Mason, P.R. Shearing, G. Hinds, D.J.L. Brett, Electrochemical pressure impedance spectroscopy applied to the study of polymer electrolyte fuel cells, Electrochem. Commun. 75 (2017) 60–63.
[69] J. Zhang, C. Song, J. Zhang, R. Baker, L. Zhang, Understanding the effects of backpressure on PEM fuel cell reactions and performance, J. Electroanal. Chem. 688 (2013) 130–136.
[70] D. Grübl, J. Janek, W.G. Bessler, Electrochemical pressure impedance spectroscopy (EPIS) as diagnostic method for electrochemical cells with gaseous reactants: a model-based analysis, J. Electrochem. Soc. 163 (2016) A599–A610.
[71] L. Schiffer, D. Grübl, W. Bessler, Model-based analysis of electrochemical pressure impedance spectroscopy (EPIS) for PEM fuel cells, in: H.A. Gasteiger, A. Bandarenka (Eds.), EFCF

2019, Proceedings of the European Fuel Cell Forum: Low-Temperature Fuel Cells, Electrolysers, Luzern, 2019.
[72] A. Kulikovsky, Analytical model for PEM fuel cell concentration impedance, J. Electroanal. Chem. 899 (2021), 115672.
[73] A. Sorrentino, T. Vidakovic-Koch, K. Sundmacher, Studying mass transport dynamics in polymer electrolyte membrane fuel cells using concentration-alternating frequency response analysis, J. Power Sources 412 (2019) 331–335.
[74] A. Sorrentino, K. Sundmacher, T. Vidakovic-Koch, Decoupling oxygen and water transport dynamics in polymer electrolyte membrane fuel cells through frequency response methods based on partial pressure perturbations, Electrochim. Acta 390 (2021), 138788.
[75] E. Engebretsen, et al., Electro-thermal impedance spectroscopy applied to an open-cathode polymer electrolyte fuel cell, J. Power Sources 302 (2016) 210–214.
[76] Y. Wu, et al., Investigation of water generation and accumulation in polymer electrolyte fuel cells using hydro electrochemical impedance imaging, J. Power Sources 414 (2019) 272–277.
[77] A. Sorrentino, K. Sundmacher, T. Vidakovic-Koch, Polymer electrolyte fuel cell degradation mechanisms and their diagnosis by frequency response analysis methods: a review, Energies 13 (2020) 5825.
[78] M. Eikerling, A.A. Kornyshev, Electrochemical impedance of the cathode catalyst layer in polymer electrolyte fuel cells, J. Electroanal. Chem. 475 (1999) 107–123.
[79] M.C. Lefebvre, R.B. Martin, P.G. Pickup, Characterization of ionic conductivity profiles within proton exchange membrane fuel cell gas diffusion electrodes by impedance spectroscopy, Electrochem. Solid St. 2 (1999) 259–261.
[80] T. Suzuki, H. Murata, T. Hatanaka, Y. Morimoto, Analysis of the Catalyst Layer of Polymer Electrolyte Fuel Cells, Vol. 39, R&D Review of Toyota, Toyota Central R&D Labs, Inc, 2003, pp. 33–38.
[81] J. Hou, W. Song, H. Yu, Y. Fu, L. Hao, Z. Shao, B. Yi, Ionic resistance of the catalyst layer after the PEM fuel cell suffered freeze, J. Power Sources 176 (2008) 118–121.
[82] M. Boillot, C. Bonnet, S. Didierjean, F. Lapicque, Investigation of the response of separate electrodes in a polymer electrolyte membrane fuel cell without reference electrode, J. Appl. Electrochem. 37 (2007) 103–110.
[83] M.L. Perry, J. Newman, J.E. Cairns, Mass transport in gas-diffusion electrodes: a diagnostic tool for fuel-cell cathodes, J. Electrochem. Soc. 145 (1998) 5–15.
[84] C.R. Tsai, F. Chen, A.C. Ruo, M.-H. Chang, H.-S. Chu, C.Y. Soong, W.M. Yan, C.H. Cheng, An analytical solution for transport of oxygen in cathode gas diffusion layer of PEMFC, J. Power Sources 160 (2006) 50–56.
[85] Y. Bultel, K. Wiezell, F. Jaouen, P. Ozil, G. Lindbergh, Investigation of mass transport in gas diffusion layer at the air cathode of a PEMFC, Electrochim. Acta 51 (2005) 474–488.
[86] G. Lin, W. He, T.V. Nguyen, Modeling liquid water effects in the gas diffusion and catalyst layers of the cathode of a PEM fuel cell, J. Electrochem. Soc. 151 (2004) A1999–A2006.
[87] R. Jiang, H.R. Kunz, J.M. Fenton, Electrochemical oxidation of H_2 and H_2/CO mixtures in higher temperature (T_{cell} > 100°C) proton exchange membrane fuel cells: electrochemical impedance spectroscopy, J. Electrochem. Soc. 152 (2005) A1329–A1340.
[88] M. Ciureanu, H. Wang, Electrochemical impedance study of electrode-membrane assemblies in PEM fuel cells I. Electro-oxidation of H_2 and H_2/CO mixtures on Pt-based gas-diffusion electrodes, J. Electrochem. Soc. 146 (1999) 4031–4040.
[89] M. Ciureanu, H. Wang, Electrochemical impedance study of anode CO-poisoning in PEM fuel cells, J. New Mater. Electrochem. Syst. 3 (2000) 107–119.

[90] M. Ciureanu, S.D. Mikhailenko, S. Kaliaguine, PEM fuel cells as membrane reactors: kinetic analysis by impedance spectroscopy, Catal. Today 82 (2003) 195–206.

[91] R. Venkataraman, H.R. Kunz, J.M. Fenton, Development of new CO tolerant ternary anode catalysts for proton exchange membrane fuel cells, J. Electrochem. Soc. 150 (2003) A278–A284.

[92] L. Bai, B.E. Conway, Complex behavior of AI dissolution in non-aqueous medium as revealed by impedance spectroscopy, J. Electrochem. Soc. 137 (1990) 3737–3747.

[93] R. de Levie, in: Delahay, Tobias (Eds.), Advances in Electrochemistry and Electrochemical Engineering, vol. 6, 1967, p. 329.

[94] S. Touhami, J. Mainka, J. Dillet, S. Ait Hammou Taleb, O. Lottin, Transmission line impedance models considering oxygen transport limitations in polymer electrolyte membrane fuel cells, J. Electrochem. Soc. 166 (2019) F1209–F1217.

[95] U. Cano-Castillo, A. Ortiz, S. Manzo, L.G. Arriaga, G. Orozco, Parameter changes during gradual flooding of a PEM fuel cell through EIS studies, ECS Trans. 3 (2006) 931–939.

[96] J. Gazzarri, M. Eikerling, Q. Wang, Z. Liu, Estimation of local relative humidity in cathode catalyst layers of PEFC, Electrochem. Solid St. 13 (2010) B58–B62.

[97] Y. Liu, M.W. Murphy, D.R. Baker, W. Gu, C. Ji, J. Jorne, H. Gasteiger, A, "proton conduction and oxygen reduction kinetics in PEM fuel cell cathodes: effects of ionomer-to-carbon ratio and relative humidity", J. Electrochem. Soc. 156 (2009) B970–B980.

[98] J. Choi, J. Sim, H. Oh, K. Min, Resistance separation of polymer electrolyte membrane fuel cell by polarization curve and electrochemical impedance spectroscopy, Energies 14 (2021) 1491.

[99] P. Agarwal, M.E. Orazem, Measurement models for electrochemical impedance spectroscopy I. Demonstration of applicability, J. Electrochem. Soc. 139 (1992) 1917–1927.

[100] P. Agarwal, M.E. Orazem, L.H. García-Rubio, Application of the Kramers Kronig relations to electrochemical impedance spectroscopy, in: Presented at the ASTM International Symposium on Electrochemical Impedance: Analysis and Interpretation, San Diego, California, November 4-5, 1991.

[101] P. Agarwal, M.E. Orazem, A. Hiser, Application of electrochemical impedance spectroscopy to metal hydrides, in: Presented at the 180th Meeting of the Electrochemical Society, Phoenix, Arizona, October 13–18, 1991.

[102] C. Montella, Voigt circuit representation model for electrochemical impedances under finite-length diffusion conditions, J. Electroanal. Chem. 879 (2020), 114785.

[103] B.A. Boukamp, A linear Kronig-Kramers transform test for immittance data validation, J. Electrochem. Soc. 142 (1995) 1885–1894.

[104] S. Cruz-Manzo, Kramers-Kronig Analysis in ZView Software for Evaluation of EIS Measurements in Polymer Electrolyte Fuel Cells, 2015, https://doi.org/10.13140/rg.2.1.1094.0887/1 (Last access June 2022).

[105] S. Cruz-Manzo, R. Chen, P. Greenwood, Evaluate the validity of electrochemical impedance measurements of polymer electrolyte fuel cells using a computational algorithm based on fast Fourier transform, Insights Anal. Electrochem. 1 (2015) 1.

[106] M. Urquidi-Macdonald, S. Real, D.D. Macdonald, Applications of Kramers-Kronig transforms in the analysis of electrochemical impedance data-III. Stability and linearity, Electrochim. Acta 35 (1990) 1559–1566.

[107] T.J. VanderNoot, Hilbert transformation of immittance data using the fast Fourier transform, J. Electroanal. Chem. 322 (1992) 9–24.

Part II

Modelling

Chapter 4

Impedance model of the cathode catalyst layer

4.1 Oscillating current and voltage in the charge double-layer

During an electrochemical impedance spectroscopy (EIS) test carried out in a polymer electrolyte membrane fuel cell (PEMFC), electrical charges in the catalyst layer are pushed and pulled away from the double-layer between dissimilar materials (electrolyte/carbon), at different frequencies. The charge distribution in the double-layer interface of the catalyst layer is analogous to the charge distribution in a capacitor. The structure of the charge double-layer in the catalyst layer has been depicted in Fig. 2.11 from Chapter 2. During the positive cycle of the oscillating current, electrical charges are accumulated in the electrode-electrolyte interface. This phenomenon attracts solvated ions from the diffusive layer. During the negative cycle of the oscillating current, solvated ions are released towards the diffusive layer. The oscillating current makes the solvated ions oscillate to and from the diffuse layer at a defined frequency. It is possible to model the distribution of electrical charges in the interface between dissimilar materials or double-layer interface of the catalyst layer in terms of capacitive elements.

Fig. 4.1 shows the relation between the oscillating voltage and the oscillating current in the charge double-layer of the catalyst layer where the oscillating current presents a 90 degrees phase shift in relation to the oscillating voltage. When the oscillating current is at its maximum amplitude (A), solvated ions from the diffusive layer are being attracted to the electrode interface, at this point the potential difference is equal to zero. When the current is passing down through zero (B), the double-layer is fully charged, the solvated ions from the diffusive layer are closer to the electrode interface and there is a maximum potential difference. When the oscillating current amplitude is at its minimum value (C), the attracted solvated ions are released back to the diffuse layer and the potential goes to zero and the double layer is fully discharged. The double layer is fully charged again when the current is zero (D) and the potential has a negative value.

Electrochemical Phenomena in the Cathode Impedance Spectrum of PEM Fuel Cells.
https://doi.org/10.1016/B978-0-323-90607-4.00001-X
Copyright © 2022 Elsevier Inc. All rights reserved.

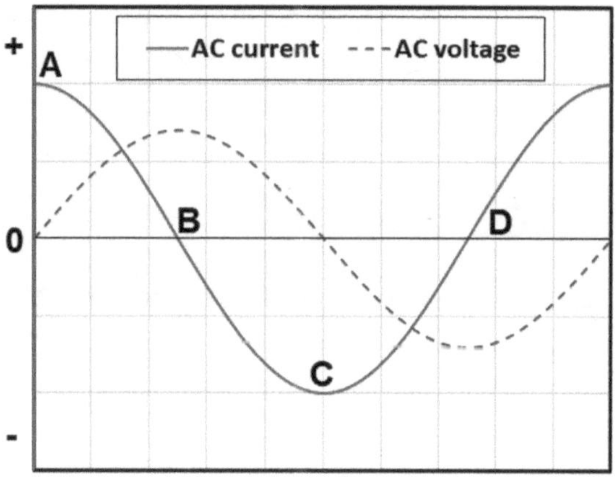

FIG. 4.1 Oscillating current and voltage in the charge double-layer of the catalyst layer.

4.2 Transmission-line model for porous electrodes

A literature review about the impedance response of porous electrodes has been carried out by Huang et al. [1]. The review of the impedance response of porous electrodes comprises the impedance theory of de Levie established in 1963 up to the application of high order models for the study of nonlinear impedance response of porous electrodes. The impedance model of the catalyst layer of the PEMFC can be simplified with the macrohomogeneous electrode theory in which the complex structure of the catalyst layer is reduced by considering average structural parameters [2]. Based on the macrohomogeneous electrode theory, the structure of the catalyst layer has been considered with a homogeneous distribution of carbon grains providing the electron conductivity, platinum Pt supported on carbon as the catalyst, polytetrafluoroethylene (PTFE) as a binder stabilising the carbon matrix and as a hydrophobizing agent, and electrolyte network of perfluorosulfonate ionomer. The transmission-line model (TLM) has been considered for the analysis of the impedance response of porous electrodes in electrochemical systems [3]. The book of Orazem and Tribollet [4] comprises a chapter to describe the TLM associated to porous electrodes. The impedance response of the catalyst layer has been considered analogous to the impedance response of the TLM [5]. Fig. 4.2 shows an analogous representation between the catalyst layer and the TLM.

The TLM shown in Fig. 4.2 represents the agglomeration of platinum particles supported on carbon and surrounded by the electrolyte for proton conduction in the catalyst layer. The TLM considers an upper rail of resistors to account for the resistance for proton H^+ conduction across the electrolyte surrounding the agglomerates in the catalyst layer. The lower rail considers resistors to account for the resistance for electron conduction e^- in the carbon structure of the catalyst layer. An equivalent impedance is connected between the upper and lower rails to account for the faradaic impedance during the redox reaction in the catalyst layer.

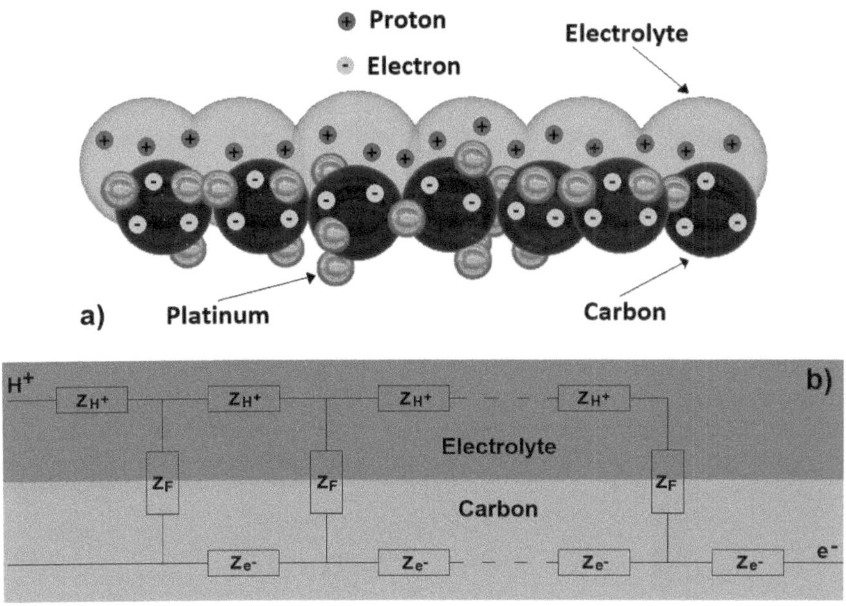

FIG. 4.2 Analogous representation between the catalyst layer and the transmission-line model, (a) proton and electron conduction, (b) transmission-line model.

4.2.1 Transmission-line model with negligible electron resistance

The resistance of electron conduction in the carbon structure of the catalyst layer has been neglected in the TLM [6] because the resistance of protons in the electrolyte is higher than the electron resistance in the carbon structure of the catalyst layer. The TLM to study the impedance response of the catalyst layer with negligible electron resistance in the carbon structure is shown in Fig. 4.3.

A resistor and a capacitor are connected between the upper and lower rails of the TLM to account for the charge transfer resistance during the redox reaction and double-layer capacitance in the catalyst layer. The TLM shown in Fig. 4.3 can be represented through Kirchhoff's current law which implies that the algebraic sum of currents flowing into a node is equal to the sum of currents flowing out of that node, as shown in Fig. 4.4.

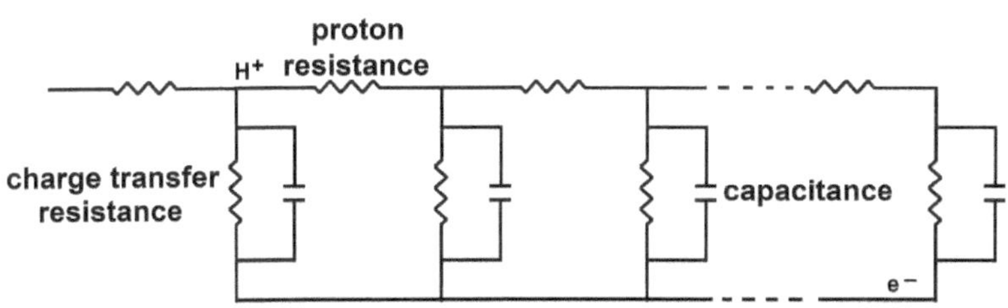

FIG. 4.3 Transmission-line model to represent the impedance response of the catalyst layer with negligible electron resistance for electron conduction in the carbon structure.

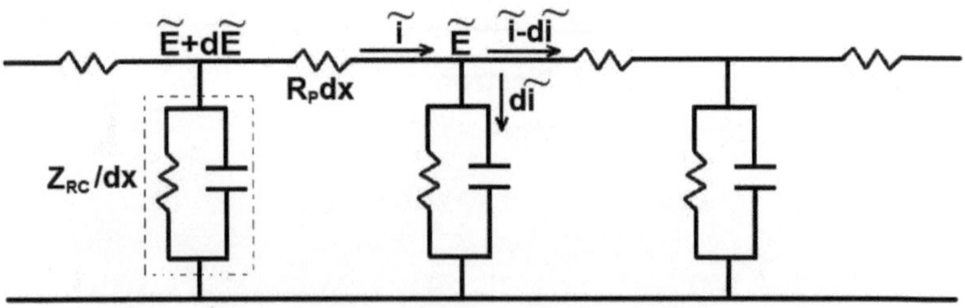

FIG. 4.4 Representation of small section *dx* of the transmission-line model with oscillating currents defined from Kirchhoff's Law.

Where dx is a small section of the TLM, $\tilde{E}$ is the oscillating potential resulting from the sinusoidal perturbation during EIS, $\tilde{i}$ is the oscillating current density, and R_P is the proton resistance. Z_{RC} is the impedance for the parallel configuration of the charge transfer resistance R_C and double-layer capacitance C_{dl}, shown in Fig. 4.4, as such:

$$Z_{RC} = \frac{R_C}{1 + R_C C_{dl} j\omega} \tag{4.1}$$

where j is the imaginary component of a complex number, $\omega = 2\pi f$ is the angular frequency, and f is the frequency.

Expressing Kirchhoff's law of currents in Fig. 4.4 yields:

$$\tilde{i} = d\tilde{i} + \tilde{i} - d\tilde{i} \tag{4.2}$$

substituting Ohm's Law defined from $d\tilde{i} = \frac{\tilde{E}}{Z_{RC}} dx$ into the first term of the right-hand side of Eq. (4.2) gives:

$$\tilde{i} = \frac{\tilde{E}}{Z_{RC}} dx + \tilde{i} - d\tilde{i} \tag{4.3}$$

Eq. (4.3) can be rearranged as:

$$\frac{d\tilde{i}}{dx} = \frac{\tilde{E}}{Z_{RC}} \tag{4.4}$$

Expressing Kirchhoff's law of voltages in Fig. 4.4 yields:

$$\tilde{E} + d\tilde{E} = \tilde{E} + \tilde{i} R_P dx \tag{4.5}$$

Eq. (4.5) can be rearranged as:

$$d\tilde{E} = \tilde{i} R_P dx \tag{4.6}$$

Substituting Eq. (4.4) into Eq. (4.6) yields:

$$\frac{d^2\tilde{E}}{dx^2} = \frac{R_P}{Z_{RC}} \tilde{E} \tag{4.7}$$

Substituting Eq. (4.1) into Eq. (4.7) gives:

$$\frac{d^2\widetilde{E}}{dx^2} = \frac{R_P}{R_C}\widetilde{E} + R_P C_{dl} j\omega \widetilde{E} \tag{4.8}$$

The solution of Eq. (4.8) can be obtained by applying the method of the *n*th-order homogeneous equation with constant coefficients and its solution is expressed as:

$$\widetilde{E} = A\exp(\lambda_1 x) + B\exp(-\lambda_1 x) \tag{4.9}$$

with

$$\lambda_1 = \sqrt{R_P\left[\frac{1}{R_C} + C_{dl} j\omega\right]}$$

Evaluating the boundary condition $\frac{d\widetilde{E}}{dx}_{x=0} = 0$ in Eq. (4.9):

$$\widetilde{E} = A[\exp(\lambda_1 x) + \exp(-\lambda_1 x)] \tag{4.10}$$

Evaluating the derivative of Eq. (4.10) as

$$\frac{d\widetilde{E}}{dx}_{x=1} = A\lambda_1[\exp(\lambda_1) - \exp(-\lambda_1)] \tag{4.11}$$

Substituting Eq. (4.11) into Eq. (4.6) to define the parameter A:

$$A = \frac{\widetilde{i} R_P}{\lambda_1[\exp(\lambda_1) - \exp(-\lambda_1)]} \tag{4.12}$$

Substituting Eq. (4.12) into Eq. (4.10) and evaluating $x=1$:

$$\widetilde{E} = \frac{\widetilde{i} R_P}{\lambda_1}\coth(\lambda_1) \tag{4.13}$$

The impedance response of the TLM shown in Fig. 4.3 can be simulated through the following equation:

$$Z_{TLM} = \frac{\widetilde{E}}{\widetilde{i}} = \frac{R_P \coth\left(\sqrt{R_P\left[\frac{1}{R_C} + C_{dl} j\omega\right]}\right)}{\sqrt{R_P\left[\frac{1}{R_C} + C_{dl} j\omega\right]}} \tag{4.14}$$

Makharia et al. [6] fitted the TLM shown in Fig. 4.3 with EIS measurements carried out in a $5\,cm^2$ PEMFC (H_2/O_2) operated at low current density with a membrane electrode assembly (MEA) containing a Nafion/carbon weight ratio of 0.8 and 0.4. The parameters of the TLM estimated from EIS measurements were obtained using ZView software (Transmission-Line Open Circuit Terminus, DX-Type 6, Scribner Associates, Inc., version 2.3). The parameters shown

TABLE 4.1 Parameters of the transmission-line model accounting for the cathode catalyst layer and reported by Makharia et al. [6].

i A/cm^2 Current density	R_P Ωcm^2 Proton resistance	C_{dl} mF/cm^2 Capacitance	R_C Ωcm^2 Charge transfer resistance
0.03	0.122	19.67	1.215

in Table 4.1 were obtained by fitting the TLM shown in Fig. 4.3 with EIS measurements carried out in a PEMFC at a current density of 0.03 A/cm^2. More details about the estimation of the parameters shown in Table 4.1 from EIS measurements of the PEMFC can be found in the study reported by Makharia et al. [6].

The parameters shown in Table 4.1 are substituted into Eq. (4.14) to simulate the impedance response of the cathode catalyst layer (CCL) and using MATLAB software. Fig. 4.5 shows the impedance response generated by Eq. (4.14) with the parameters shown in Table 4.1. The impedance response was simulated in MATLAB software considering a range of frequencies from 10 MHz to 0.1 Hz. The diameter of the semicircle shown in the spectrum from Fig. 4.5 is related to the charge transfer resistance R_C during the oxygen reduction reaction (ORR) in the CCL.

Fig. 4.6 shows the high frequency region of the impedance response shown in Fig. 4.5. A 45-degree straight line at the end of the impedance spectrum at

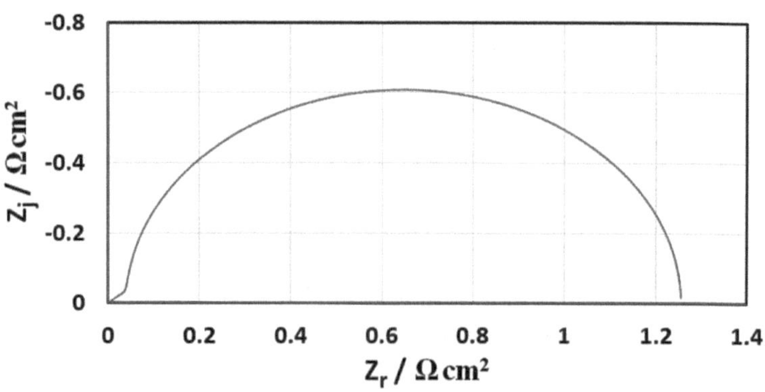

FIG. 4.5 Simulated impedance response of the cathode catalyst layer with negligible electron resistance. The impedance response was simulated in MATLAB using Eq. (4.14) and using parameters $R_P = 0.122$ Ωcm^2, $R_C = 1.215$ Ωcm^2, $C_{dl} = 19.67 \times 10^{-3}$ F/cm^2 reported in the study by Makharia et al. [6].

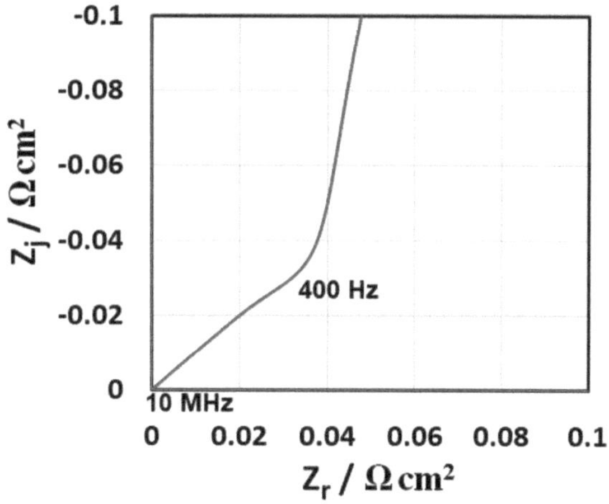

FIG. 4.6 High frequency simulated impedance response of the cathode catalyst layer with negligible electron resistance. The impedance response was simulated in MATLAB using Eq. (4.14) and using parameters $R_P = 0.122\ \Omega cm^2$, $R_C = 1.215\ \Omega cm^2$, $C_{dl} = 19.67 \times 10^{-3}\ F/cm^2$ reported in the study by Makharia et al. [6].

frequencies from 10 MHz to 400 Hz has been associated with the resistance of proton conduction through the electrolyte of the CCL.

4.2.2 Transmission-line model with no faradaic resistance

Some studies [7,8] have opted to estimate the proton resistance in the CCL by carrying out EIS measurements in PEMFCs operated with H_2/N_2 as reactants. Eq. (4.14) can only be used to characterise the impedance response of the CCL in a PEMFC supplied with either H_2/air or H_2/O_2 at low current operation (no oxygen transport limitations). If a PEMFC is operated with H_2/N_2; no faradaic reaction during the ORR in the cathode will take place, and under this operating condition the charge transfer resistance R_C in Eq. (4.14) becomes infinite. If $R_C = \infty$ (no faradaic reaction) is considered in Eq. (4.14), the impedance of the TLM can be expressed as:

$$Z_{TLM, R_C=\infty} = \frac{\widetilde{E}}{\widetilde{i}} = \frac{R_P \coth\left(\sqrt{R_P C_{dl} j\omega}\right)}{\sqrt{R_P C_{dl} j\omega}} \tag{4.15}$$

Eq. (4.15) can be represented through a TLM neglecting the charge transfer resistance connected between the upper and lower rails and also neglecting the electron resistance in the lower rail as shown in Fig. 4.7. The TLM shown in Fig. 4.7 can be considered to estimate the proton resistance of the CCL in EIS measurements carried out in H_2/N_2 PEMFCs [8].

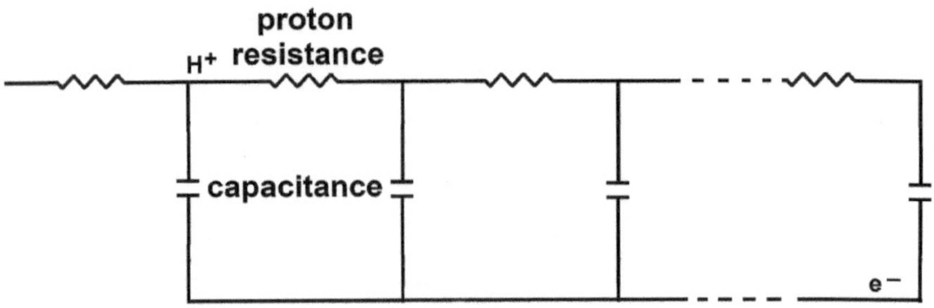

FIG. 4.7 Transmission-line model to represent the impedance response of catalyst layers with negligible electron resistance for electron conduction in the carbon structure and negligible charge transfer resistance (no faradaic reaction).

It is possible to separate the real and imaginary components in the impedance response generated by Eq. (4.15) through the use of trigonometric identities [9]. The real component of Eq. (4.15) can be expressed as:

$$Re\_Z_{\mathrm{TLM},R_\mathrm{C}=\infty} = R_\mathrm{P}\frac{\sinh(\varphi) - \sin(\varphi)}{\varphi[\cosh(\varphi) - \cos(\varphi)]} \tag{4.16}$$

with

$$\varphi = \sqrt{2R_\mathrm{P}C_\mathrm{dl}\omega}$$

Fig. 4.8 shows a script in MATLAB software to solve the Taylor series expansion of Eq. (4.16) evaluated at $\omega=0$.

Based on the solution of the Taylor series expansion of Eq. (4.16), if the frequency tends to zero $\omega\rightarrow 0$, the real component of the impedance response of the TLM represented in Eq. (4.16) results in

$$Re\_Z_{\mathrm{TLM},R_\mathrm{C}=\infty,\omega=0} = \frac{R_\mathrm{P}}{3} \tag{4.17}$$

Fig. 4.9 shows the impedance response generated by Eq. (4.15) considering the parameters from Table 4.1. The simulation was carried out in MATLAB software and the considered frequency range was from 10 MHz to 0.1 Hz. As there is no faradaic reaction, the impedance response of the capacitive effect between dissimilar materials (electrode/electrolyte interface) is represented through a vertical line and is comprised of imaginary components which become more negative as the frequency is reduced. The impedance response of a capacitor was discussed in Section 1.4.1 from Chapter 1. The proton resistance R_P can be graphically estimated by projecting the 45-degree straight line onto the real axis Z_r and dividing it by 3 ($R_\mathrm{P}/3$), as shown in Fig. 4.9. Eq. (4.17) demonstrated that the value of the real component Z_r of the impedance response as the frequency tends to zero $\omega\rightarrow 0$ results in $R_\mathrm{P}/3$.

Taylor_series.m

```
% Definition of symbolic parameters
syms Rp C w
% Parameter defined in Eq. 4.16
q=sqrt(2*Rp*C*w);
% Real component of TLM, Eq. 4.16
Re_Z=Rp*(sinh(q)-sin(q))/(q*(cosh(q)-cos(q)));
% Taylor series expansion
Ts=taylor(Re_Z);
% Real component evaluated at w=0
Re_w_zero=(subs(Ts,w,0))
```

Command Window

```
>> Taylor_series

Re_w_zero =

Rp/3

fx >>
```

FIG. 4.8 Script in MATLAB software for the solution of Taylor series expansion of Eq. (4.16) and evaluated at $\omega = 0$.

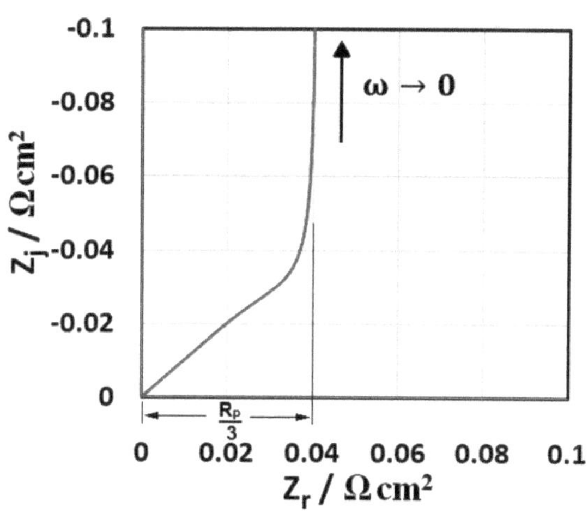

FIG. 4.9 Simulated impedance spectrum generated through Eq. (4.15) with $R_P = 0.122\ \Omega cm^2$ and $C_{dl} = 19.67 \times 10^{-3}$ F/cm^2.

4.3 Impedance model of the cathode catalyst layer with low electron conductivity

Physics-based impedance models of the CCL have been reported by Kulikovsky [10–17]. The impedance models consider proton, electron, and oxygen transport through the CCL. The impedance model of the CCL considering low electron conductivity in the carbon structure reported by Kulikovsky [18] is described in this section. The TLM shown in Fig. 4.2b can also simulate the impedance response of the CCL with low electron conductivity and will be demonstrated later. The impedance model of the CCL reported by Kulikovsky [18] during low current operation is expressed as:

$$\widetilde{Z}_{\text{ct+pe}} = \frac{jq\sin(p) + [1+j]\left[2k_\sigma + \left[1+k_\sigma^2\right]\cos(p)\right]}{k_\sigma\left[jq - [1+j]p\right]\sin(p)} \tag{4.18}$$

with

$$p = \sqrt{-\left[\widetilde{i} + j\widetilde{\Omega}\right]\left[1 + \frac{1}{k_\sigma}\right]}$$

$$q = \sqrt{2k_\sigma[1+k_\sigma]\left[j\widetilde{i} - \widetilde{\Omega}\right]}$$

$$\widetilde{\Omega} = \frac{\omega C_{\text{dl}} \ell_{\text{CCL}}{}^2}{\sigma_{\text{H}^+}}$$

$$k_\sigma = \frac{\sigma_{\text{e}^-}}{\sigma_{\text{H}^+}}$$

where ℓ_{CCL} is the thickness of the CCL and k_σ is the ratio between electron σ_{e^-} and proton σ_{H^+} conductivities in the CCL. More details about the nomenclature in Eq. (4.18) can be found in the study from Kulikovsky [18].

If the frequency is reduced towards a very low value $\widetilde{\Omega} \to 0$ in Eq. (4.18), the DC resistance in the low frequency limit can be expressed as:

$$\widetilde{R}_{\text{ct+pe}} = \frac{1}{\widetilde{i}} + \frac{1}{3} + \frac{1}{3k_\sigma} \tag{4.19}$$

If the frequency is increased towards a very high value $\widetilde{\Omega} \to \infty$ in Eq. (4.18), a high frequency resistance is predicted and expressed as:

$$\widetilde{R}_{\text{HF}} = \frac{1}{1+k_\sigma} \tag{4.20}$$

Fig. 4.10 shows the impedance response of the CCL generated through Eq. (4.18) and reported in the study by Kulikovsky [18]. The effect of the ratio between electron and proton conductivity k_σ on the impedance response of the CCL can be observed. The high and low frequency limits predicted by Eqs (4.19) and (4.20) can also be observed in Fig. 4.10a. If the electron conduction

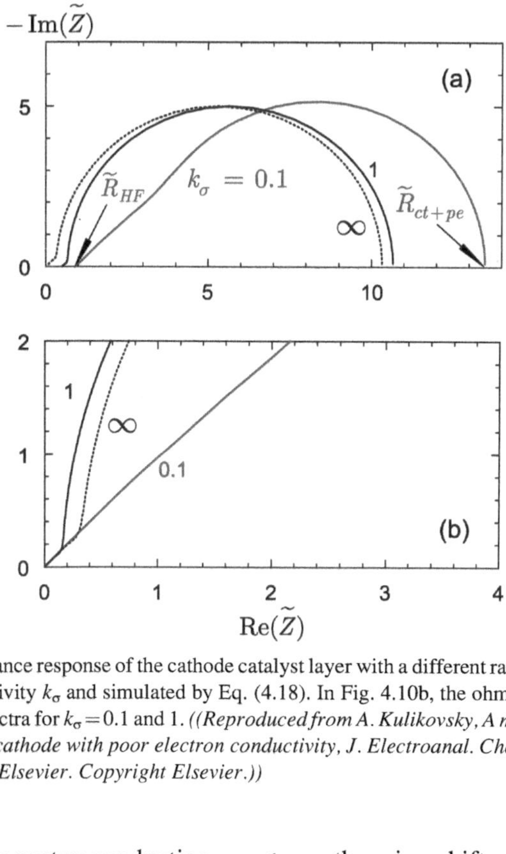

FIG. 4.10 Impedance response of the cathode catalyst layer with a different ratio between electron and proton conductivity k_σ and simulated by Eq. (4.18). In Fig. 4.10b, the ohmic resistivity is subtracted from the spectra for $k_\sigma = 0.1$ and 1. *((Reproduced from A. Kulikovsky, A model for impedance of a PEM fuel cell cathode with poor electron conductivity, J. Electroanal. Chem. 801 (2017) 125, with permission of Elsevier. Copyright Elsevier.))*

is lower than the proton conduction $\sigma_{e^-} < \sigma_{H^+}$, there is a shift of the impedance spectrum towards the right or towards the upper end of the real component of the complex-impedance plot as shown in Fig. 4.10a.

The impedance model of the CCL considering oxygen diffusion during low current operation was also reported by Kulikovsky [18] as:

$$Z_{ccl} = \frac{\ell_{CCL}\widetilde{Z}_{ct+pe}}{\sigma_{H^+}} + Z_{ox} \tag{4.21}$$

with

$$Z_{ox} = \frac{b[1 - W]}{i\left[W - \frac{\omega^2}{\omega_0^* \omega_{ct}^*} + j\omega\left[\frac{1}{\omega_{ct}^*} + \frac{1}{\omega_0^*}\right]\right]\left[1 + \frac{j\omega}{\omega_{ct}^*}\right]} \tag{4.22}$$

and the DC resistance in the low frequency limit of the impedance response at $\omega \to 0$ is represented as:

$$R_{ccl,\omega\to 0} = \frac{b}{i} + \frac{\ell_{CCL}}{3\sigma_{H^+}} + \frac{\ell_{CCL}}{3\sigma_{e^-}} + \frac{b\ell_{CCL}}{12FD_{ox}c_1} \tag{4.23}$$

The impedance model reported by Kulikovsky [18] can simulate the impedance response of the CCL with different proton and electron conductivities whilst also considering oxygen diffusion in the CCL. Nevertheless, the impedance model is limited to low current operation. In addition, the proton and electron resistances associated with the ORR were defined as $R_P = \ell_{CCL}/3\sigma_{H^+}$ and $R_e = \ell_{CCL}/3\sigma_{e^-}$ and the proton and electron resistances associated with proton and electron conduction in the carbon and Nafion clusters of the CCL were defined as $R_{HF,p} = \ell_{CCL}/\sigma_{H^+}$ and $R_{HF,e} = \ell_{CCL}/\sigma_{e^-}$ which are related to the total ohmic high frequency resistance expressed in Eq. (4.20).

4.4 Electrochemical phenomena in the cathode catalyst layer

The electrochemical reaction occurring at the electrode considering a simple reaction or a single-step electron transfer takes the following form:

$$\text{Ox} + \text{e}^- \overset{k}{\leftrightarrow} \text{Red} \tag{4.24}$$

where Ox is the oxidised form of the chemical species (reactants), and Red is the reduced form of the chemical species (products). In equilibrium both reactions take place simultaneously, and the reactant rate (i.e. the amount of a reactant utilised or the amount of a product formed per unit time) is proportional to the surface concentration. However, the chemical reaction proceeds at different rates. The factors that affect reaction rates are: the surface area of a catalyst site, temperature, the phase of the reactants (i.e. gas or liquid) and the catalyst loading. A change in one or more of these factors may alter the rate of a reaction. When a catalyst is used (e.g. platinum in PEMFCs) the reaction rate can be considered on a catalyst weight or surface area basis. In this analysis, catalyst layer thickness is also considered.

For reaction (4.24), the rate of change for forward reaction flux of the chemical species across the nondimensional thickness x of the catalyst layer is given by:

$$\frac{dv_f}{dx} = k_f c_{Ox} \tag{4.25}$$

and the rate of change for the backward reaction is given by:

$$\frac{dv_b}{dx} = k_b c_{Red} \tag{4.26}$$

where $v_{f,b}$ is the flux of the species with units of mol/scm^2, $k_{f,b}$ is the reaction rate coefficient with units cm/s and $c_{Ox,Red}$ is the surface concentration of the chemical species with units mol/cm^3. Substituting Faraday's law $v = i/zF$ into the forward reaction flux Eq. (4.25) yields an expression for the rate of change for the forward current density:

$$\frac{d\vec{i}_f}{dx} = zFk_f c_{Ox} \tag{4.27}$$

and substituting Faraday's law $v = i/zF$ into the backward reaction flux Eq. (4.26) yields the backward current density:

$$\frac{d\overleftarrow{i_b}}{dx} = zFk_b c_{Red} \tag{4.28}$$

where z is the number of electrons released or consumed and F is the Faraday constant. Another factor altering the rate of the reaction is temperature. This can be expressed by the Arrhenius equation:

$$k_{f,b} = A_a \exp\left(\frac{-\Delta G_{f,b}}{RT}\right) \tag{4.29}$$

where $\Delta G_{f,b}$ is the standard free energy of activation (Gibbs free energy) related to the forward or backward reaction, R is the ideal gas constant, T is the temperature and A_a is called frequency factor. If the potential in the electrode is changed to a new value, the Gibbs free energy will be changed as well. The effect of change in Gibbs free energy for the forward and backward reaction is shown in Fig. 4.11.

It is possible to relate a parameter defined as the charge transfer coefficient α which is a measure of the symmetry of the energy barrier during the electrochemical reaction for a change of potential, as shown in Fig. 4.11.

Taking into account the change in potential represented in Fig. 4.11, the change in Gibbs free energy for forward and back reaction rates can be represented as:

$$\Delta G_f = \Delta G_{f_{eq}} + zF\alpha\left[E - E_{eq}\right] \tag{4.30}$$

$$\Delta G_b = \Delta G_{b_{eq}} - zF[1 - \alpha]\left[E - E_{eq}\right] \tag{4.31}$$

where $\Delta G_{f_{eq},b_{eq}}$ is the activation energy at equilibrium conditions, E_{eq} is the potential at equilibrium conditions and E is the potential at nonequilibrium conditions. Substituting Eq. (4.30) into Eq. (4.29) yields the forward reaction rate

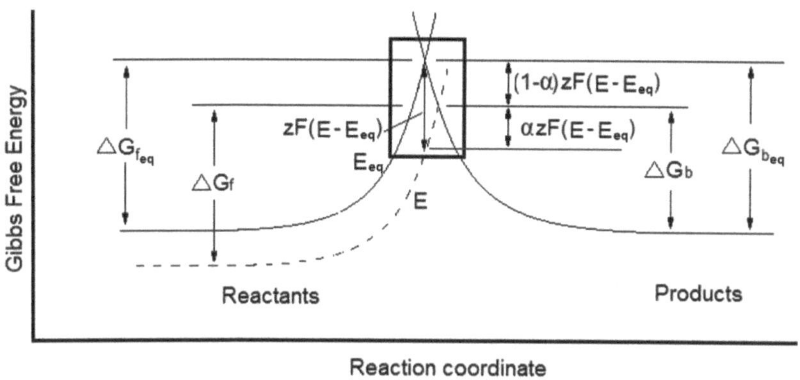

FIG. 4.11 Effect of the potential change on the free energy of activation.

coefficient considering the change in equilibrium potential on the free energy of activation.

$$k_{\mathrm{f}} = A_{\mathrm{a}} \exp\left(\frac{-\Delta G_{\mathrm{f_{eq}}}}{RT}\right) \exp\left(-\frac{zF\alpha\left[E - E_{\mathrm{eq}}\right]}{RT}\right) \tag{4.32}$$

Substituting Eq. (4.31) into Eq. (4.29) yields the backward reaction rate coefficient:

$$k_{\mathrm{b}} = A_{\mathrm{a}} \exp\left(\frac{-\Delta G_{\mathrm{b_{eq}}}}{RT}\right) \exp\left(\frac{zF[1-\alpha]\left[E - E_{\mathrm{eq}}\right]}{RT}\right) \tag{4.33}$$

Substituting Eq. (4.32) into the forward current density Eq. (4.27) yields:

$$\frac{d\vec{i}_{\mathrm{f}}}{dx} = zFc_{\mathrm{Ox}}k_{\mathrm{f_{eq}}} \exp\left(-\frac{\alpha zF\left[E - E_{\mathrm{eq}}\right]}{RT}\right) \tag{4.34}$$

and substituting Eq. (4.33) into the backward current density Eq. (4.28) yields:

$$\frac{d\overleftarrow{i}_{\mathrm{b}}}{dx} = zFc_{\mathrm{Red}}k_{\mathrm{b_{eq}}} \exp\left(\frac{zF[1-\alpha]\left[E - E_{\mathrm{eq}}\right]}{RT}\right) \tag{4.35}$$

where $k_{\mathrm{f_{eq}},\mathrm{b_{eq}}} = A_{\mathrm{a}} \exp\left(-\Delta G_{\mathrm{f_{eq}},\mathrm{b_{eq}}}/RT\right)$ is called the standard rate constant which is also called the intrinsic rate constant in the electrochemical literature.

Under equilibrium conditions, the reaction in Eq. (4.24) proceeds in both directions simultaneously at the same rate. Therefore the equilibrium potential is equal to the potential of the electrode at nonequilibrium conditions, $E_{\mathrm{eq}}=E$. In addition, the concentration (denominated bulk concentration under equilibrium conditions) is the same for the chemical species, $c^*_{\mathrm{Ox,Red}}$. Substituting $E_{\mathrm{eq}}=E$ and $c_{\mathrm{Ox,Red}}=c^*_{\mathrm{Ox,Red}}$ into Eqs (4.34) and (4.35) gives the rate at which the forward and backward transfer of charge occurs at equilibrium conditions which is known as the exchange current density, i_0, as such:

$$i_0 = zFc^*_{\mathrm{Ox}}k_{\mathrm{f_{eq}}} = zFc^*_{\mathrm{Red}}k_{\mathrm{b_{eq}}} \tag{4.36}$$

When an external load is connected to a PEMFC a current flows through the electrodes. The total faradaic current density is the difference between the forward and backward current densities and can be expressed as:

$$\frac{di_{\mathrm{F}}}{dx} = \frac{d\vec{i}_{\mathrm{f}}}{dx} - \frac{d\overleftarrow{i}_{\mathrm{b}}}{dx} \tag{4.37}$$

Substituting Eqs (4.34) and (4.35) into Eq. (4.37) gives:

$$\begin{aligned}\frac{di_{\mathrm{F}}}{dx} = {} & zFc_{\mathrm{Ox}}k_{\mathrm{f_{eq}}} \exp\left(-\frac{\alpha zF\left[E - E_{\mathrm{eq}}\right]}{RT}\right) \\ & - zFc_{\mathrm{Red}}k_{\mathrm{b_{eq}}} \exp\left(\frac{zF[1-\alpha]\left[E - E_{\mathrm{eq}}\right]}{RT}\right)\end{aligned} \tag{4.38}$$

A relationship between the potential far from equilibrium with regard to the equilibrium potential can be expressed as:

$$\eta = E - E_{\text{eq}} \tag{4.39}$$

where η is a deviation from both potentials and is known as overpotential. Substituting Eq. (4.39) into the expression of total current Eq. (4.38) gives:

$$\frac{di_{\text{F}}}{dx} = zFc_{\text{Ox}}k_{\text{f}_{\text{eq}}}\exp\left(-\frac{\alpha zF\eta}{RT}\right) - zFc_{\text{Red}}k_{\text{b}_{\text{eq}}}\exp\left(\frac{[1-\alpha]zF\eta}{RT}\right) \tag{4.40}$$

Multiplying both terms in Eq. (4.40) by the bulk concentration $c^*_{\text{Ox,Red}}$, and substituting the exchange current density i_0 defined in Eq. (4.36), gives:

$$\frac{di_{\text{F}}}{dx} = \frac{c_{\text{Ox}}}{c^*_{\text{Ox}}}i_0\exp\left(-\frac{\alpha zF\eta}{RT}\right) - \frac{c_{\text{Red}}}{c^*_{\text{Red}}}i_0\exp\left(\frac{[1-\alpha]zF\eta}{RT}\right) \tag{4.41}$$

At large negative overpotential η (Tafel approximation), the contribution of the second term on the right-hand side of Eq. (4.41) becomes small, and therefore it can be neglected. The final equation for the rate of current density generation across the CCL is:

$$\frac{di_{\text{F}}}{dx} = i_0\frac{c_{\text{Ox}}}{c^*_{\text{Ox}}}\exp\left(-\frac{\alpha zF\eta}{RT}\right) \tag{4.42}$$

4.4.1 Faradaic current and current in double-layer capacitance

The CCL is formed by a double-layer structure for the interface between the dissimilar materials, i.e., the electrode/electrolyte interface. The electrode represents the electron-conducting network of carbon-supported platinum agglomerates while the electrolyte represents the ion-conducting dispersion of Nafion. At this interface, an electric field is present and has a determinant role in the charge distribution within the reactants as well as in the position and orientation of the reactant to form the desired products. This double layer can behave like a capacitor C_{dl} that is in parallel with the electrode reactions, the current passing from the electrode to the electrolyte either can take part in the charge transfer reaction or can contribute to the charge in the capacitive effect. The current density due to double-layer capacitance between dissimilar materials in the CCL is expressed as:

$$\frac{i_{C_{\text{dl}}}}{dx} = C_{\text{dl}}\frac{dE}{dt} \tag{4.43}$$

where C_{dl} is the double-layer capacitance between electrode and electrolyte with units F/cm^2 and E is the potential difference between the solution phase and the solid phase in the CCL structure.

Both faradaic and nonfaradaic processes are present when the ORR takes place. The total current density is defined as the sum between faradaic current density and nonfaradaic current density (charge capacitance current density):

$$\frac{di}{dx} = \frac{di_{\mathrm{F}}}{dx} + \frac{di_{C_{\mathrm{dl}}}}{dx} \tag{4.44}$$

where i_{F} is the current density related to the faradaic process due to the charge transfer in the ORR, and $i_{C_{\mathrm{dl}}}$ is the current density related to the nonfaradic process between the electrode/electrolyte interface in the CCL. Substituting Eqs (4.42) and (4.43) into Eq. (4.44) yields:

$$\frac{di}{dx} = i_0 \frac{c_{\mathrm{Ox}}}{c_{\mathrm{Ox}}^*} \exp\left(-\frac{\alpha z F \eta}{RT}\right) + C_{\mathrm{dl}} \frac{dE}{dt} \tag{4.45}$$

4.4.2 Proton and electron resistances in the cathode catalyst layer

A potential difference in the CCL is present because of the charge distribution between electrons and protons at the electrode/electrolyte interface. The potential in the solution phase E_{Φ} across the nondimensional thickness x of the CCL can be defined as:

$$\frac{dE_{\Phi}}{dx} = i_{\mathrm{p}} R_{\mathrm{P}} \tag{4.46}$$

where R_{P} is the proton resistance in the electrolyte network of the CCL. The potential in the solid phase E_{ϕ} across the nondimensional thickness x of the CCL can be defined as:

$$\frac{dE_{\phi}}{dx} = -i_{\mathrm{e}} R_{\mathrm{e}} \tag{4.47}$$

where R_{e} is the electron resistance in the carbon-structure network in the CCL. The negative sign in Eq. (4.47) implies that electron conduction occurs from the gas diffusion layer (GDL)–CCL interface to the polymer electrolyte membrane (PEM)–CCL interface and is opposite to the proton conduction that occurs from the CCL–PEM interface to the CCL–GDL interface as shown in Fig. 4.12. The impedance model of the CCL reported by Kulikovsky [18] and presented in the previous Section 4.3 considered charge conservation $i_{\mathrm{m}} = i_{\mathrm{p}} + i_{\mathrm{e}}$ for proton and electron current density as shown in Fig. 4.12.

The electron resistance in the carbon structure of the CCL can be neglected where $R_{\mathrm{e}} = 0$ because it is commonly considered that the resistance for proton conduction is higher than the resistance for electron conduction $R_{\mathrm{P}} \gg R_{\mathrm{e}}$ by several orders of magnitude [5,19]. The potential across the nondimensional thickness of the CCL considering the potential difference at the electrode/electrolyte

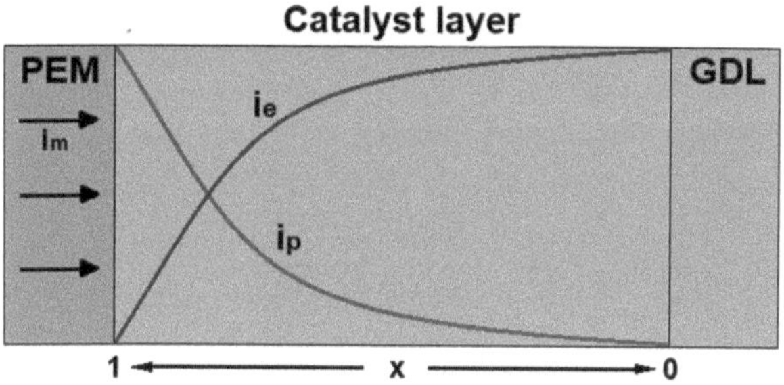

FIG. 4.12 Proton and electron current density distribution across the cathode catalyst layer. Polymer electrolyte membrane (PEM), gas diffusion layer (GDL).

interface $E = E_\Phi - E_\phi$ and neglecting electron resistance $R_e = 0$ can be represented as:

$$\frac{dE}{dx} = iR_P \tag{4.48}$$

Eq. (4.48) implies that the current density across the CCL thickness is only a function of the protonic current density $i = i_p$.

4.5 Electrochemical impedance of the cathode catalyst layer at low current

At low current operation the gradient of reactant concentration across the CCL can be considered negligible $\frac{c_{Ox}}{c^*_{Ox}} = 1$ in Eq. (4.45) as such:

$$\frac{di}{dx} = i_0 \exp\left(-\frac{\alpha z F \eta}{RT}\right) + C_{dl}\frac{dE}{dt} \tag{4.49}$$

The PEMFC demonstrates nonlinear current–voltage behaviour, as the current has an exponential dependence on potential. This can be seen in the polarisation curve representing the performance of the PEMFC as demonstrated in Fig. 2.5 from Chapter 2. The use of a low amplitude value in the sinusoidal signal during EIS measurements allows application of a linear model in order to interpret the impedance spectra. A linear model can be derived using the Taylor series expansion about the steady-state value as such:

$$\tilde{i} = \left(\frac{\partial i}{\partial E}\right)_{ss} \tilde{E} \tag{4.50}$$

Eq. (4.50) has been reported by Orazem and Tribollet [4] and can be applied to the faradaic current from any electrochemical reaction. $\tilde{i}$ and $\tilde{E}$ represent the oscillating current density and potential. Applying Eq. (4.50) into Eq. (4.42) and considering $\frac{c_{Ox}}{c^*_{Ox}} = 1$ yields:

$$\frac{d\widetilde{i_{\mathrm{F}}}}{dx}=\frac{\widetilde{E}}{R_{\mathrm{C}}} \tag{4.51}$$

where R_C represents the charge transfer resistance during the ORR and is defined as

$$R_{\mathrm{C}}=\frac{|b_{\mathrm{c}}|}{i_0 \exp\left(\eta_{\mathrm{SS}}/b_{\mathrm{c}}\right)} \tag{4.52}$$

where η_{SS} is the overpotential at steady state and b_c is the Tafel slope defined as

$$b_{\mathrm{c}}=-\frac{RT}{\alpha zF} \tag{4.53}$$

The oscillating charge capacitance current has been derived in Chapter 1. This derivation has been demonstrated through Eqs (1.49)–(1.51) from Chapter 1. Eq. (4.43) can be expressed in the frequency domain considering the Laplace transform and the same derivation process of Eq. (1.51) from Chapter 1:

$$\frac{d\widetilde{i_{C_{\mathrm{dl}}}}}{dx}=C_{\mathrm{dl}}j\omega\widetilde{E} \tag{4.54}$$

The total oscillating current density can be defined as the sum of the oscillating faradaic current density (Eq. 4.51) and the oscillating charge capacitance current density (Eq. 4.54) as such:

$$\frac{d\widetilde{i}}{dx}=\frac{\widetilde{E}}{R_{\mathrm{C}}}+C_{\mathrm{dl}}j\omega\widetilde{E} \tag{4.55}$$

Noting that Eq. (4.55) is similar to Eq. (4.4) that is related to the equivalent impedance connected between the upper and lower rails of the TLM shown in Fig. 4.4. At this stage, it could be deduced that the impedance response of the CCL would be analogous to the impedance response of the TLM shown in Fig. 4.3.

The electrochemical reaction in a PEMFC results in an inhomogeneous distribution of charge in the catalyst layer. As a result, a nonideal capacitive behaviour has to be considered in the theoretical treatment. In order to account for this inhomogeneity, a constant phase element (CPE) is considered instead of an ideal capacitor [20]. A CPE improves the quality of fitting between an impedance model and the impedance response of an electrochemical system. The book of Orazem and Tribollet [4] has published a full chapter to describe the physical meaning of the CPE when applying it to interpret the impedance response of electrochemical systems with time-constant dispersion.

The impedance response of a CPE has been introduced in Chapter 1 Section 1.4.5. The oscillating current for a CPE can be defined as:

$$\frac{d\widetilde{i_{\mathrm{CPE}}}}{dx}=Y[j\omega]^{P}\widetilde{E} \tag{4.56}$$

where Y represents a parameter related to CPE with units $s^P/\Omega cm^2$, and the superscript P represents a parameter to correct the inhomogeneity in the distribution of charge. Eq. (4.56) represents the oscillating current density attributed to the difference of potential between dissimilar materials and nonhomogeneous charge distribution in the carbon/electrolyte interface in the CCL. For the specific case if superscript P is equal to 1 in Eq. (4.56); Eq. (4.56) would represent the oscillating current density through an ideal capacitor with Y having units F/cm^2. The second term on the right-hand side of Eq. (4.55) can be replaced by Eq. (4.56) to account for the nonhomogeneous charge distribution at the electrolyte/carbon interface of the CCL, as such:

$$\frac{d\tilde{i}}{dx} = \frac{\tilde{E}}{R_C} + Y[j\omega]^P \tilde{E} \tag{4.57}$$

Expressing Eq. (4.48) as a function of the oscillating current density and potential resulting from the sinusoidal waveform during EIS yields:

$$\frac{d\tilde{E}}{dx} = \tilde{i}R_P \tag{4.58}$$

4.5.1 Effect of proton resistance on steady-state overpotential

It is important to mention that the steady-state potential E_{SS}, related to the oscillating potential $\tilde{E}$ (oscillating deviation of the potential in the steady-state) in Eq. (4.58), will change across the nondimensional thickness x of the CCL. Therefore, if Eq. (4.57) is substituted into Eq. (4.58), the steady-state overpotential η_{SS} defined in the charge transfer resistance R_C (Eq. 4.52) will change across the coordinate x as well. It is possible to consider a constant value for the steady-state overpotential η_{SS} in R_C across x only if the condition $R_P/R_C < 1$ is complied. The relation $R_P/R_C < 1$ is encountered in catalyst layers of PEMFCs and has been considered for the derivation of the impedance model of the CCL as reported by Makharia et al. [6]. This is demonstrated as follows:

Eq. (4.58) can be represented in terms of steady-state potential and current density as such:

$$\frac{dE_{SS}}{dx} = i_{SS}R_P \tag{4.59}$$

The ratio between charge transfer resistance R_C expressed in Eq. (4.52) and proton resistance R_P expressed in Eq. (4.59) can be expressed as:

$$\frac{R_P}{R_C} = \frac{dE_{SS}}{dx} \frac{i_0 \exp\left(\eta_{SS}/b_c\right)}{|b_c| i_{SS}} \tag{4.60}$$

Separating variables in Eq. (4.60) gives:

$$\exp\left(\frac{\eta_{SS}}{b_c}\right) dE_{SS} = \frac{R_P}{R_C} \frac{|b_c| i_{SS}}{i_0} dx \tag{4.61}$$

Integrating Eq. (4.61) with limits at $x=0$ in the CCL–GDL interface where $E_{SS}=E_{SS0}$, and at $x=1$ in the CCL–PEM interface where $E_{SS}=E_{SS1}$

$$\frac{b_c}{\exp\left(E_{eq}/b_c\right)}\int_{E_{SS0}}^{E_{SS1}}\exp\left(\frac{E_{SS}}{b_c}\right)dE_{SS}=\frac{R_P}{R_C}\frac{|b_c|i_{SS}}{i_0}\int_0^1 dx \tag{4.62}$$

yields:

$$\exp\left(\frac{\eta_{SS1}}{b_c}\right)-\exp\left(\frac{\eta_{SS0}}{b_c}\right)=-\frac{R_P}{R_C}\frac{i_{SS}}{i_0} \tag{4.63}$$

Rearranging Eq. (4.63) yields:

$$\frac{i_0}{i_{SS}}\exp\left(\frac{\eta_{SS0}}{b_c}\right)=\frac{i_0}{i_{SS}}\exp\left(\frac{\eta_{SS1}}{b_c}\right)+\frac{R_P}{R_C} \tag{4.64}$$

If the relation $R_P/R_C<1$ is considered in Eq. (4.64), the contribution of the second term on the right-hand side of Eq. (4.64) is reduced and can be considered negligible and the difference between the overpotential η_{SS0} at $x=0$ in the CCL–GDL interface and the overpotential η_{SS1} at $x=1$ in the CCL–PEM interface is small $\eta_{SS0}\approx\eta_{SS1}$.

4.5.2 Impedance model of the cathode catalyst layer

As demonstrated previously in Section 4.5.1 the value of the steady-state overpotential η_{SS} in the charge transfer resistance R_C can be considered constant along the nondimensional thickness x of the CCL if the relation $R_P/R_C<1$ is considered. Substituting Eq. (4.57) into Eq. (4.58) and considering that the relation $R_P/R_C<1$ is complied yields:

$$\frac{d^2\tilde{i}}{dx^2}=R_P\left[\frac{\tilde{i}}{R_C}+Y[j\omega]^P\tilde{i}\right] \tag{4.65}$$

Eq. (4.65) is a second order homogeneous equation and its solution can be obtained by applying the method of the nth-order homogeneous equation with constant coefficients. The solution of a homogeneous equation with different roots can be obtained as follows:

$$\tilde{i}=A\exp\left(\beta_1 x\right)+B\exp\left(-\beta_1 x\right) \tag{4.66}$$

with

$$\beta_1=\sqrt{R_P\left[\frac{1}{R_C}+Y[j\omega]^P\right]}$$

Evaluating boundary conditions in Eq. (4.66), at the CCL–GDL interface $x=0$ and $\tilde{i}=0$, while at the CCL–PEM interface $x=1$ and $\tilde{i}=\tilde{i}_m$ yields:

$$\tilde{i} = \tilde{i}_{\mathrm{m}} \frac{\sinh(\beta_1 x)}{\sinh(\beta_1)} \tag{4.67}$$

The oscillating potential in the frequency domain can be obtained by arranging Eq. (4.57) as such:

$$\tilde{E} = \frac{R_{\mathrm{C}}}{1 + Y[j\omega]^P R_{\mathrm{C}}} \frac{d\tilde{i}}{dx} \tag{4.68}$$

Differentiating Eq. (4.67) with respect to x and substituting into Eq. (4.68) gives the oscillating potential:

$$\tilde{E} = \frac{R_{\mathrm{C}} \tilde{i}_{\mathrm{m}} \beta_1 \cosh(\beta_1 x)}{\left[1 + Y[j\omega]^P R_{\mathrm{C}}\right] \sinh(\beta_1)} \tag{4.69}$$

The impedance model of the CCL is defined as the ratio between the oscillating potential, Eq. (4.69), and the oscillating current density, Eq. (4.67), in the frequency domain:

$$Z_{\mathrm{CCL}} = \frac{\tilde{E}}{\tilde{i}} = \frac{R_{\mathrm{P}} \coth(\beta_1 x)}{\beta_1} \tag{4.70}$$

with

$$\beta_1 = \sqrt{R_{\mathrm{P}} \left[\frac{1}{R_{\mathrm{C}}} + Y[j\omega]^P\right]}$$

where x represents the nondimensional thickness of the CCL and for EIS measurements in PEMFCs (lumped parameters) is equal to 1. Eq. (4.70) represents the opposition to the flow of an alternating current AC within the CCL which contains physical processes that dissipate energy and store energy. The opposition to the AC load can be measured over a range of frequencies, thereby revealing the impedance response of the CCL. By understanding the magnitude of the electrochemical mechanisms represented in Eq. (4.70), it is possible to generate a deeper understanding of how and where the chemical energy that is released from the ORR is being dissipated and retained within the CCL. The solution of Eq. (4.70) over a range of frequencies will contain real and imaginary components, which can be presented on a complex-impedance plot. Eq. (4.70) can only be applied during operating conditions when oxygen diffusion does not limit the performance of the CCL. Eq. (4.70) is similar to the equation that represents the impedance response generated by a TLM (Eq. 4.14), but Eq. (4.70) considers distributed CPEs instead of ideal capacitors.

Based on the impedance of the CCL represented in Eq. (4.70), Fig. 4.13 shows a TLM to represent the impedance of the CCL. The TLM shown in Fig. 4.13 differs from the TLM shown in Fig. 4.3 in which the distributed CPEs instead of capacitors are connected between the upper and lower rails of the

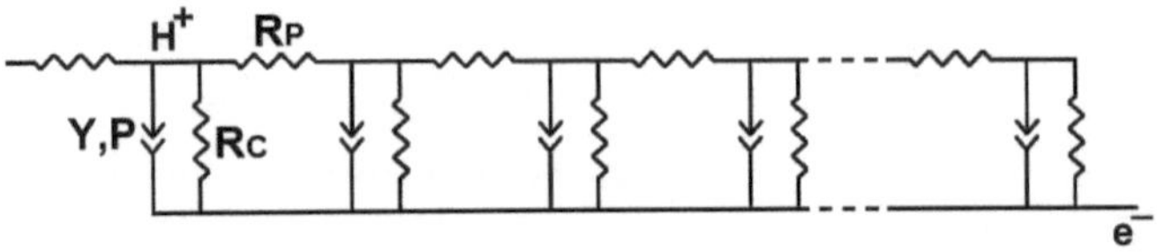

FIG. 4.13 Transmission-line model to represent the impedance response of the cathode catalyst layer with nonhomogeneous charge distribution between the electrolyte/carbon interface.

TLM. The TLM shown in Fig. 4.13 considers a nonhomogeneous charge distribution between the electrolyte/carbon interface in the CCL. It is also assumed that the nonhomogeneous charge distribution is constant (the CPEs in the TLM shown in Fig. 4.13 have the same value) along the length of the CCL thickness. The TLM shown in Fig. 4.3 considers that there is a constant and homogeneous charge distribution between the electrolyte/carbon interface across the thickness of the CCL. The nonhomogeneous charge distribution effect in the CCL along the CCL thickness could be simulated through the TLM shown in Fig. 4.3, but considering different values of the distributed capacitors along the TLM structure. Different values of the capacitors distributed along the TLM yield different time constants represented in the impedance of the CCL. A time constant dispersion can represent the behaviour of the impedance response of a CPE [4]. In Chapter 1 Section 1.5.1, it was demonstrated that the Voigt circuit comprising a time constant dispersion (a resistor-capacitor parallel configuration repeated a finite number of times with different values) was able to reproduce the impedance response of a Randles circuit considering a CPE. Eq. (4.70) simulates a depressed semicircle associated with the nonhomogeneous charge distribution between the electrolyte/carbon interface in the CCL and also simulates a 45-degree straight line at high frequencies associated with the opposition for proton conduction in the electrolyte surrounding the agglomerates of the CCL. The impedance model of the CCL has been developed from electrode theory considering a single-step electron transfer reaction and no multistep electron transfer mechanisms during the ORR. The mathematical treatment could be extended to consider the potential dependence of all steps and surface concentration of all intermediate species during the ORR reaction.

4.5.3 External capacitance for high frequency analysis

In Chapter 3 Section 3.4.1, it was demonstrated that the inductance of the electrical cables of the measurement system deforms the straight-line feature of the PEMFC impedance spectrum at high frequencies. Reshetenko and Kulikovsky [21] investigated the origin of the high frequency featured in the PEMFC impedance spectrum and concluded that a variation in slope of the straight-line feature represented in the Nyquist plot at high frequencies can be attributed to an ‘external capacitance’ due to the electric double-layer at the carbon/Nafion and carbon/water interface in the CCL. Reshetenko and Kulikovsky [21] concluded that the 'external capacitance' does not contribute in the ORR mechanisms. Reshetenko and Kulikovsky [21] connected the external capacitance in parallel to the impedance model of the CCL previously developed by the same authors [22] for a better prediction of the high frequency feature in the impedance response of the CCL. In this section, the impedance of an external capacitance is connected in parallel to the impedance of the CCL represented in Eq. (4.70) to demonstrate that it is possible to change the slope of the straight-line high frequency feature of the impedance spectrum of the CCL.

$$Z_{CCL} \| Z_{C_{ext}} = \left[Z_{CCL}^{-1} + Z_{C_{ext}}^{-1} \right]^{-1} \quad (4.71)$$

Z_{CCL} is the impedance of the CCL represented in Eq. (4.70) and $Z_{C_{ext}}$ is the impedance of a capacitor where the impedance of a capacitor has been introduced in Section 1.4.1 of Chapter 1. Parameters reported in the literature have been considered to simulate the effect of the external capacitance on the CCL impedance spectrum.

In the study by Cruz-Manzo and Chen [23], a cathodic impedance spectrum was measured in a 25 cm^2 H_2/Air PEMFC and measured at 8 mA/cm^2 using a three-electrode configuration (reference electrode) to separate the cathodic impedance spectrum from the overall PEMFC impedance spectrum. The TLM shown in Fig. 4.13 was fitted to the cathodic impedance measurements to estimate the electrochemical parameters of the CCL. The estimated parameters of the TLM reported in the study by Cruz-Manzo and Chen [23] are represented in Table 4.2.

The simulation exercise will be carried out considering ideal capacitors rather than CPEs, therefore it is necessary to convert the CPE into a capacitor in Eq. (4.70). Noting that if capacitors are considered in Eq. (4.70) instead of CPEs, the impedance of the CCL is analogous to the impedance of the TLM

TABLE 4.2 Parameters of the transmission-line model reported in the study by Cruz-Manzo and Chen [23] to simulate the impedance response of the cathode catalyst layer.

R_P Ωcm^2 Proton resistance	Y $s^P/\Omega cm^2$ Constant phase element	R_C Ωcm^2 Charge transfer resistance	P dimensionless Constant phase element
0.1825	16.1×10^{-3}	5.065	0.9264

Parameters estimated from cathodic EIS measurements at 0.008 A/cm^2.

represented in Eq. (4.14). The double-layer capacitance has been calculated from a CPE using the following relation [20]:

$$C_{\mathrm{dl}} = Y[\omega^*]^{P-1} \tag{4.72}$$

where ω^* is the angular frequency at which the imaginary component of the EIS measurements shows a minimum value or becomes more negative $-Z_{\mathrm{j}}$. Orazem and Tribollet [4] reported different approaches [20,24–26] to calculate a capacitance from the parameters of the CPE. The value of the double-layer capacitance considering the parameters of the CPE shown in Table 4.2, ω^* of the measured impedance spectrum [23] and Eq. (4.72) resulted in $C_{\mathrm{dl}}=0.0134$ F/cm^2.

It was possible to consider the value of the double-layer capacitance $C_{\mathrm{dl}}=0.0134$ F/cm^2 in Eq. (4.70) by considering $P=1$ and $Y=0.0134$ F/cm^2. Fig. 4.14a shows the simulated impedance response generated by Eq. (4.70) and considering $R_{\mathrm{C}}=5.065$ Ωcm^2, $R_{\mathrm{P}}=0.1825$ Ωcm^2, $Y=0.0134$ F/cm^2 and $P=1$. The simulations shown in Fig. 4.14 were carried out in MATLAB software in the frequency range from 10 MHz to 0.1 Hz. Two values of the external capacitance $C_{\mathrm{ext}}=0.01$ F/cm^2 and $C_{\mathrm{ext}}=0.003$ F/cm^2 have been considered in Eq. (4.71) to simulate their effect on the CCL impedance response at high frequencies. Fig. 4.14a also shows the simulated impedance response generated by Eq. (4.71) and considering two different values of the external capacitance C_{ext}. The external capacitance does not have an effect on the impedance spectrum at medium-low frequencies. Therefore, as concluded by Reshetenko and Kulikovsky [21], the external capacitance does not contribute in the charge transfer mechanisms during the ORR. The simulation results demonstrate that the external capacitance has only an effect at high frequencies as shown in Fig. 4.14b where the external capacitance can change the slope of the straight-line feature at high frequencies. The proton resistance is graphically represented in the CCL impedance spectrum through the projection of the 45-degree straight line on the real component Z_{r} at high frequencies as $R_{\mathrm{P}}/3$, although the external capacitance could mislead the graphical estimation of the proton resistance on the impedance spectrum at high frequencies. The external capacitance coupled with the impedance of the CCL as represented in Eq. (4.71) could provide a better prediction of the EIS measurements represented at the high frequency region of the complex-impedance plot.

4.5.4 Impedance response of the cathode catalyst layer considering proton and electron resistivities

The impedance response of the CCL considering proton and electron resistivities can be analogous to the TLM shown in Fig. 4.2b. The impedance of the TLM shown in Fig. 4.2b has been derived in the study by Illig et al. [27] to study the impedance response of porous electrodes in lithium-ion batteries. It is

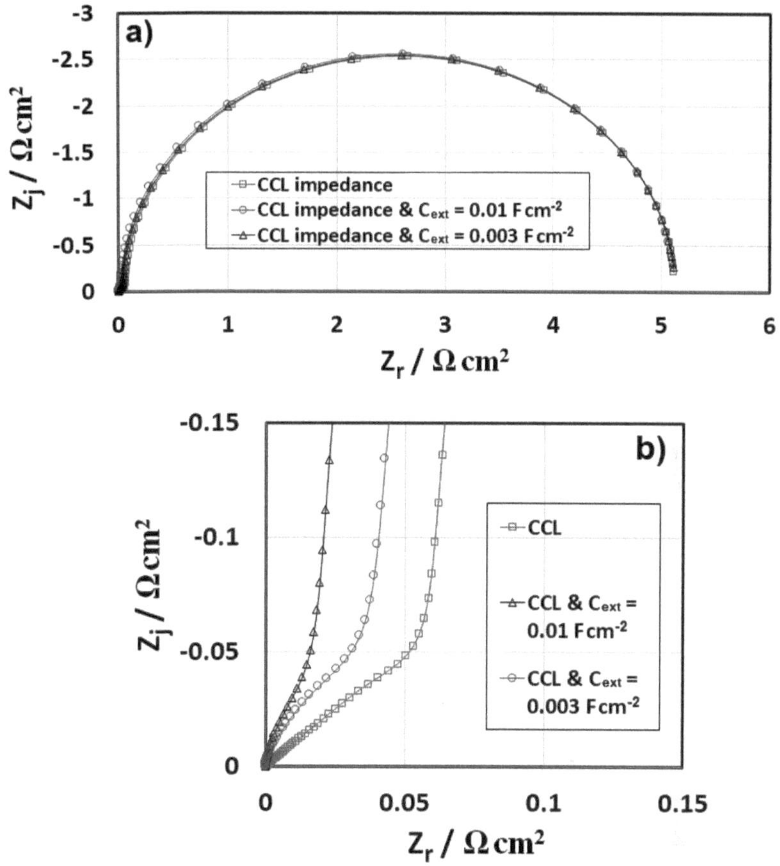

FIG. 4.14 Effect of an external capacitance C_{ext} on the impedance response of the cathode catalyst layer (CCL). The external capacitance is attributed to the electric double-layer at the carbon/Nafion and carbon/water interface in the CCL, (a) full impedance spectrum, (b) high frequency region.

possible to relate the equation of the TLM reported by Illig et al. [27] with the impedance response of the CCL as such:

$$Z_{TLM} = \frac{Z_{e^-} Z_{H^+}}{Z_{e^-} + Z_{H^+}} \left[\ell_{CCL} + \frac{2\zeta}{\sinh\left(\frac{\ell_{CCL}}{\zeta}\right)} \right] + \zeta \frac{Z_{H^+}^{\ 2} + Z_{e^-}^{\ 2}}{Z_{e^-} + Z_{H^+}} \coth\left(\frac{\ell_{CCL}}{\zeta}\right) \tag{4.73}$$

with $\zeta = \sqrt{\frac{Z_F}{Z_{e^-} + Z_{H^+}}}$ and $Z_F = \ell_{CCL} Z_{RC}$ and $Z_{RC} = \frac{R_C}{1 + R_C C_{dl} j\omega}$.

Eq. (4.73) is the solution of the TLM shown in Fig 4.2b with the parameters Z_{H^+}, Z_{e^-}, and Z_F. Z_{H^+} is the proton resistivity with units Ωcm, Z_{e^-} is the electron resistivity with units Ωcm, ℓ_{CCL} is the CCL thickness in cm, Z_{RC} is the impedance of the parallel connection between the charge transfer resistance R_C with units Ωcm^2 and C_{dl} double-layer capacitance with units F/cm^2.

For the specific case that an infinite electron conductivity $\sigma_{e^-}=\infty$ or negligible electron resistivity $Z_{e^-}=0$ is considered in Eq. (4.73), the impedance model of the CCL represented in Eq. (4.73) will resemble the impedance of the TLM represented in Eq. (4.14). The impedance model of the CCL considering electron resistance will not be considered in the next section which is related to the impedance of the CCL with oxygen transport limitations because contrary to PEM water electrolysers, PEMFCs commonly have catalyst layers with proton conductivity lower than electron conductivity by several orders of magnitude $\sigma_{e^-} \gg \sigma_{H^+}$ [5,19]. However, low electron conductivity in the CCL of PEMFCs can be attributed to carbon corrosion during start-up and shut-down events during PEMFC operation [28]. In this section, the effect of electron resistivity on the CCL impedance response will be simulated with parameters of the CCL reported by Kosakian and Secanell [29] and using Eq. (4.73). The authors simulated the effect of proton and electron conductivities on the impedance response of the CCL using a numerical one-dimensional catalyst layer model.

Parameters reported in the literature [29] as shown in Table 4.3 have been considered to demonstrate the effect of the proton and electron resistivities on the impedance response of the CCL. Table 4.3 shows electron and proton resistivities calculated from proton and electron conductivities reported in the study by Kosakian and Secanell [29]. A charge transfer resistance $R_C=0.6\ \Omega cm^2$, double-layer capacitance $C_{dl}=0.0243\ F/cm^2$, thickness of the CCL of $\ell_{CCL}=4.5\ \mu m$ and the proton and electron resistivities shown in Table 4.3 are considered in Eq. (4.73) to simulate the impedance response of the CCL.

Fig. 4.15 shows the simulated impedance spectra of the CCL resulting from Eq. (4.73) and the parameters shown in Table 4.3. The simulations shown in Fig. 4.15 were carried out in MATLAB software in the frequency range from 10 MHz to 0.1 Hz to extend the high frequency straight-line close to the real axis Z_r of the complex-impedance plot. Similar results were reported by Kosakian and Secanell [29] (Fig. 6a from the authors' study). If the proton resistivity is increased and the electron resistivity is very low for case 2, there is an increase in length of the high frequency straight-line which increases the real component of the impedance response of the CCL at low frequencies as shown in Fig. 4.15. The high frequency straight-line becomes negligible with decreasing proton and electron resistivities as demonstrated in case 3. The script written in MATLAB to carry out the simulations shown in Fig. 4.15 can be found in Appendix A.

Fig. 4.16 shows the high frequency region of the impedance spectra shown in Fig. 4.15. If the electron resistivity increases, the impedance spectrum is shifted towards the upper end of the real axis Z_r or to the right of the complex-impedance plot as shown in Fig. 4.16 for case 4. The straight-line at high frequencies becomes negligible when the proton and electron conductivities are high as shown in Figs 4.15 and 4.16 for case 3. The results shown in

TABLE 4.3 Proton and electron conductivities reported in the study by Kosakian and Secanell [29].

	σ_{H^+} [29] mS/cm	σ_{e^-} [29] mS/cm	Z_{H^+} Ωcm	Z_{e^-} Ωcm
Case 1	2	3788	500	0.2640
Case 2	1	3788	1000	0.2640
Case 3	2000	3788	0.5	0.2640
Case 4	2	37.9	500	26.385
Case 5	3788	2	0.2640	500

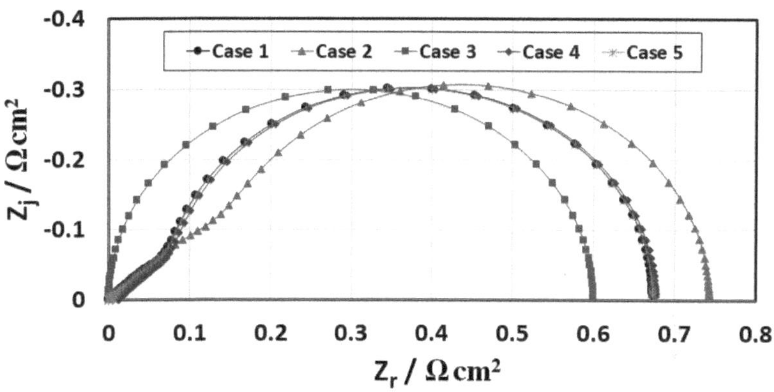

FIG. 4.15 Simulated impedance response of the cathode catalyst layer at different proton and electron resistivities. The impedance spectra generated by Eq. (4.73) agree with the results reported by Kosakian and Secanell [29] (Fig. 6a from the authors' study).

Fig. 4.16 are also similar to the results reported by Kosakian and Secanell [29] (Fig. 6c from the authors' study).

The impedance model of the CCL reported by Kulikovsky [18] and previously discussed in Section 4.3 predicted that the straight-line at high frequencies is not only associated with proton resistance but also to the electron resistance in the CCL. In addition, the author predicted a shift of the impedance spectrum of the CCL towards the right or towards the upper end of the real axis Z_r of the complex-plot with decreasing electron conductivity in the CCL. Fig. 4.17 shows the impedance response of the CCL generated by Eq. (4.73) and considering similar electron and proton conductivities $\sigma_{e^-} = 5$ mS/cm and $\sigma_{H^+} = 5$ mS/cm and considering $k_\sigma = \sigma_{e^-}/\sigma_{H^+} = 1$. In addition, the impedance response of the CCL considering the relation $k_\sigma = 0.1$ by decreasing the electron conductivity

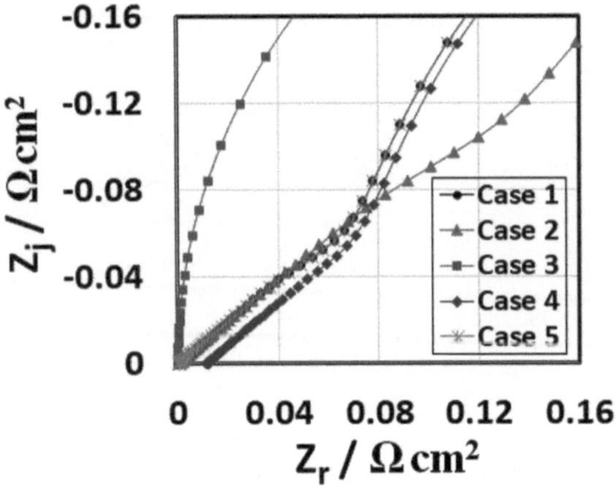

FIG. 4.16 High frequency region of the simulated impedance response of the cathode catalyst layer at different proton and electron resistivities. The impedance spectra generated by Eq. (4.73) agree with the results reported by Kosakian and Secanell [29] (Fig. 6c from the authors' study).

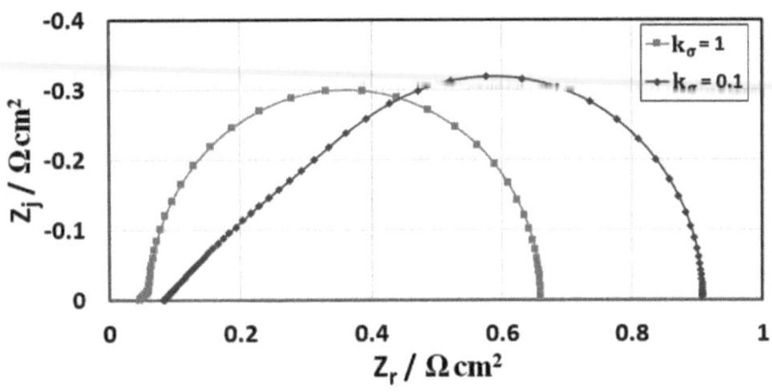

FIG. 4.17 Impedance response of the cathode catalyst layer generated by Eq. (4.73) and considering similar electron and proton conductivity $k_\sigma = \sigma_{e^-}/\sigma_{H^+} = 1$ and reduction in electron conductivity $k_\sigma = 0.1$. The simulations generated by Eq. (4.73) also considered charge transfer resistance $R_C = 0.6\ \Omega cm^2$, double-layer capacitance $C_{dl} = 0.0243\ F/cm^2$ and thickness of the cathode catalyst layer of $\ell_{CCL} = 4.5\ \mu m$.

by one order of magnitude in Eq. (4.73) is shown in Fig. 4.17 as well. The simulations shown in Fig. 4.17 were carried out in MATLAB software in the frequency range from 10 MHz to 0.1 Hz. The simulations generated by Eq. (4.73) also considered charge transfer resistance $R_C = 0.6\ \Omega cm^2$, double-layer capacitance $C_{dl} = 0.0243\ F/cm^2$ and thickness of the CCL of $\ell_{CCL} = 4.5\ \mu m$. The impedance results shown in Fig. 4.17 are qualitatively similar to the impedance results reported by Kulikovsky [18] as depicted in Fig. 4.10 from the previous Section 4.3. A shift of the impedance spectrum and increase in the length of the straight-line at high frequencies with decreasing electron conductivity $k_\sigma = 0.1$ is shown in Fig. 4.17.

4.6 Electrochemical impedance of the cathode catalyst layer at high current

The performance of a PEMFC depends on the concentration of the reactants supplied to the catalyst layers. During high current operation, a gradient for oxygen concentration in the CCL is established because the rate of oxygen consumption increases and can become greater than the rate of oxygen transport into the active sites of the CCL. There are three modes of transport of oxygen in the CCL which can influence oxygen transport limitations: gas-diffusion in the electrode-pore, dissolved oxygen diffusion in the liquid water surrounding the agglomerate, and dissolved oxygen diffusion in the ionomer phase. The finite diffusion distance for oxygen to reach the active sites in the CCL forms a complicated network of multiphase parallel and serial paths. The finite diffusion distance could change in dimension for different CCL structure (e.g. Nafion loading, porosity, tortuosity) and at different fuel cell operating conditions (current density, temperature, relative humidity (RH), etc.). Water produced by the ORR can form an additional double-layer capacitance effect between carbon/water interface in the CCL. Fig. 4.18 shows the multiphase structure of the CCL considering charge carriers during the ORR where the Pt-C structure is interfaced with the solution and void structures.

The study of the reactant transport in the electrode is complex, yet has been treated in the literature with certain simplifications and approximations. Studies in the literature [30–34] consider that the drop in reactant concentration is linear with the diffusion distance, as shown in Fig. 4.19.

In Eq. (4.57) from Section 4.5, the oscillating current density resulting from the sinusoidal perturbation during EIS and through the nondimensional thickness x of the CCL was defined as:

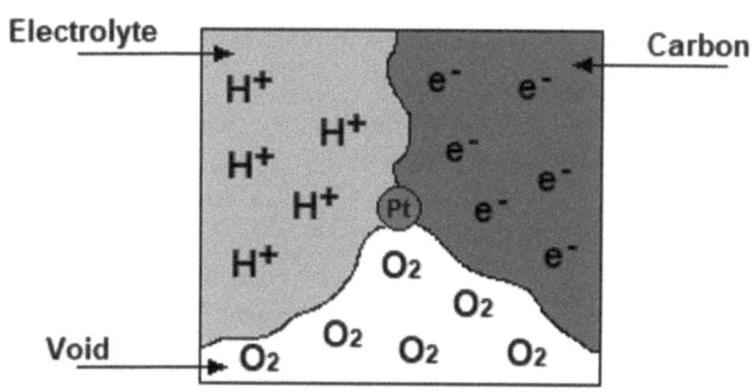

FIG. 4.18 Multiphase network interface of the cathode catalyst layer.

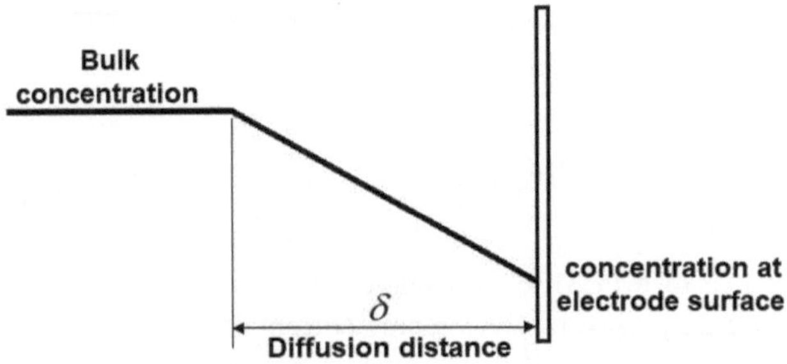

FIG. 4.19 Linear drop in reactant concentration.

$$\frac{d\widetilde{i}}{dx} = \widetilde{\varsigma_{Ox}} \frac{\widetilde{E}}{R_C} + Y[j\omega]^P \widetilde{E} \tag{4.74}$$

with

$$\widetilde{\varsigma_{Ox}} = \frac{\widetilde{c_{Ox}}}{\widetilde{c^*_{Ox}}} = 1$$

where $\widetilde{\varsigma_{Ox}}$ is the oscillating ratio between reactant concentration at the CCL surface $\widetilde{c_{Ox}}$ and bulk concentration c^*_{Ox}. Eq. (4.74) allowed the derivation of an impedance model of the CCL during low current operation as demonstrated in Section 4.5. An oxygen concentration gradient across the CCL is considered in Eq. (4.74) to derive an impedance model of the CCL for high current operation.

The following considerations based on the theory of EIS will be considered to develop the oxygen transport theory of the CCL in the frequency domain.

- EIS only measures bulk parameters in the total CCL thickness and reflects a total oxygen transport resistance and a total effective diffusion coefficient for the three modes of oxygen transport in the CCL.
- In the mathematical treatment for the oxygen transport analysis, the finite diffusion distance y for oxygen to diffuse through the CCL will be considered to be independent from the thickness x of the CCL, as shown in Fig. 4.20, to derive the mathematical model of the CCL impedance at high currents.
- The change in oxygen concentration in the CCL during the ORR will be considered from the CCL–GDL interface $c^*_{O_2}$ at $y=0$ to the CCL–PEM interface c'_{O_2} at $y=\delta_{CCL}$, as shown in Fig. 4.20.
- Even though this mathematical treatment results in an oversimplification for the oxygen diffusion in the total CCL; this will simplify the mathematical analysis. It will allow the derivation of a model with parameters commonly known in the electrochemical area such as the finite-length Warburg

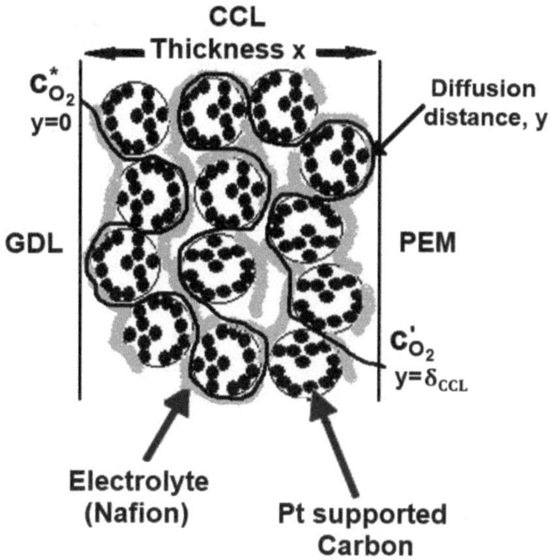

FIG. 4.20 Oxygen transport in the cathode catalyst layer (CCL). Polymer electrolyte membrane (PEM), gas diffusion layer (GDL).

component which has been considered to simulate the low frequency impedance response of PEMFCs operated at high currents.

The experimental EIS technique only measures the change in oxygen concentration in the total CCL thickness (bulk parameters). Based on the modelling considerations previously mentioned, first the oscillating ratio between oxygen concentration at the CCL–PEM interface and oxygen concentration at the CCL–GDL interface is considered in Eq. (4.74). The oxygen concentration at the CCL–GDL interface and at the CCL–PEM interface was depicted in Fig. 4.20. Second, this ratio between oxygen concentrations is considered to be less than 1, as the oxygen concentration is reduced from the CCL–GDL interface to the CCL–PEM interface. The oxygen diffusion distance is independent from the thickness x of the CCL. These considerations simplify the mathematical treatment for oxygen transport through the CCL and yield the derivation of an impedance model of the CCL during high current operation. Considering the aforementioned assumptions, Eq. (4.74) can be re-defined as:

$$\frac{d\widetilde{i}}{dx} = \widetilde{\vartheta_{O_2}}\frac{\widetilde{E}}{R_C} + Y[j\omega]^P\widetilde{E} \tag{4.75}$$

with

$$\widetilde{\vartheta_{O_2}} = \frac{\widetilde{c'_{O_2}}}{\widetilde{c^*_{O_2}}} < 1$$

where $\widetilde{c'_{O_2}}$ is the oscillating oxygen concentration at the CCL–PEM interface and $\widetilde{c^*_{O_2}}$ is the oscillating oxygen concentration at the CCL–GDL interface as shown in Fig. 4.20.

The mathematical treatment for the derivation of the impedance response of the CCL assumes that a simple reaction or single-step electron transfer describes the ORR. A concentration profile of chemical species during a simple electron-transfer reaction can be derived from the general theory of controlled-current methods [35]. This theory solves Fick's Second Law of diffusion and relates its solution with the flux of chemical species and Faraday's Law to account for the change in concentration of chemical species during an electrochemical reaction at a fixed current. The same procedure will be considered for the derivation of the change of oxygen concentration during the ORR in the CCL.

4.6.1 Solution of Fick's second law in Laplace domain

Fick's Second Law is considered to model the effect of diffusion on the distribution of oxygen concentration in the CCL with respect to time, as such:

$$D_{\mathrm{CCL}} \frac{\partial^2 c_{O_2}(y,t)}{\partial y^2} = \frac{\partial c_{O_2}(y,t)}{\partial t} \tag{4.76}$$

where D_{CCL} is the effective diffusion coefficient of oxygen and y is the distance for oxygen to diffuse in the CCL from the CCL–GDL interface $y=0$ to the CCL–PEM interface $y=\delta_{\mathrm{CCL}}$, as shown in Fig. 4.20.

EIS measurements are carried out in PEMFCs operated at steady state to comply with Kramers–Kronig transformations. At steady state, the oxygen concentration is independent of time; hence, Fick's Second Law can be expressed in Laplace domain s. The solution of Eq. (4.76) in the Laplace domain s requires an initial condition at $t=0$ as $c_{O_2}(y,0)=c^*_{O_2}$. The initial condition at $t=0$ states that oxygen concentration at the CCL–GDL interface is available before the ORR takes place in the CCL.

The Laplace transform of a derivative of a function using the Real Differentiation Theorem was represented in Eq. (1.49) from Chapter 1. Transforming Eq. (4.76) in the Laplace domain using the Real Differentiation Theorem and considering the initial condition at $t=0$ as $c_{O_2}(y,0)=c^*_{O_2}$ gives:

$$D_{\mathrm{CCL}} \frac{\partial^2 \overline{c_{O_2}}(y,s)}{\partial y^2} = s\overline{c_{O_2}}(y,s) - c^*_{O_2} \tag{4.77}$$

where the bar located over a letter represents the Laplace transform of a variable. The solution of Eq. (4.77) in the Laplace domain s through the method of undetermined coefficients for a nonhomogeneous linear differential equation is solved as follows:

$$\overline{c_{O_2}}(y,s) = \overline{Hc_{O_2}}(y,s) + \overline{Pc_{O_2}}(y,s) \tag{4.78}$$

where $\overline{Hc_{O_2}}(y,s)$ is the solution of the homogeneous equation expressed in Eq. (4.77), and $\overline{Pc_{O_2}}(y,s)$ is a parameter associated with the method of the undetermined coefficients. The coefficients expressed on the right-hand side of Eq. (4.78) can be estimated through the following steps:

First step. The characteristic equation for the homogeneous part in Eq. (4.77) is $D_{CCL}\psi^2 - s = 0$ which has two distinct roots $\psi_{1,2} = \pm\sqrt{s/D_{CCL}}$. Hence the solution of the homogeneous equation is:

$$\overline{Hc_{O_2}}(y,s) = A\exp(\psi_1 y) + B\exp(\psi_2 y) \tag{4.79}$$

Second step. The modification rule calls for

$$\overline{Pc_{O_2}}(y,s) = \overline{c_{O_2}}(y,s) = K \tag{4.80}$$

The first and second derivatives of Eq. (4.80) yield:

$$\frac{d\overline{c_{O_2}}(y,s)}{dy} = 0 \tag{4.81}$$

$$\frac{d^2\overline{c_{O_2}}(y,s)}{dy^2} = 0 \tag{4.82}$$

Substituting Eqs (4.80) and (4.82) into Eq. (4.77) yields:

$sK - c^*_{O_2} = 0$ hence $K = \frac{c^*_{O_2}}{s}$ this gives

$$\overline{Pc_{O_2}}(y,s) = \frac{c^*_{O_2}}{s} \tag{4.83}$$

Third step. Substituting Eqs (4.79) and (4.83) into Eq. (4.78) yields:

$$\overline{c_{O_2}}(y,s) - \overline{c^*_{O_2}}(s) = A\exp(\psi_1 y) + B\exp(\psi_2 y) \tag{4.84}$$

with $\overline{c^*_{O_2}}(s) = c^*_{O_2}/s$ and $\psi_{1,2} = \pm\sqrt{s/D_{CCL}}$

Evaluating boundary conditions in Eq. (4.84) at the CCL–GDL interface $\overline{c_{O_2}}(0,s) = \overline{c^*_{O_2}}(s)$ while at the CCL–PEM interface, $\overline{c_{O_2}}(\delta_{CCL},s) = \overline{c'_{O_2}}(s)$ gives:

$$A = \frac{\overline{c^*_{O_2}}(s) - \overline{c'_{O_2}}(s)}{\exp(\psi_2\delta_{CCL}) - \exp(\psi_1\delta_{CCL})} \tag{4.85}$$

$$B = \frac{\overline{c'_{O_2}}(s) - \overline{c^*_{O_2}}(s)}{\exp(\psi_2\delta_{CCL}) - \exp(\psi_1\delta_{CCL})} \tag{4.86}$$

Substituting Eqs (4.85) and (4.86) back into Eq. (4.84) gives:

$$\overline{c_{O_2}}(y,s) - \overline{c^*_{O_2}}(s) = \frac{\left[\overline{c^*_{O_2}}(s) - \overline{c'_{O_2}}(s)\right]\exp(\psi_1 y)}{\exp(\psi_2\delta_{CCL}) - \exp(\psi_1\delta_{CCL})} + \frac{\left[\overline{c'_{O_2}}(s) - \overline{c^*_{O_2}}(s)\right]\exp(\psi_2 y)}{\exp(\psi_2\delta_{CCL}) - \exp(\psi_1\delta_{CCL})} \tag{4.87}$$

4.6.2 Finite-length Warburg impedance and charge transfer resistance

Eq. (4.75) represents the sum between the faradaic current density and the non-faradaic current density related to the charge double-layer in the CCL. The Faradic current density represents the current density at which oxygen is consumed during the ORR and is expressed in the first term on the right-hand side of Eq. (4.75). It can be expressed in the Laplace domain s considering the transformation between the Laplace space and Fourier space $s = j\omega$ as such:

$$\frac{d\bar{i}_F(s)}{dx} = \overline{\vartheta_{O_2}}(s)\frac{\overline{E}(s)}{R_C} \tag{4.88}$$

where $\overline{\vartheta_{O_2}}(s) = \overline{c'_{O_2}}(s)/\overline{c^*_{O_2}}(s)$ is ratio between oxygen concentration at the CCL–PEM interface and CCL–GDL interface in the Laplace domain s.

From Faraday's Law, it is possible to establish that the current density in the CCL in the Laplace domain is proportional to the charge transferred and the consumption of reactant:

$$\frac{d\bar{i}_F(s)}{dx} = zF\overline{v}(s) \tag{4.89}$$

where $\overline{v}(s)$ is the flux of reactant, z is the number of electrons consumed during the ORR, and F is the Faraday constant.

From Fick's First Law it is possible to establish that the flux of reactant is proportional to concentration gradient

$$\overline{v}(s) = -D_{CCL}\left.\frac{d\overline{c_{O_2}}(y,s)}{dy}\right|_{y=\delta_{CCL}} \tag{4.90}$$

In the steady state, the current density at which the oxygen is consumed in the ORR from Faraday's Law is equal to the diffusion flux from Fick's First Law. Combining Eqs (4.88), (4.89), and (4.90) yields:

$$\overline{\vartheta_{O_2}}(s)\frac{\overline{E}(s)}{zFR_C} = -D_{CCL}\left.\frac{d\overline{c_{O_2}}(y,s)}{dy}\right|_{y=\delta_{CCL}} \tag{4.91}$$

Differentiating Eq. (4.87) with respect to the finite diffusion distance y yields:

$$\frac{d\overline{c_{O_2}}(y,s)}{dy} = \frac{\left[\overline{c^*_{O_2}}(s) - \overline{c'_{O_2}}(s)\right]\exp(\psi_1 y)}{\exp(\psi_2 \delta_{CCL}) - \exp(\psi_1 \delta_{CCL})}\psi_1 + \frac{\left[\overline{c'_{O_2}}(s) - \overline{c^*_{O_2}}(s)\right]\exp(\psi_2 y)}{\exp(\psi_2 \delta_{CCL}) - \exp(\psi_1 \delta_{CCL})}\psi_2 \tag{4.92}$$

Rearranging Eq. (4.92) considering hyperbolic trigonometric identities for exponentials and considering $\psi_2 = -\psi_1$ yields:

$$\frac{d\overline{c_{O_2}}(y,s)}{dy} = \frac{\psi_1\left[\overline{c'_{O_2}}(s) - \overline{c^*_{O_2}}(s)\right]\cosh(\psi_1 y)}{\sinh(\psi_1 \delta_{CCL})} \tag{4.93}$$

Substituting Eq. (4.93) into Eq. (4.91) with the boundary of the CCL–PEM interface $y = \delta_{CCL}$ yields:

$$\overline{\vartheta_{O_2}}(s)\frac{\overline{E}(s)}{zFR_C} = -D_{CCL}\psi_1 \frac{\overline{c'_{O_2}}(s) - \overline{c^*_{O_2}}(s)}{\tanh(\psi_1 \delta_{CCL})} \tag{4.94}$$

Rearranging Eq. (4.94) yields:

$$\overline{\vartheta_{O_2}}(s) = \left[1 + \frac{\overline{E}(s)\tanh(\psi_1 \delta_{CCL})}{R_C zFD_{CCL}\psi_1 \overline{c^*_{O_2}}(s)}\right]^{-1} \tag{4.95}$$

The ratio between oxygen concentration at the CCL–PEM interface and CCL–GDL interface $\overline{\vartheta_{O_2}}(s)$ can be expressed as a function of oxygen transport resistance and charge transfer resistance. The use of a small amplitude in the sinusoidal perturbation through the EIS technique allows the use of a linear model to represent the impedance of the CCL. Therefore, it is possible to consider a linearized relation of the potential by considering the Laplace form $\overline{E}(s) = E/s$ and considering $E = RT/zF$. In addition, considering the Laplace form $\overline{c^*_{O_2}}(s) = c^*_{O_2}/s$, and considering $\psi_1 = \sqrt{s/D_{CCL}}$ in the term on the right-hand side of Eq. (4.95) and considering the transformation between the Laplace space and Fourier space $s = j\omega$ in Eq. (4.95) yields:

$$\widetilde{\vartheta_{O_2}} = \frac{R_C}{R_C + Z_{O_2\_CCL}} \tag{4.96}$$

where

$$Z_{O_2\_CCL} = R_{O_2\_CCL}\frac{\tanh\left(\sqrt{j\omega\tau_{O_2\_CCL}}\right)}{\sqrt{j\omega\tau_{O_2\_CCL}}} \tag{4.97}$$

is known as the finite-length Warburg Impedance [36,37] and represents the impedance response for oxygen transport limitation of the CCL in the frequency domain, with

$$R_{O_2\_CCL} = \frac{RT\delta_{CCL}}{z^2F^2c^*_{O_2}D_{CCL}} \tag{4.98}$$

defined as resistance for the oxygen diffusion process and

$$\tau_{O_2\_CCL} = \frac{{\delta_{CCL}}^2}{D_{CCL}} \tag{4.99}$$

defined as the time constant to diffuse oxygen through the CCL. The oxygen diffusion resistance $R_{O_2\_CCL}$ from the finite-length Warburg impedance is a function of the operating temperature and gas constant [37,38]. Some studies [39,40] have represented the oxygen diffusion resistance in the finite-length Warburg impedance representing the diffusion impedance of the GDL as a function of the Tafel slope b. Kulikovsky [39] reported an impedance equation to characterise oxygen transport limitations in GDLs from PEMFCs using EIS. The author also reported that the finite-length Warburg impedance which is represented through Eq. (4.97) is not appropriate for oxygen transport analysis in GDLs of PEMFCs. In this chapter, it is considered that the finite-length Warburg impedance represented in Eq. (4.97) accounts for the analysis of the oxygen diffusion in the impedance of the CCL.

4.6.3 Impedance response of the cathode catalyst layer

Substituting Eq. (4.96) into Eq. (4.75) yields:

$$\frac{d\widetilde{i}}{dx} = \frac{\widetilde{E}}{R_C + Z_{O_2\_CCL}} + Y[j\omega]^P\widetilde{E} \tag{4.100}$$

Eq. (4.100) represents the oscillating current density through the nondimensional thickness x of the CCL considering oxygen transport limitations at the frequency domain.

Substituting Eq. (4.100) into Eq. (4.58) yields the current density as a function of the proton resistance as such:

$$\frac{d^2\widetilde{i}}{dx^2} = R_P\left[\frac{\widetilde{i}}{R_C + Z_{O_2\_CCL}} + Y[j\omega]^P\widetilde{i}\right] \tag{4.101}$$

The same mathematical procedure described in Section 4.5.2 can be applied to solve Eq. (4.101) to represent the oscillating current density.

Rearranging Eq. (4.100) yields the oscillating potential as such:

$$\widetilde{E} = \frac{R_C + Z_{O_2\_CCL}}{1 + Y[j\omega]^P[R_C + Z_{O_2\_CCL}]}\frac{d\widetilde{i}}{dx} \tag{4.102}$$

It is possible to define the impedance response of the CCL during high current operation considering the solution of Eq. (4.101) with Eq. (4.102) as such:

$$Z_{\mathrm{CCL}} = \frac{\widetilde{E}}{\widetilde{i}} = \frac{R_{\mathrm{P}}\coth(\gamma_1 x)}{\gamma_1} \tag{4.103}$$

with

$$\gamma_1 = \sqrt{R_{\mathrm{P}}\left[\frac{1}{R_{\mathrm{C}} + Z_{\mathrm{O_2\_CCL}}} + Y[j\omega]^P\right]}$$

where x represents the nondimensional thickness of the CCL and for EIS measurements in PEMFCs (lumped parameters) is equal to 1. For the specific case where $Z_{\mathrm{O_2\_CCL}}$ is considered to be negligible either due to a high diffusion coefficient D_{CCL} or high bulk oxygen concentration $c^*_{\mathrm{O_2}}$ in Eq. (4.97) (reduction of oxygen transport limitations), Eq. (4.103) will represent the impedance response of the CCL with equilibrium boundary conditions in terms of oxygen concentration. These conditions can occur for low current operation and as such, Eq. (4.103) reduces to Eq. (4.70) represented in Section 4.5.2 for impedance of the CCL at low current density. The solution of Eq. (4.103) over a range of frequencies will contain real and imaginary components, which can be presented on a complex-impedance plot. Eq. (4.103) is analogous to the impedance of a TLM considering distributed Warburg components as shown in Fig. 4.21. The TLM shown in Fig. 4.21 ignores the electron resistance in the lower rail which is related to the conduction of electrons in the carbon structure in the CCL. The TLM shown in Fig. 4.21 has been considered in some studies [23,41,42] to study the impedance of the CCL using EIS measurements of PEMFCs.

4.7 Electrochemical phenomena of the cathode catalyst layer at low frequencies

The mathematical treatment developed in this chapter and expressed through Eq. (4.103) simulates the impedance response of the CCL at high current density considering charge transfer resistance during the ORR, nonhomogeneous charge distribution between dissimilar materials (carbon/electrolyte interface), oxygen diffusion resistance, and proton resistance in the CCL. Parameters

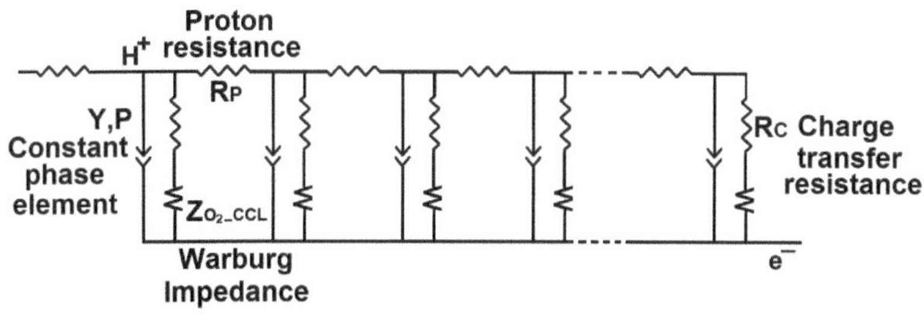

FIG. 4.21 Transmission-line model for analysis of the impedance response of the cathode catalyst layer considering oxygen transport limitations.

reported in the literature [23] and extracted by fitting the TLM shown in Fig. 4.21 with cathodic EIS measurements in a PEMFC are considered to study the phenomenological processes in the CCL.

In the study by Cruz-Manzo and Chen [23], a cathodic impedance spectrum was measured in a 25 cm^2 H_2/Air PEMFC and measured at 1200 mA/cm^2 and using a three-electrode configuration (reference electrode) to separate the cathodic impedance spectrum from the overall PEMFC impedance spectrum. The temperature of the PEMFC was controlled to 50°C by a test rig acquired from Baltic Fuel Cells and the gas back pressure was held to 0.9 bar(g) for both anode and cathode. The stoichiometry of hydrogen was 2 and for air was 2.5. The RH of hydrogen entering the PEMFC was 98% and the RH of the air supplied to the cathode was 55%. The thickness of the CCL was 12 μm. In the study by Cruz-Manzo and Chen [23], the TLM shown in Fig. 4.21 was connected in series with a resistor R_{ohm} to account for the ohmic resistance in the PEM, GDL, and flow field plate (FFP). An inductor L was also connected in series with the TLM to account for the inductance of the measurement system. The impedance of the electrical circuit reported by Cruz-Manzo and Chen [23] to estimate the parameters of the CCL from cathodic impedance measurements is represented through the following equation:

$$Z_C = Lj\omega + R_{ohm} + Z_{CCL} \tag{4.104}$$

Eq. (4.104) represents the sum of the impedance of the inductor, ohmic resistance, and impedance of the TLM shown in Fig. 4.21. In the study by Cruz-Manzo and Chen [23], Eq. (4.104) was fitted to the cathodic experimental impedance spectrum as shown in Fig. 4.22 to estimate charge transfer resistance, parameters of the CPE, proton resistance and oxygen diffusion impedance in the CCL. The fitting process was carried out in MATLAB software.

The parameters of the impedance of the electrical circuit represented by Eq. (4.104) and extracted from the cathodic experimental impedance spectrum shown in Fig. 4.22 and reported in the study by Cruz-Manzo and Chen [23] will be considered for the simulation of the CCL impedance model. As discussed in the literature [43], once the impedance of the electrical circuit represented in Eq. (4.104) is fitted to the measured impedance spectrum of the cathodic PEMFC, the ohmic resistor R_{ohm}, and inductor L are removed in order to generate the impedance response of just the CCL in isolation. The parameters of the TLM representing electrochemical mechanisms in the CCL and estimated in the study by Cruz-Manzo and Chen [23] are shown in Table 4.4.

MATLAB software allowed the simulation of the impedance response of the CCL generated by Eq. (4.103) and considering the parameters from Table 4.4. The range of frequencies was considered from 10 MHz to 0.1 Hz. Fig. 4.23 shows the spectrum generated from Eq. (4.103) and using the parameters from Table 4.4. The impedance response of the CCL considering the charge transfer resistance, charge distribution effect, proton resistance, and oxygen diffusion impedance is comprised of two overlapped semicircles as shown in

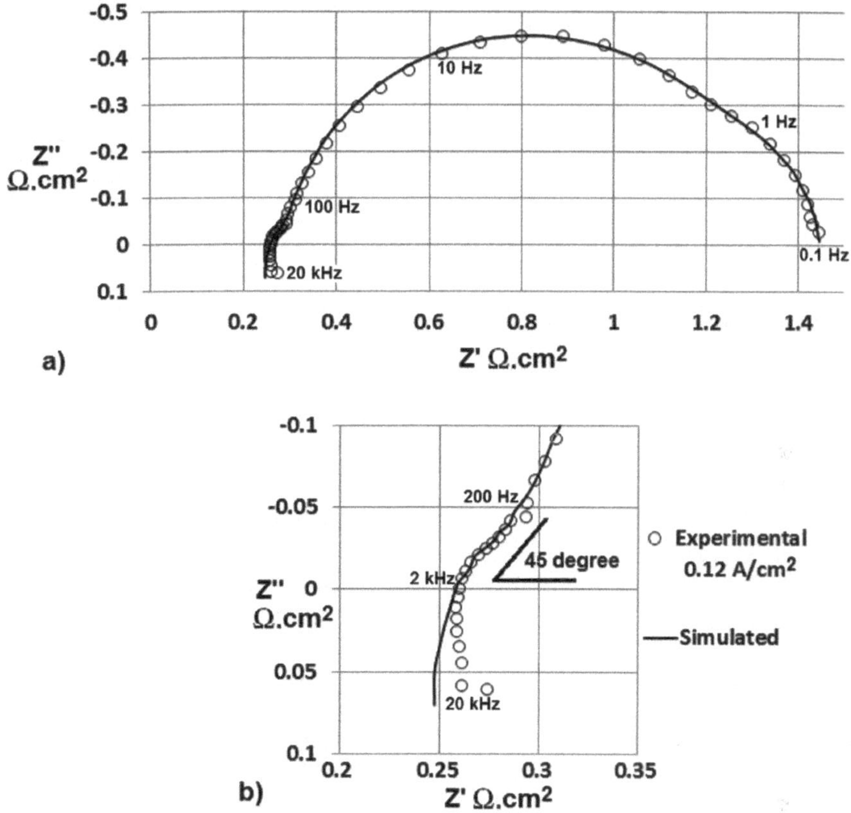

FIG. 4.22 Comparison between measured cathodic impedance spectrum and simulated cathodic impedance response, (a) full impedance spectrum, (b) high frequency region. *((Reproduced from S. Cruz-Manzo, R. Chen, A generic electrical circuit for performance analysis of the fuel cell cathode catalyst layer through electrochemical impedance spectroscopy, J. Electroanal. Chem. 694 (2013) 54, with permission of Elsevier. Copyright Elsevier.))*

TABLE 4.4 Parameters of the transmission-line model reported in the study by Cruz-Manzo and Chen [23] to simulate the impedance of the cathode catalyst layer.

R_P Ωcm^2 Proton resistance	R_C Ωcm^2 Charge transfer resistance	Y $s^P/\Omega cm^2$ Constant phase element	P Constant phase element	$R_{O_2\_CCL}$ Ωcm^2 Oxygen transport resistance	$\tau_{O_2\_CCL}$ s Oxygen diffusion time constant
0.18	0.9975	8.836×10^{-3}	0.9112	0.16	0.0920

Parameters estimated from cathodic EIS measurements at 0.12 A/cm^2.

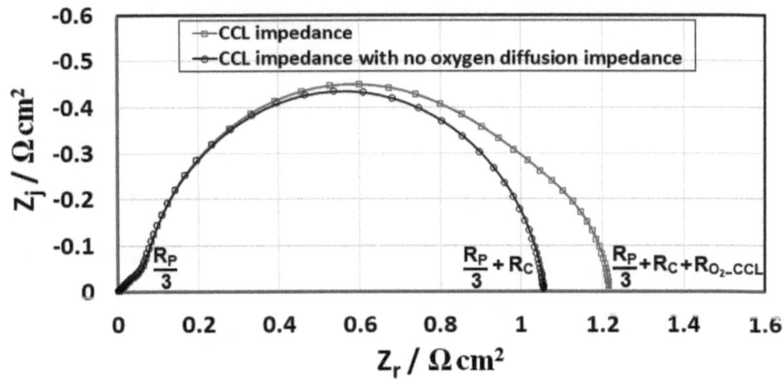

FIG. 4.23 Simulated impedance response of the cathode catalyst layer considering and neglecting oxygen diffusion impedance.

Fig. 4.23. The high frequency straight-line is related to the proton resistance in the CCL and can be projected onto the real axis Z_r of the complex-impedance plot through the relation $\frac{R_P}{3}$. The low frequency resistance of the impedance spectrum labelled as "CCL impedance" where the imaginary component converges closer to the real axis Z_r of the Nyquist plot at low frequencies $\omega \rightarrow 0$ is calculated through the relation $\frac{R_P}{3} + R_C + R_{O_2\_CCL}$.

The impedance spectrum of the CCL neglecting the oxygen diffusion impedance $Z_{O_2\_CCL}=0$ in Eq. (4.103) was simulated as well. The semicircle of the impedance spectrum of the CCL that neglects the oxygen diffusion impedance $Z_{O_2\_CCL}=0$ shown in Fig. 4.23 considers only the charge distribution effect, charge transfer resistance during the ORR, and proton resistance. Its low frequency resistance where the imaginary component converges closer to the real axis Z_r of the Nyquist plot $\omega \rightarrow 0$ is calculated through the relation $\frac{R_P}{3} + R_C$ as shown in Fig. 4.23.

The Tafel slope can be calculated through the relation $R_C = |b_c|/i_{cell}$ as expressed in Eq. (4.52). The charge transfer resistance R_C is shown in Table 4.4. The Tafel slope resulted in $|b_c| = 0.1197$ V or $|b_c| = 0.275$ V/dec at a current density of $0.12\,A/cm^2$. The parameters shown in Tables 4.2 and 4.4 were estimated from the same PEMFC at different current densities. The Tafel slope calculated from the charge transfer resistance R_C shown in Table 4.2 at a current density of $8\,mA/cm^2$ resulted in $|b_c| = 0.041$ V or $|b_c| = 0.0943$ V/dec. An increase of the Tafel slope with increasing current density can be attributed to oxygen transport limitations [44]. In the study by Mainka et al. [45], the charge transfer resistance was estimated by fitting the Randles circuit with EIS measurements of a PEMFC. Mainka et al. [45] calculated the Tafel slope from the charge transfer resistance corresponding to the Randles circuit. The calculated Tafel slope reported by Mainka et al. [45] increased from 23 mV/dec at a current density of $0.063\,A/cm^2$ to 384 mV/dec at a current density of $1.08\,A/cm^2$. The increase of the Tafel slope attributed to oxygen transport limitations has been discussed in Section 2.3.1 from Chapter 2. The estimated value of the Tafel slope $|b_c| = 0.275\ \text{V/dec}$ from the charge transfer resistance R_C at a current density of $0.12\ \text{A/cm}^2$ expressed in Table 4.4 may be overestimated because the analysis of the impedance spectrum shown in Fig. 4.22 neglected oxygen diffusion in the GDL and oxygen transport in the channels of the FFP. Bultel et al. [46] however, reported a Tafel slope of 0.22 V/dec in a PEMFC operated at 80°C. A decrease in the value of the charge transfer resistance and hence a decrease in the value of the Tafel slope may result if the electrical circuit fitted to the cathodic impedance spectrum considers not only oxygen diffusion in the CCL, but also oxygen diffusion in the GDL and oxygen transport in the channels of the FFP.

4.7.1 Effect of oxygen diffusion parameters on the impedance spectrum

The oxygen diffusion impedance $Z_{O_2\_CCL}$ in the CCL which is represented as a finite-length Warburg impedance is represented in Eq. (4.97). It comprises the oxygen diffusion coefficient D_{CCL} and oxygen concentration in the CCL–GDL interface $c^*_{O_2}$. A change in these parameters can be a consequence of different operating conditions. The parameters represented in Table 4.4 will be considered as a base-line to simulate the effect of decreasing oxygen diffusion coefficient D_{CCL} and decreasing oxygen concentration in the CCL–GDL interface $c^*_{O_2}$ on the impedance response of the CCL. The parameters represented in Table 4.4 were estimated from EIS measurements carried out in a PEMFC with CCL thickness of 12 μm [23]. The modelling assumptions for the development of the impedance model of the CCL considered that the diffusion distance of oxygen through the multiphase structure of the CCL is different from the thickness of the CCL as shown in Fig. 4.20. The diffusion distance through the multiphase structure of the CCL could change at different operating conditions. For instance, the electrolyte in the CCL is swelled when it is in contact with water vapour during PEMFC operation [47]. This effect can increase the diffusion distance of dissolved oxygen in the electrolyte of the CCL. The calculation of the diffusion parameters in the CCL is not straightforward because these can be related to three modes of transport of oxygen in the CCL which can influence oxygen transport limitations: gas-diffusion in the electrode-pore, dissolved oxygen diffusion in the liquid water surrounding the agglomerate, and dissolved oxygen diffusion in the ionomer phase [5]. In this section, the analysis of the effect of decreasing oxygen diffusion coefficient D_{CCL} and oxygen concentration in the CCL–GDL interface $c^*_{O_2}$ on the impedance response of the CCL is only qualitative. Therefore, the estimation of the oxygen diffusion coefficient D_{CCL} will be calculated using the diffusion time constant $\tau_{O_2\_CCL}$ represented in Eq. (4.99) and diffusion distance considered as the CCL thickness $\delta_{CCL} = 12 \times 10^{-6}$ m. The oxygen diffusion coefficient resulted in

$D_{CCL} = 1.5652 \times 10^{-9}\,m^2/s$ considering $\tau_{O_2\_CCL} = 0.092\,s$, and $\delta_{CCL} = 12 \times 10^{-6}$ m in Eq. (4.99). A diffusion coefficient with order of magnitude of $10^{-9}\,m^2/s$ has been related to dissolved oxygen diffusion in the liquid water surrounding the agglomerates of the CCL [36].

The oxygen concentration at the CCL–GDL interface $c^*_{O_2}$ can be calculated from the oxygen diffusion resistance $R_{O_2\_CCL}$ represented in Eq. (4.98) as a first approximation. The equation of the oxygen diffusion resistance $R_{O_2\_CCL}$ comprises the operating temperature T of the PEMFC. The oxygen concentration at the CCL–GDL interface resulted in $c^*_{O_2} = 8.6434 \times 10^{-6}\,mol/cm^3$ considering $T = 50\,°C$, $D_{CCL} = 1.5652 \times 10^{-9}\,m^2/s$ and $\delta_{CCL} = 12 \times 10^{-6}$ m in Eq. (4.98). The true value of the oxygen concentration at the CCL–GDL interface $c^*_{O_2}$ could only be calculated by considering the real oxygen diffusion distance δ_{CCL} through the CCL which is higher than the CCL thickness as oxygen is transported though the multiphase structure of the CCL as shown in Fig. 4.20. In addition, the true value of the oxygen concentration at the CCL–GDL interface $c^*_{O_2}$ should also be calculated considering the depletion of oxygen concentration along the gas channels of the FFP and oxygen diffusion through the GDL.

Fig. 4.24 shows the simulation results considering the parameters shown in Table 4.4 in Eq. (4.103). The simulation was carried out in MATLAB software and considered a range of frequencies from 10 MHz to 0.1 Hz. In addition, Fig. 4.24 shows the effect of decreasing the oxygen diffusion coefficient by 60% as $D_{CCL} = 6.2608 \times 10^{-10}\,m^2/s$ on the impedance spectrum of the CCL. A diffusion coefficient with order of magnitude of $10^{-10}\,m^2/s$ has been related to dissolved oxygen diffusion in the ionomer phase of the CCL [36]. Decreasing the diffusion coefficient by 60% is equivalent to an increase by a factor of 2.5 in the oxygen diffusion resistance $R_{O_2\_CCL}$ represented in Eq. (4.98) and the oxygen diffusion time constant $\tau_{O_2\_CCL}$ represented in

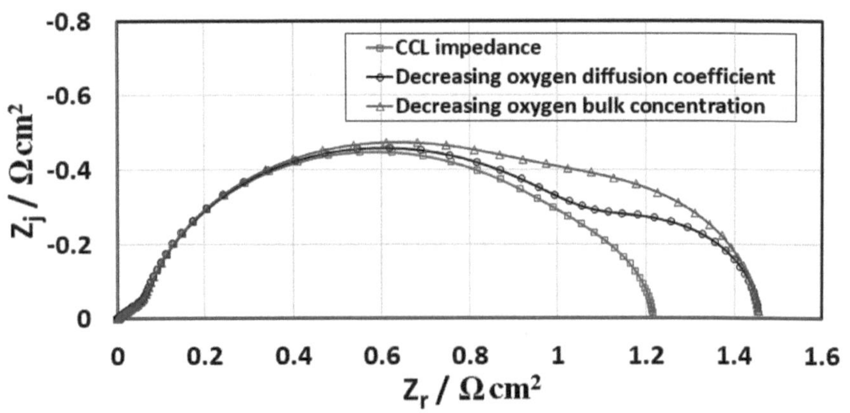

FIG. 4.24 Simulated impedance responses with decreasing oxygen diffusion coefficient and oxygen concentration at the CCL–GDL interface (oxygen bulk concentration) in the impedance model of the cathode catalyst layer.

Eq. (4.99). Fig. 4.24 shows that the low frequency limit of the impedance spectrum where the imaginary component converges closer to the real axis Z_r of the Nyquist plot $\omega \rightarrow 0$ increases with decreasing oxygen diffusion coefficient. A second semicircle is apparent at frequencies lower than 3 Hz and is related to the oxygen diffusion impedance $Z_{O_2\_CCL}$ and attributed to an increase in the oxygen diffusion time constant with decreasing oxygen diffusion coefficient. The diameter of the second semicircle at low frequencies represents the value of oxygen diffusion resistance $R_{O_2\_CCL}$ with decreasing oxygen diffusion coefficient. The increase in the oxygen diffusion time constant can also be simulated with increasing diffusion distance. Increasing the diffusion distance is equivalent to increasing water film surrounding the agglomerate for dissolved oxygen diffusion in the liquid water. Fig. 4.24 also shows a simulation considering a reduction by 60% of the oxygen concentration at the CCL–GDL interface (oxygen bulk concentration). Under this consideration, only the oxygen diffusion resistance $R_{O_2\_CCL}$ represented in Eq. (4.98) is increased by a factor of 2.5. An increase of the overall impedance spectrum with decreasing oxygen bulk concentration is observed in Fig. 4.24. The decrease in oxygen concentration at the CCL–GDL interface can be related to oxygen transport limitation in the layers adjacent to the CCL. The oxygen concentration within the adjacent layers has an influence on the CCL performance. Therefore, any change in oxygen concentration across the channels of the FFP and GDL will have an effect on the oxygen concentration at the CCL–GDL interface. The separation of the semicircle representing oxygen diffusion impedance from the semicircle representing charge transfer resistance and double-layer capacitance with increasing oxygen diffusion time constant in the CCL has been demonstrated in the study by Cruz-Manzo et al. [48] and can also be observed in Fig. 3.26 from Chapter 3.

4.8 Further application of the impedance model of the cathode catalyst layer

The TLM shown in Fig. 4.2b can represent the impedance of the CCL considering proton and electron resistances in the electrolyte and carbon structures. Finite-length Warburg impedances could be implemented in the TLM that considers proton and electron resistances as shown in Fig. 4.25. For the specific case that the electron resistance is neglected in the TLM shown in Fig. 4.25, the TLM will be similar to the TLM shown previously in Fig. 4.21.

The impedance of the electrical components connected between the upper and lower rails of the TLM shown in Fig. 4.25 is represented as:

$$Z_{RCW} = \frac{R_C + Z_{O_2\_CCL}}{1 + [R_C + Z_{O_2\_CCL}]Y[j\omega]^P} \tag{4.105}$$

Eq. (4.105) can be considered in Eq. (4.73) with $Z_F = \ell_{CCL} Z_{RCW}$ to simulate the impedance response of the CCL considering proton and electron resistances,

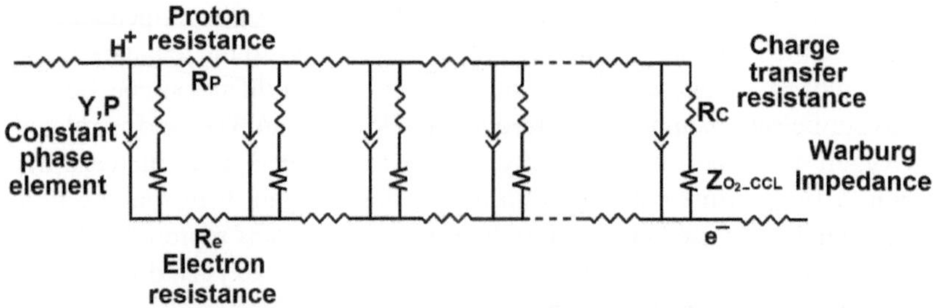

FIG. 4.25 Transmission-line model to represent the impedance response of the cathode catalyst layer considering proton and electron resistances, charge transfer resistance, and oxygen transport resistance during the oxygen reduction reaction.

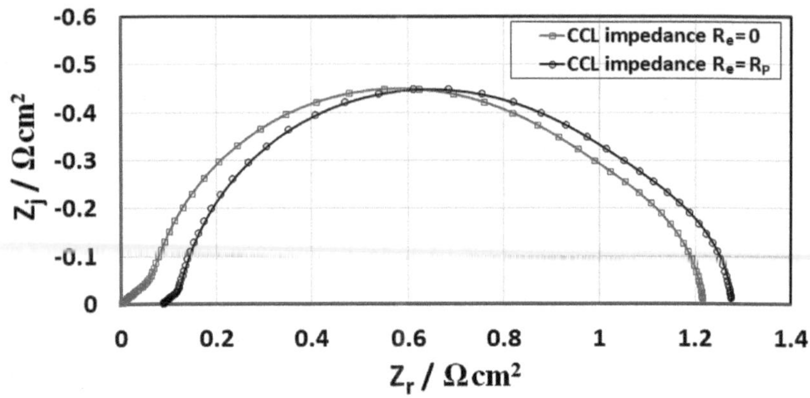

FIG. 4.26 Simulated impedance response of the cathode catalyst layer considering electron and proton resistances, charge transfer and oxygen transport limitations during the oxygen reduction reaction.

charge transfer resistance, and oxygen transport resistance during the ORR. The parameters shown in Table 4.4 can be applied to Eq. (4.73) considering Eq. (4.105) with $Z_F = \ell_{CCL} Z_{RCW}$. The proton resistivity Z_{H^+} with units Ωcm represented in Eq. (4.73) can be calculated from the proton resistance $R_P = \ell_{CCL} Z_{H^+}$ with units Ωcm^2 shown in Table 4.4 and CCL thickness $\ell_{CCL} = 12 \times 10^{-4}$ cm. It is also considered that $Z_{e^-} = Z_{H^+}$. Fig. 4.26 shows the simulated impedance response of the CCL generated by Eq. (4.73) considering Eq. (4.105) with $Z_F = \ell_{CCL} Z_{RCW}$. The simulation was carried out in MATLAB software over a range of frequencies from 10 MHz to 0.1 Hz. The impedance spectrum that neglects the electron resistance $R_e = 0$ is similar to the impedance spectrum previously shown in Fig. 4.23. A shift of the impedance spectrum of the CCL across the real axis Z_r of the complex-impedance plot by considering $R_e = R_P$ is shown in Fig. 4.26.

The TLM shown in Fig. 4.25 can be applied to cathodic EIS measurements of PEMFCs with low electron conductivity in the CCL. For instance, it has been considered that carbon corrosion in the catalyst layer yields oxygen starvation during accelerated stress tests in PEMFCs [49]. The TLM shown in Fig. 4.25 applied to EIS measurements carried out during accelerated stress tests in PEMFCs, could provide a correlation between the electron resistance and oxygen transport resistance as reported by Meyer et al. [49].

References

[1] J. Huang, Y. Gao, J. Luo, S. Wang, C. Li, S. Chen, J. Zhang, Impedance response of porous electrodes: theoretical framework, physical models and applications, J. Electrochem. Soc. 167 (2020), 166503.

[2] M. Eikerling, Water management in cathode catalyst layers of PEM fuel cells, J. Electrochem. Soc. 153 (2006) E58–E70.

[3] Z. Siroma, N. Fujiwara, S. Yamazaki, M. Asahi, T. Nagai, T. Ioroi, Multi-rail transmission-line model as an equivalent circuit for electrochemical impedance of a porous electrode, J. Electroanal. Chem. 878 (2020), 114622.

[4] M.E. Orazem, B. Tribollet, Electrochemical Impedance Spectroscopy, second ed., Wiley, Hoboken, NJ, 2017.

[5] M. Eikerling, A.A. Kornyshev, Electrochemical impedance of the cathode catalyst layer in polymer electrolyte fuel cells, J. Electroanal. Chem. 475 (1999) 107–123.

[6] R. Makharia, M.F. Mathias, D.R. Baker, Measurement of catalyst layer electrolyte resistance in PEFCs using electrochemical impedance spectroscopy, J. Electrochem. Soc. 152 (2005) A970–A977.

[7] J. Gazzarri, M. Eikerling, Q. Wang, Z. Liu, Estimation of local relative humidity in cathode catalyst layers of PEFC, Electrochem. Solid-State Lett. 13 (2010) B58–B62.

[8] Y. Liu, M.W. Murphy, D.R. Baker, W. Gu, C. Ji, J. Jorne, H. Gasteiger, A, "proton conduction and oxygen reduction kinetics in PEM fuel cell cathodes: effects of ionomer-to-carbon ratio and relative humidity", J. Electrochem. Soc. 156 (2009) B970–B980.

[9] S. Cruz-Manzo, P. Greenwood, Frequency transition from diffusion to capacitive response in the blocked-diffusion Warburg impedance for EIS analysis in modern batteries, J. Electrochem. Soc. 167 (2020), 140507.

[10] A. Kulikovsky, A physical model for catalyst layer impedance, J. Electroanal. Chem. 669 (2012) 28–34.

[11] A. Kulikovsky, M. Eikerling, Analytical solutions for impedance of the cathode catalyst layer in PEM fuel cell: layer parameters from impedance spectrum without fitting, J. Electroanal. Chem. 691 (2013) 13–17.

[12] A. Kulikovsky, Analytical solutions for polarization curve and impedance of the cathode catalyst layer with fast oxygen transport in a PEM fuel cell, J. Electrochem. Soc. 161 (2014) E3171–E3179.

[13] A. Kulikovsky, Impedance of a PEM fuel cell cathode with nonuniform ionomer loading: analytical and numerical study, J. Electroanal. Chem. 789 (2017) 174–180.

[14] A. Kulikovsky, Exact low–current analytical solution for impedance of the cathode catalyst layer in a PEM fuel cell, Electrochim. Acta 147 (2014) 773–777.

[15] A. Kulikovsky, Analytical physics–based impedance of the cathode catalyst layer in a PEM fuel cell at typical working currents, Electrochim. Acta 225 (2017) 559–565.

[16] A. Kulikovsky, A simple physics–based equation for low–current impedance of a PEM fuel cell cathode, Electrochim. Acta 196 (2016) 231–235.

[17] A. Kulikovsky, Analytical low-current impedance of the cathode side of a PEM fuel cell, J. Electroanal. Chem. 823 (2018) 335–341.

[18] A. Kulikovsky, A model for impedance of a PEM fuel cell cathode with poor electron conductivity, J. Electroanal. Chem. 801 (2017) 122–128.

[19] T. Suzuki, H. Murata, T. Hatanaka, Y. Morimoto, Analysis of the Catalyst Layer of Polymer Electrolyte Fuel Cells, vol. 39, R&D Review of Toyota, Toyota Central R&D Labs, Inc, 2003, pp. 33–38.

[20] C.H. Hsu, F. Mansfeld, Concerning the conversion of the constant phase element parameter Y_0 into a capacitance, Corrosion 57 (2001) 747–748.

[21] T. Reshetenko, A. Kulikovsky, On the origin of high frequency impedance feature in a PEM fuel cell, J. Electrochem. Soc. 166 (2019) F1253–F1257.

[22] T. Reshetenko, A. Kulikovsky, Nafion film transport properties in a low-Pt PEM fuel cell: impedance spectroscopy study, RSC Adv. 9 (2019) 38797–38806.

[23] S. Cruz-Manzo, R. Chen, A generic electrical circuit for performance analysis of the fuel cell cathode catalyst layer through electrochemical impedance spectroscopy, J. Electroanal. Chem. 694 (2013) 45–55.

[24] G.J. Brug, A.L.G. van den Eeden, M. Sluyters-Rehbach, J.H. Sluyters, The analysis of electrode impedances complicated by the presence of a constant phase element, J. Electroanal. Chem. 176 (1984) 275–295.

[25] B. Hirschorn, M.E. Orazem, B. Tribollet, V. Vivier, I. Frateur, M. Musiani, Constant-phase-element behavior caused by resistivity distributions in films: I. Theory, J. Electrochem. Soc. 157 (2010) C452–C457.

[26] B. Hirschorn, M.E. Orazem, B. Tribollet, V. Vivier, I. Frateur, M. Musiani, Constant-phase-element behavior caused by resistivity distributions in films: II. Applications, J. Electrochem. Soc. 157 (2010) C458–C463.

[27] J. Illig, M. Ender, A. Weber, E. Ivers-Tiffee, Modeling graphite anodes with serial and transmission line models, J. Power Sources 282 (2015) 335–347.

[28] M. Baghalha, J. Stumper, D. Harvey, M. Eikerling, Modeling the effect of low carbon conductivity of the cathode catalyst layer on PEM fuel cell performance, ECS Trans. 28 (2010) 113–123.

[29] A. Kosakian, M. Secanell, Estimating charge-transport properties of fuel-cell and electrolyzer catalyst layers via electrochemical impedance spectroscopy, Electrochim. Acta 367 (2021), 137521.

[30] C. Berger, Handbook of Fuel Cell Technology, Prentice-Hall, Inc., Englewood Cliffs, NJ, New York, 1968.

[31] J.C. Amphlett, R.M. Baumert, R.F. Mann, B.A. Peppley, P.R. Roberge, Performance modeling of the Ballard mark IV solid polymer electrolyte fuel cell I. mechanistic model development, J. Electrochem. Soc. 142 (1995) 1–8.

[32] J.E. Larminie, A. Dicks, Fuel Cell Systems Explained, John Wiley & Sons, LTD, Chichester, 2000.

[33] A.R. Balkin, Modelling a 500 Watt Polymer Electrolyte Membrane Fuel Cells (thesis), Faculty of Engineering, University of Technology, Sidney, 2002.

[34] J.M. Correa, F.A. Farret, L.N. Canha, An analysis of the dynamic performance of proton exchange membrane fuel cells using an electrochemical model, Ind. Electron. Soc. Annu. Conf. 1 (2001) 141–146.

[35] A.J. Bard, L.R. Faulkner, Electrochemical Methods, Fundamentals and Applications, John Wiley & Sons, Inc, New York, 2001.

[36] D. Malevich, E. Halliop, B.A. Peppley, J.G. Pharoah, K. Karan, Investigation of charge-transfer and mass-transport resistances in PEMFCs with microporous layer using electrochemical impedance spectroscopy, J. Electrochem. Soc. 156 (2009) B216–B224.

[37] N. Fouquet, C. Doulet, C. Nouillant, D. Dauphin-Tanguy, B. Ould-Bouamama, Model based PEM fuel cell state-of-health monitoring via AC impedance measurements, J. Power Sources 159 (2006) 905–913.

[38] B.A. Boukamp, Derivation of a distribution function of relaxation times for the(fractal) finite length Warburg, Electrochim. Acta 252 (2017) 154–163.

[39] A. Kulikovsky, Why impedance of the gas diffusion layer in a PEM fuel cell differs from the Warburg finite-length impedance? Electrochem. Commun. 84 (2017) 28–31.

[40] J. Mainka, G. Maranzana, J. Dillet, S. Didierjean, O. Lottin, Effect of oxygen depletion along the air channel of a PEMFC on the Warburg diffusion impedance, J. Electrochem. Soc. 157 (2010) B1561–B1568.

[41] S. Touhami, J. Mainka, J. Dillet, S. Ait Hammou Taleb, O. Lottin, Transmission line impedance models considering oxygen transport limitations in polymer electrolyte membrane fuel cells, J. Electrochem. Soc. 166 (2019) F1209–F1217.

[42] J. Choi, J. Sim, H. Oh, K. Min, Resistance separation of polymer electrolyte membrane fuel cell by polarization curve and electrochemical impedance spectroscopy, Energies 14 (2021) 1491.

[43] U. Cano-Castillo, A. Ortiz, S. Manzo, L.G. Arriaga, G. Orozco, Parameter changes during gradual flooding of a PEM fuel cell through EIS studies, ECS Trans. 3 (2006) 931–939.

[44] M. Chatenet, J. Benziger, M. Inaba, S. Kjelstrup, T. Zawodzinski, R. Raccichini, Good practice guide for papers on fuel cells and electrolysis cells for the journal of power sources, J. Power Sources 451 (2020), 227635.

[45] J. Mainka, G. Maranzana, J. Dillet, S. Didierjean, O. Lottin, On the estimation of high frequency parameters of proton exchange membrane fuel cells via electrochemical impedance spectroscopy, J. Power Sources 253 (2014) 381–391.

[46] Y. Bultel, K. Wiezell, F. Jaouen, P. Ozil, G. Lindbergh, Investigation of mass transport in gas diffusion layer at the air cathode of a PEMFC, Electrochim. Acta 51 (2005) 474–488.

[47] L. Xing, X. Liu, T. Alaje, R. Kumar, M. Mamlouk, K. Scott, A two-phase flow and non-isothermal agglomerate model for a proton exchange membrane (PEM) fuel cell, Energy 73 (2014) 618–634.

[48] S. Cruz-Manzo, R. Chen, P. Rama, Study of current distribution and oxygen diffusion in the fuel cell cathode catalyst layer through electrochemical impedance spectroscopy, Int. J. Hydrog. Energy 38 (2013) 1702–1713.

[49] Q. Meyer, I. Pivac, F. Barbir, C. Zhao, Detection of oxygen starvation during carbon corrosion in proton exchange membrane fuel cells using low-frequency electrochemical impedance spectroscopy, J. Power Sources 470 (2020), 228285.

Chapter 5

Impedance model of the gas diffusion layer and air channel

5.1 Oxygen transport through the gas diffusion layer and air channel

Oxygen transport limitation reduces the performance of the polymer electrolyte membrane fuel cell (PEMFC). Fig. 5.1 shows a simplified representation of oxygen and water transport through the cathodic side of the PEMFC. Oxygen is travelling through the channels of the flow field plate (FFP) and diffuses through the gas diffusion layer (GDL) until reaching the cathode catalyst layer (CCL). Although PEMFCs are commonly operated with air, only oxygen travelling along the channels of the FFP and through the GDL is considered in Fig. 5.1 for simplicity. Water produced from the oxygen reduction reaction (ORR) in the CCL travels towards the FFP and passes through the GDL. An excess of water concentration hinders the transport of oxygen in the GDL to reach the active sites in the CCL. Oxygen transport limitations represented in electrochemical impedance spectroscopy (EIS) measurements of PEMFCs have been mainly attributed to the resistance of oxygen diffusion in the GDL. Resistance of oxygen diffusion in the GDL is commonly related to EIS measurements of PEMFCs at low frequencies [1,2]. In addition, oxygen depletion during the ORR and along the length of the air channels of the FFP can be manifested in EIS measurements of PEMFCs at low frequencies [3,4]. Studies in the literature [4–6] have demonstrated experimentally and analytically that the oxygen transport through the air channels of PEMFCs operated at low oxygen stoichiometry can also contribute to the total oxygen diffusion impedance due to oscillations of oxygen concentration along the air channel.

The analysis of oxygen transport through the GDL and air channel of the FFP can be complicated because liquid water and water vapour can travel from the CCL to the GDL and air channel, and can also travel towards the anode side. In addition, different operating conditions can change the water phase diffusion where the molar fraction of water vapour changes with increasing pressure [7]. The derivation of impedance models based on the oxygen transport through the GDL and air channel can be simplified by considering modelling assumptions.

Electrochemical Phenomena in the Cathode Impedance Spectrum of PEM Fuel Cells.
https://doi.org/10.1016/B978-0-323-90607-4.00005-7
Copyright © 2022 Elsevier Inc. All rights reserved.

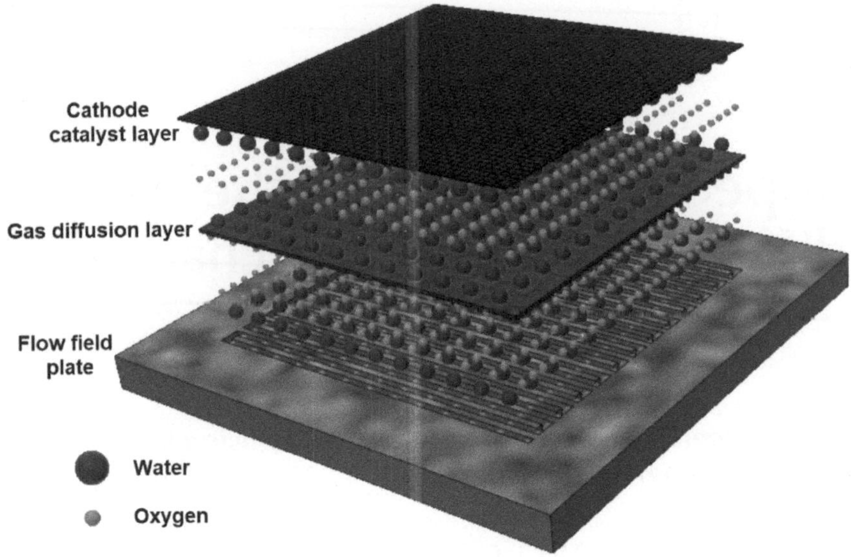

FIG. 5.1 Oxygen and water transport across the cathode.

Mainka et al. [3] derived an impedance model of the GDL-channel considering that only half of the water produced by the ORR in the CCL is transported towards the air channel. An impedance model of the GDL-channel considering water produced by the ORR in the CCL and transported towards the air channel could provide valuable information about oxygen transport limitations in GDLs with clogging effect in the air channel.

5.2 Finite-length Warburg impedance

The finite-length Warburg impedance has been the most conventional mathematical expression to analyse EIS measurements of PEMFCs at low frequencies with oxygen transport limitations present. The finite-length Warburg impedance has been commonly considered in the Randles electrical circuit [1,2,8] to estimate the oxygen diffusion resistance in the GDL that is present in EIS measurements of PEMFCs. The impedance response of the Randles circuit comprising the finite-length Warburg impedance has been described in Section 1.4.6 from Chapter 1. The finite-length Warburg impedance to represent the oxygen diffusion impedance in the GDL is defined as:

$$Z_{W\_O_2} = R_{W\_O_2} \frac{\tanh\left(\sqrt{j\omega\tau_{O_2\_GDL}}\right)}{\sqrt{j\omega\tau_{O_2\_GDL}}} \tag{5.1}$$

with

$$\tau_{O_2\_GDL} = \frac{\delta_{GDL}{}^2}{D_{GDL}} \quad (5.2)$$

where $R_{W\_O_2}$ represents the resistance for oxygen diffusion in the GDL, $\tau_{O_2\_GDL}$ is the oxygen diffusion time constant in the GDL, δ_{GDL} is the diffusion distance for oxygen to diffuse in the GDL, D_{GDL} is the oxygen diffusion coefficient in the GDL, j is the imaginary component of a complex number, and ω is the angular frequency. Eq. (5.1) can be derived from the Fick's second law for diffusion and has been applied to EIS measurements of PEMFCs to study oxygen transport limitations in GDLs at different operating conditions [1] and with different GDL morphology [2]. The impedance response of the finite-length Warburg impedance was described in Chapter 1 Section 1.4.3. Malevich et al. [2] applied the finite-length Warburg impedance represented in Eq. (5.1) with EIS measurements carried out in a PEMFC with microporous layers (MPL) attached to the GDLs. The oxygen diffusion resistance in the GDL considering and neglecting the MPL was estimated using the finite-length Warburg impedance.

5.2.1 Effect of capacitance of the cathode catalyst layer on impedance response of the gas diffusion layer

Kulikovsky [9] reported that the finite-length Warburg impedance represented in Eq. (5.1) is not appropriate for oxygen transport analysis in GDLs of PEMFCs. Kulikovsky [9] reported a new mathematical expression to represent the impedance response of the GDL as such:

$$Z_{K\_GDL} = \frac{Z_{W\_O_2}}{1 + j\omega C_{dl}\left[b/i_{cell}\right]} \quad (5.3)$$

where $Z_{W\_O_2}$ is similar to Eq. (5.1), C_{dl} is the double-layer capacitance in the CCL which is attached to the GDL, b is the Tafel slope, and i_{cell} is the current density of the PEMFC. Kulikovsky [9] reported that Eq. (5.3) should be considered instead of Eq. (5.1) to study oxygen transport limitations in GDLs attached to CCLs with high capacitance effect.

Fig. 5.2 shows the impedance response of the finite-length Warburg generated by Eq. (5.1). Fig. 5.2 also shows the impedance response generated by Eq. (5.3). The parameters for the simulation of the impedance spectra are the same as reported by Kulikovsky [9] for $R_{W\_O_2}=0.027\ \Omega cm^2$, $\delta_{GDL}=0.025$ cm, $D_{GDL}=0.01\ cm^2/s$, $b=0.03$ V (expressed as $b=0.069$ V/dec if b with units V is multiplied by a factor of 2.3), $C_{dl}=0.02\ F/cm^2$, $i_{cell}=0.1\ A/cm^2$ where the range of frequencies was from 10 kHz to 0.01 Hz. The simulation of the impedance spectra was carried out in MATLAB software. The impedance spectra shown Fig. 5.2 are similar to the impedance results reported by Kulikovsky [9] (Fig. 2 from the author's study). The difference in the impedance spectra

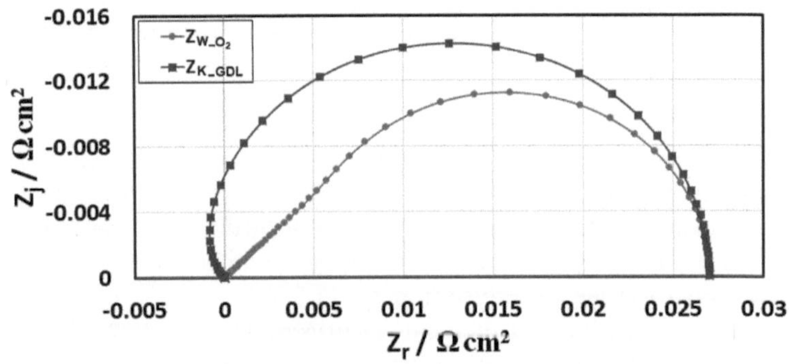

FIG. 5.2 Comparison between impedance spectrum generated by Eq. (5.1) (finite-length Warburg $Z_{W\_O_2}$) and impedance spectrum generated by Eq. (5.3) (equation reported by Kulikovsky [9] Z_{K_GDL}). The parameters for the simulation of the impedance spectra are $R_{W\_O_2}=0.027\ \Omega\text{cm}^2$, $\delta_{GDL}=0.025$ cm, $D_{GDL}=0.01\ \text{cm}^2/\text{s}$, $b=0.03$ V, $C_{dl}=0.02\ \text{F/cm}^2$, $i_{cell}=0.1\ \text{A/cm}^2$. The range of frequencies was from 10kHz to 0.01Hz. The impedance spectra are similar to the impedance results reported by Kulikovsky [9] (Fig. 2 from the author's study).

predicted by Eqs (5.1) and (5.3) is mainly attributed to the value of the capacitance C_{dl} considered in the denominator of Eq. (5.3). If the value of the capacitance C_{dl} is reduced to a very small value, the value of the denominator of Eq. (5.3) becomes close to 1 and the impedance spectrum generated by Eq. (5.3) resembles the impedance spectrum generated by Eq. (5.1) as such $Z_{K\_GDL} \approx Z_{W\_O_2}$.

In the study by Cruz-Manzo and Greenwood [10], an impedance model of the PEMFC considering the impedance model of the GDL reported by Kulikovsky [9] was fitted to EIS measurements of a PEMFC. The PEMFC was operated with 20% O_2/80% N_2 (nitrox) as gas reactant at 30% relative humidity (RH) in the cathode side. The impedance model of the GDL reported by Kulikovsky [9] was considered in terms of oxygen diffusion resistance, charge transfer resistance, constant phase element (CPE), and oxygen diffusion time constant as follows:

$$Z_{GDL} = \frac{Z_{W\_O_2}}{1 + Y[j\omega]^P R_C} \tag{5.4}$$

where $Z_{W\_O_2}$ is the finite-length Warburg impedance defined in Eq. (5.1). A CPE has been considered instead of an ideal capacitor C_{dl} in the denominator of Eq. (5.4) to account for the nonhomogeneous charge distribution between dissimilar materials (e.g. Nafion/carbon) in the CCL. Fig. 5.3 shows the impedance response of the GDL generated by Eq. (5.4) considering parameters extracted from EIS measurements carried out in a PEMFC operated with 20% O_2/80% N_2 (nitrox) as gas reactant and 30% RH at the cathode. The parameters for the simulation of the impedance responses are $R_{W\_O_2}=0.1197\ \Omega\text{cm}^2$, $\tau_{O_2\_GDL}=0.1692$ s, $R_C=0.9571\ \Omega\text{cm}^2$, $Y=0.0118\ \text{s}^P/\Omega\text{cm}^2$, $P=0.7669$.

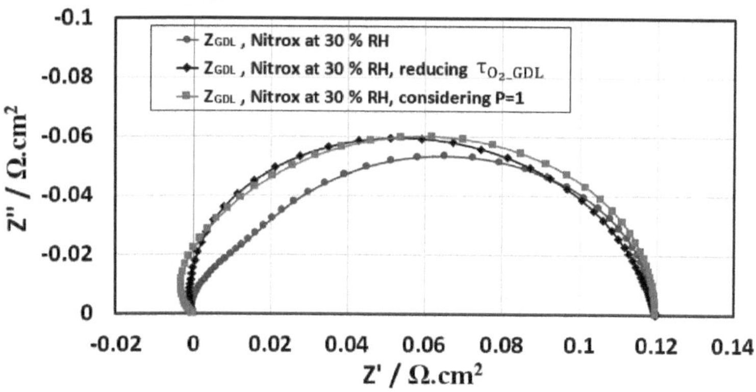

FIG. 5.3 Simulated impedance response of the gas diffusion layer generated by Eq. (5.4). The parameters for the simulation of the impedance responses are $R_{W\_O_2}=0.1197\ \Omega cm^2$, $\tau_{O_2\_GDL}=0.1692$ s, $R_C=0.9571\ \Omega cm^2$, $Y=0.0118\ s^P/\Omega cm^2$ and $P=0.7669$. Simulated effect of reducing oxygen diffusion time constant $\tau_{O_2\_GDL}=0.01692$ s and increasing capacitance effect $P=1$ on the impedance response of the gas diffusion layer. *(Reproduced from S. Cruz-Manzo, P. Greenwood, Study of oxygen diffusion in the cathode catalyst layer and gas diffusion layer for polymer electrolyte fuel cells with EIS, J. Electroanal. Chem. 892 (2021) 115270–115279, with permission of Elsevier. Copyright Elsevier.)*

Further information about the extraction of the parameters from the EIS-PEMFC measurements can be found in the study by Cruz-Manzo and Greenwood [10]. The impedance response of the GDL resembled the impedance response of the finite-length Warburg impedance $Z_{W\_O_2}$ because the term of the denominator of Eq. (5.4) had a small effect on the overall impedance of the GDL. If the oxygen diffusion time constant $\tau_{O_2\_GDL}$ is reduced or the parameter P is considered to be equal to 1, the impedance response of the GDL resembled the impedance response reported by Kulikovsky [9] (Fig. 2 from the author's study) as shown in Fig. 5.3. The results from Fig. 5.3 indicate that the impedance response of the GDL generated by Eq. (5.4) differs from the impedance response of the finite-length Warburg impedance $Z_{W\_O_2}$ at low oxygen diffusion time constant $\tau_{O_2\_GDL}$ and high double-layer capacitance of the CCL which is attached to the GDL.

5.2.2 Local Warburg impedance

Mainka et al. [3] reported a mathematical expression, Eq. (5.5), which is a modification of the finite-length Warburg impedance represented in Eq. (5.1) to simulate the impedance response of the GDL considering oxygen depletion along the channel.

$$Z_{M\_GDL-ch} = R_d^{2D} \frac{\tanh\left(\sqrt{j\omega\tau_d}\right)}{\sqrt{j\omega\tau_d}} \tag{5.5}$$

with

$$R_{\text{d}}^{\text{2D}} = \frac{b\delta_{\text{GDL}}}{4FD_{\text{GDL}}\frac{I_{\text{cell}}S_{\text{O}_2} - d\int_0^y i_{\text{F}}(y)dy}{I_{\text{cell}}S_{\text{O}_2}[1+H]}c_{\text{ch\_in}} - i_{\text{F}}(y)\delta_{\text{GDL}}} \quad (5.6)$$

where τ_{d} is the characteristic diffusion time constant, S_{O_2} is the oxygen stoichiometry, I_{cell} is the current of the cell, $c_{\text{ch\_in}}$ is the inlet oxygen concentration in the channel, and $i_{\text{F}}(y)$ is the local faradaic current density. Detailed information about the derivation of Eq. (5.6) can be found in the study by Mainka et al. [3]. Eq. (5.5) does not consider the oscillating oxygen concentration along the channel attributed to the sinusoidal perturbation superimposed during an EIS test and only considers the time average oxygen depletion along the channel. In the study by Mainka et al. [3], Eq. (5.5) was implemented with the Randles circuit and fitted to local EIS measurements of a segmented PEMFC to estimate the oxygen transport resistance R_{d}^{2D} along the air channel. The EIS results from the segmented cell along the air channel and reported by Mainka et al. [3] featured two semicircles in the complex-impedance plot. The semicircle at low frequencies attributed to the GDL-channel impedance response increased from the inlet to the outlet of the air channel. The impedance model of the GDL expressed in Eq. (5.5) could reproduce the second semicircle at low frequencies attributed to oxygen transport limitations in the GDL considering oxygen depletion along the channel. Nevertheless, the effect of the oscillating oxygen concentration along the channel during EIS measurements cannot be analysed using Eq. (5.5).

Maranzana et al. [4] also reported a physics-based impedance model of the PEMFC to simulate the local impedance response of a segmented PEMFC along the channel. The impedance model of the PEMFC considered a local Warburg impedance with oxygen concentration at the electrode and along the channel as such:

$$Z_{\text{W}}(x) = \frac{b}{c_{\text{O}_2}(x, y=\delta)_{\text{t}}} \frac{\tanh\left(\sqrt{\delta^2[j\omega/D_{\text{O}_2}]}\right)}{4F\varepsilon_{\text{eff}}\sqrt{j\omega D_{\text{O}_2}}} \quad (5.7)$$

where D_{O_2} is the oxygen diffusion coefficient, δ is the thickness of the diffusion layer, ε_{eff} is the diffusion media porosity, $c_{\text{O}_2}(x, y=\delta)$ is the oxygen concentration at the electrode along the length x of the channel. Detailed information about the derivation of Eq. (5.7) can be found in the study by Maranzana et al. [4]. The impedance model was able to consider the oscillating oxygen concentration along the channel. A second semicircle at low frequencies of the local impedance spectrum of the PEMFC was attributed to the oxygen depletion along the air channel as oxygen is consumed during the ORR and the amplification of the oscillating oxygen concentration along the air channel. The impedance model demonstrated that at low oxygen stoichiometry, the low-frequency

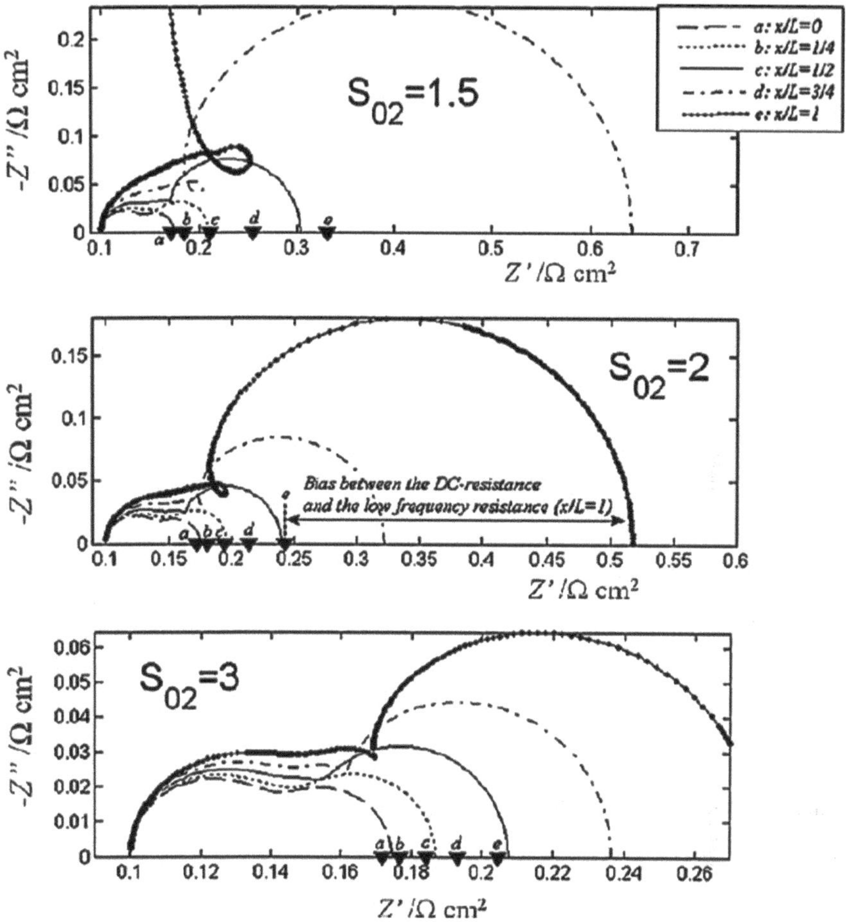

FIG. 5.4 Simulated local impedance spectra along the channel considering different oxygen stoichiometry S_{O_2}. The nondimensional coordinate $x/L=0$ corresponds to the inlet channel. The black triangles correspond to the DC polarisation resistance from the inlet (a) to the outlet (e) of the channel. *(Reproduced from G. Maranzana, J. Mainka, O. Lottin, J. Dillet, A. Lamibrac, A. Thomas, S. Didierjean, A proton exchange membrane fuel cell impedance model taking into account convection along the air channel: on the bias between the low frequency limit of the impedance and the slope of the polarization curve, Electrochim. Acta 83 (2012) 13–27, with permission of Elsevier. Copyright Elsevier.)*

resistance of the EIS measurements is not in agreement with the DC polarisation resistance calculated through the slope of the polarisation curve. Fig. 5.4 shows that the second semicircle of the impedance spectrum at low frequencies, attributed to oscillations in oxygen concentration and oxygen depletion along the channel, decreases with increasing oxygen stoichiometry S_{O_2} towards the coordinate $x/L=0$ at the inlet of the channel. Fig. 5.4 also shows a discrepancy between the low-frequency resistance of the impedance spectra and the DC polarisation resistance calculated through the slope of the polarisation curve. This discrepancy increases with decreasing oxygen stoichiometry S_{O_2}.

Other physics-based impedance models of the PEMFC considering oxygen transport in the air channel have been developed [6,11,12]. Among these models, Kulikovsky and Shamardina [6] reported the same results shown in Fig. 5.4 reported by Maranzana et al. [4] in which oxygen flow in the channel triggers spatial and frequency oscillations of the local impedance of the PEMFC. Kulikovsky [13] has also extended his previous impedance model [11] to study the effect of air flow velocity oscillations in phase with the AC potential perturbation during EIS. The results demonstrated that the second semicircle of the PEMFC impedance spectrum attributed to the oscillating oxygen concentration along the channel decreases with the air flow velocity oscillations.

5.3 Impedance model of the segmented cathode

The global impedance in a sectioned PEMFC along the air channel has been represented through the following equation:

$$\frac{1}{\widehat{Z}_{\mathrm{T}}} = \int_0^1 \frac{d\widehat{z}}{\widehat{Z}_{\mathrm{local}}} \tag{5.8}$$

Eq. (5.8) calculates the integral of the inverse of the local impedance. Eq. (5.8) is also analogous to the integral of the current distribution along the air channel as the impedance is the relation between potential and current in the frequency domain. An equation for the current distribution along the air channel has to be defined [14,15] for the derivation of the equation of the local impedance of the PEMFC. Eq. (5.8) is analogous to the total impedance of individual local impedances connected in parallel as shown in Fig. 5.5. Kulikovsky and Shamardina [6] reported a similar equation as Eq. (5.8) to calculate the global impedance of the segmented cathodic side of the PEMFC represented in Fig. 5.5.

In the next section, an analytical mathematical equation coined as analytical-Warburg impedance model will be developed to simulate the

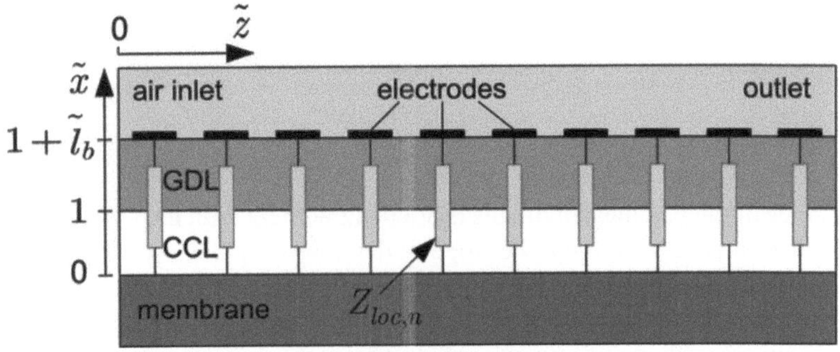

FIG. 5.5 Segmented cathodic side of a PEMFC. *(From A. Kulikovsky, O. Shamardina, A model for PEM fuel cell impedance: oxygen flow in the channel triggers spatial and frequency oscillations of the local impedance, J. Electrochem. Soc. 162 (2015) F1068. Figure published under license of Creative Commons Attribution 4.0 License (CC BY).)*

impedance response of the GDL considering oxygen depletion along the air channel. The analytical mathematical expression considers oxygen diffusion through the GDL and convective oxygen transport through the air channel. It will be possible to simulate the contribution of the GDL and channel impedance responses on the low frequency EIS spectrum of the PEMFC. In addition, similar to the simulation results shown in Fig. 5.4 and reported by Maranzana et al. [4], it will be possible to simulate the effect of the oscillating oxygen concentration along the channel on the low frequency impedance spectrum of the GDL-channel.

5.4 Mathematical model for oxygen transport in the gas diffusion layer and air channel

Impedance models considering oxygen transport in the GDL and air channel have been derived by considering oxygen transport in the coordinate x (thickness) of the GDL and the coordinate y (length) of the channel [4,16,17]. A similar mathematical approach will be considered for the derivation of the impedance model of the GDL and air channel presented in this chapter. The structure of a PEMFC comprises a much larger length of air channel compared to the GDL thickness. The coordinates for oxygen transport through the length of the air channel and thickness of the GDL are represented in Fig. 5.6. The mathematical

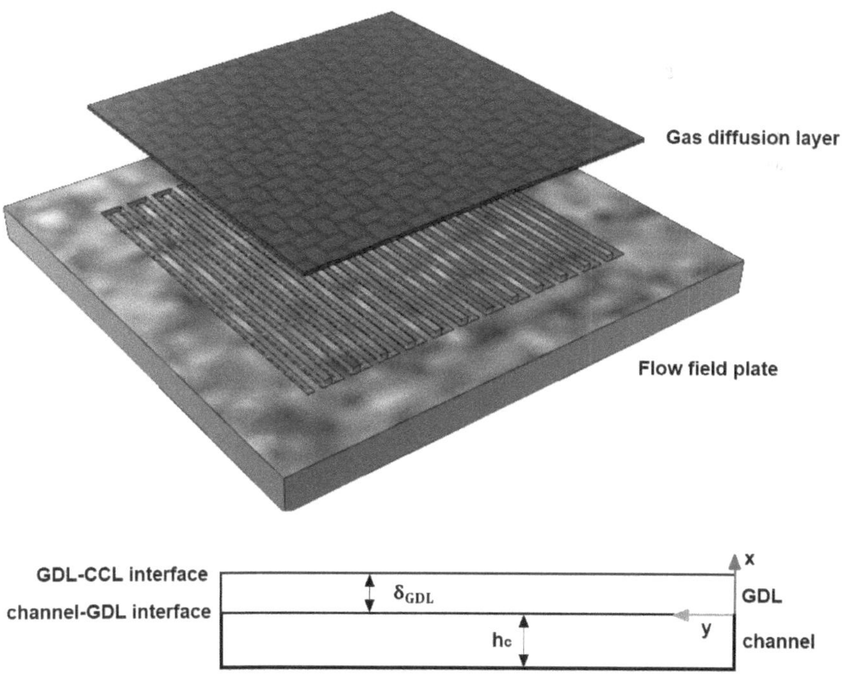

FIG. 5.6 Oxygen transport through the gas diffusion layer (GDL) and air channel, oxygen is transported through coordinate y (channel length) and coordinate x (thickness of GDL).

framework presented in this chapter considers that the oxygen transport in the CCL is fast. The transport of oxygen in the GDL-channel is a convective-diffusion process.

5.4.1 Oxygen diffusion in the gas diffusion layer

The oxygen diffusion in the GDL can be represented through Fick's second law:

$$\frac{\partial c_{GDL}(x,y,t)}{\partial t} - D_{GDL}\frac{\partial^2 c_{GDL}(x,y,t)}{\partial x^2} \tag{5.9}$$

where c_{GDL} is the oxygen concentration in the GDL. Laplace transform of Eq. (5.9) with oxygen concentration at the inlet of the air channel c_{ch_in} as initial condition $t=0$ yields:

$$D_{GDL}\frac{\partial^2 \bar{c}_{GDL}(x,y,s)}{\partial x^2} = s\bar{c}_{GDL}(x,y,s) - c_{ch\_in} \tag{5.10}$$

where the bar located over a letter represents the Laplace transform of a variable.

The general solution of Eq. (5.10) yields [18]:

$$\bar{c}_{GDL}(x,y,s) = A(y,s)\exp(\lambda_1 x) + B(y,s)\exp(-\lambda_1 x) + \bar{c}_{ch\_in}(s) \tag{5.11}$$

with $\lambda_1 = \sqrt{\frac{s}{D_{GDL}}}$ and $\bar{c}_{ch\_in}(s) = c_{ch\_in}/s$.

Solving Eq. (5.11) with boundary conditions considering coordinates from Fig. 5.6:

$x=0$ at the air channel/GDL interface $\bar{c}_{GDL}(0,y,s) = \bar{c}_{ch}(y,s)$

$x=\delta_{GDL}$ at the GDL/CCL interface $-D_{GDL}\left.\frac{\partial \bar{c}_{GDL}(x,y,s)}{\partial x}\right|_{x=\delta_{GDL}} = \frac{\bar{i}_F(y,s)}{4F}$ where $\bar{i}_F(y,s)$ is the Laplace space representation of the faradaic current density along the coordinate y.

$$\bar{c}_{GDL}(x,y,s) = \frac{[\bar{c}_{ch}(y,s) - \bar{c}_{ch\_in}(s)]4FD_{GDL}\lambda_1\exp(-\lambda_1\delta_{GDL}) - \bar{i}_F(y,s)}{4FD_{GDL}\lambda_1[\exp(\lambda_1\delta_{GDL}) + \exp(-\lambda_1\delta_{GDL})]}\exp(\lambda_1 x)$$
$$+ \frac{[\bar{c}_{ch}(y,s) - \bar{c}_{ch\_in}(s)]4FD_{GDL}\lambda_1\exp(\lambda_1\delta_{GDL}) + \bar{i}_F(y,s)}{4FD_{GDL}\lambda_1[\exp(\lambda_1\delta_{GDL}) + \exp(-\lambda_1\delta_{GDL})]}\exp(-\lambda_1 x) + \bar{c}_{ch\_in}(s) \tag{5.12}$$

5.4.2 Oxygen transport in the air channel

The oxygen transport through the air channel can be represented as follows [12]:

$$\frac{\partial c_{ch}(y,t)}{\partial t} + u_{ch}\frac{\partial c_{ch}(y,t)}{\partial y} = \frac{D_{GDL}}{h_c}\left.\frac{\partial c_{GDL}(x,y,t)}{\partial x}\right|_{x=0} \tag{5.13}$$

where c_{ch} is the oxygen concentration in the channel, y is the coordinate along the channel from 0 to ℓ_c (total length), u_{ch} is the velocity of oxygen through the channel, h_c is the depth of the channel. The Laplace transform of Eq. (5.13) with initial condition $t=0$ in $c_{ch}(y,0)=c_{ch\_in}$ yields:

$$s\bar{c}_{ch}(y,s) - c_{ch\_in} + u_{ch}\frac{\partial \bar{c}_{ch}(y,s)}{\partial y} = \frac{D_{GDL}}{h_c}\left.\frac{\partial \bar{c}_{GDL}(x,y,s)}{\partial x}\right|_{x=0} \tag{5.14}$$

where c_{ch_in} is the oxygen concentration at the inlet of the air channel. The term on the right-hand side of Eq. (5.14) can be estimated from Eq. (5.12) (oxygen concentration in GDL) as follows:

$$\left.\frac{\partial \bar{c}_{GDL}(x,y,s)}{\partial x}\right|_{x=0} = [\bar{c}_{ch\_in}(s) - \bar{c}_{ch}(y,s)]\lambda_1 \tanh(\lambda_1 \delta_{GDL}) - \frac{\bar{i}_F(y,s)}{4FD_{GDL}\cosh(\lambda_1 \delta_{GDL})} \tag{5.15}$$

The current density distribution along the channel is considered from the study of Chevalier et al. [12]:

$$\bar{i}_F(y,s) = \bar{i}_{F\_in}(s)\ \exp\left(-\frac{y}{\ell_c \Lambda}\right) \tag{5.16}$$

where $\bar{i}_{F\_in}(s)$ is the Laplace form of the inlet current density and Λ is the inlet oxygen stoichiometry ratio $\Lambda = 4F\bar{c}_{ch\_in}(s)u_{ch}h_c/\left[\bar{i}_{F\_in}(s)\ell_c\right]$, the oxygen velocity along the channel u_{ch} is considered to be constant and is defined as:

$$u_{ch} = \frac{i_{cell}\ell_c}{4Fh_c c_{ch\_in}} S_{O_2} \tag{5.17}$$

where i_{cell} is the cell current density or the average current density distribution along the channel.

The mathematical expression of the current density distribution could also be considered from the study reported by Kulikovsky [15]. A discussion about different approaches for the current distribution will be considered later.

The oxygen concentration in the channel can be derived by solving Eq. (5.14) and substituting Eqs (5.15) and (5.16) into Eq. (5.14) as well as considering trigonometric identities and the humidity molar ratio H:

$$\bar{c}_{ch}(y,s) = \frac{p_a}{5RT_a}\frac{\bar{\varphi}_{ch}(y,s)}{\left[\frac{\bar{\varphi}_{ch}(y,s)}{5} + 0.8 + H\right]} \tag{5.18}$$

with

$$\bar{\varphi}_{ch}(y,s) = 1 + \frac{u_{ch}}{\cosh(\lambda_1 \delta_{GDL})}\frac{1}{\beta \ell_c \Lambda - u_{ch}}\left[\frac{1}{\gamma} - \exp\left(-\frac{y}{\ell_c \Lambda}\right)\right] \tag{5.19}$$

and

$$\beta = s + D_{GDL}\lambda_1 \tanh(\lambda_1 \delta_{GDL}) h_c^{-1} \tag{5.20}$$

and

$$\gamma = \exp\left(y\frac{\beta}{u_{ch}}\right) \tag{5.21}$$

where p_a and T_a is the pressure and temperature of air at the inlet of the channel where for dry conditions the humidity molar ratio is neglected $H=0$. A detailed derivation of Eq. (5.18) can be found in the study by Cruz-Manzo and Greenwood [19].

In the derivation of Eq. (5.18) it is assumed that the amount of water flowing through the channel is only the amount of water mixed in the air entering the channel. The water produced by the ORR and transported from the CCL to the air channel is ignored because the mathematical approach does not consider the two-phase water transported towards the air channel and the liquid water transported towards the polymer electrolyte membrane (PEM) and anode.

The humidity molar ratio H is calculated as:

$$H = \frac{\varphi p_{vs}}{p_a - \varphi p_{vs}} \tag{5.22}$$

where p_{vs} is the saturation pressure of vapour and φ is the RH. The saturation pressure can be calculated through the relation [20]:

$$p_{vs} = e^{c1 {T_a}^{-1} + c2 + c3 T_a + c4 {T_a}^2 + c5 {T_a}^3 + c6 \ln(T_a)} \tag{5.23}$$

where $c1$, $c2$, $c3$, $c4$, $c5$, and $c6$ are coefficients defined as:

$$c1 = -5800.22$$

$$c2 = 1.3915$$

$$c3 = -0.0486$$

$$c4 = 0.4176 \times 10^{-4}$$

$$c5 = -0.1445 \times 10^{-7}$$

$$c6 = 6.5459$$

5.4.3 Steady-state oxygen transport in the channel

At steady-state the transport of oxygen through the air channel which is defined in Eq. (5.13) is expressed as:

$$u_{ch}\frac{\partial c_{ch}(y)}{\partial y} = \frac{D_{GDL}}{h_c}\left.\frac{\partial c_{GDL}(x,y)}{\partial x}\right|_{x=0} \tag{5.24}$$

The term on the right-hand side of Eq. (5.24) can be represented through Fick's first law. The Fick's first law can be expressed as:

$$-D_{\mathrm{GDL}} \frac{\partial c_{\mathrm{GDL}}(x,y)}{\partial x}\bigg|_{x=0} = \frac{i_{\mathrm{F}}(y)}{4F} \tag{5.25}$$

with the current density distribution expressed from the study of Chevalier et al. [12]:

$$i_{\mathrm{F}}(y) = i_{\mathrm{F\_in}} \exp\left(-\frac{y}{\ell_{\mathrm{c}}\Lambda}\right) \tag{5.26}$$

where Λ is the inlet oxygen stoichiometry ratio $\Lambda = 4Fc_{\mathrm{ch\_in}}u_{\mathrm{ch}}h_{\mathrm{c}}/[i_{\mathrm{F\_in}}\ell_{\mathrm{c}}]$. The mathematical expression of the current density distribution could also be considered from the study reported by Kulikovsky [15]. Substituting Eqs (5.25) and (5.26) into Eq. (5.24) and solving Eq. (5.24) with the boundary conditions at $y=0$, $c_{\mathrm{ch}}(0)=c_{\mathrm{ch\_in}}$ and considering the humidity molar ratio H yields:

$$c_{\mathrm{ch}}(y) = \frac{p_{\mathrm{a}}}{5RT_{\mathrm{a}}} \frac{\varphi_{\mathrm{ch}}(y)}{\left[\frac{\varphi_{\mathrm{ch}}(y)}{5} + 0.8 + H\right]} \tag{5.27}$$

with the molar fraction of oxygen concentration $\varphi_{\mathrm{ch}}(y) = c_{\mathrm{ch}}(y)/c_{\mathrm{ch\_in}}$ as such:

$$\varphi_{\mathrm{ch}}(y) = \exp\left(-\frac{y}{\ell_{\mathrm{c}}\Lambda}\right) \tag{5.28}$$

A detailed derivation of Eq. (5.27) can be found in the study by Cruz-Manzo and Greenwood [19].

5.5 Impedance model of the GDL-channel

Two models for the impedance of the GDL-channel can be defined from the equations of the aforementioned Section 5.4. An impedance model of the GDL-channel considering nonstationary oxygen depletion along the channel caused by the sinusoidal perturbation (AC conditions) during EIS measurements and another considering steady-state oxygen depletion along the channel. The impedance response of the GDL-channel considering nonstationary oxygen depletion during AC conditions should converge with the impedance response of the GDL-channel considering steady-state oxygen depletion at the low frequency limit of the impedance spectrum as the frequency is reduced to a value near zero and neglecting the sinusoidal current amplitude effect during EIS. The impedance model of the GDL-channel will not consider the high-frequency effect of the charge double-layer of the CCL on the GDL-channel impedance because the analysis in this chapter is mainly focussed on the low-frequency effect of oxygen concentration in the channel on the GDL-channel impedance. In the last section of this chapter, the effect of the double-layer capacitance of the CCL on the GDL-channel impedance spectrum will be demonstrated.

5.5.1 Impedance model considering nonstationary oxygen depletion along the air channel during AC conditions

Considering the general form of the Tafel equation in Laplace domain s

$$\bar{i}_F(y,s) = i_0 \frac{\bar{c}_{GDL}(\delta_{GDL},y,s)}{\bar{c}_{ch\_in}(s)} \exp\left(\frac{\bar{\eta}(y,s)}{b_{O_2}}\right) \tag{5.29}$$

where $\bar{c}_{GDL}(\delta_{GDL},y,s)$ is the oxygen concentration at the GDL/CCL interface, $\bar{c}_{ch\_in}(s)$ is the oxygen concentration at the inlet of the channel (reference concentration), i_0 is the exchange current density, and b_{O_2} is the absolute value of the Tafel slope defined in Eq. (4.53) from Chapter 4 and considers $\bar{\eta}(y,s) > 0$.

Eq. (5.29) can be rearranged and differentiated with respect to $\bar{i}_F\ (y,s)$ as:

$$\frac{\partial\bar{\eta}(y,s)}{\partial\bar{i}_F(y,s)} = \frac{b_{O_2}}{\bar{i}_F(y,s)} - \frac{b_{O_2}}{\bar{c}_{GDL}(\delta_{GDL},y,s)} \frac{\partial\bar{c}_{GDL}(\delta_{GDL},y,s)}{\partial\bar{i}_F(y,s)} \tag{5.30}$$

The first term on the right-hand side of Eq. (5.30) represents the charge transfer resistance during the ORR and the second term corresponds to the oxygen transport resistance in the GDL-channel. The impedance of oxygen transport in the GDL-channel represented in the second term of Eq. (5.30) can be expressed as follows:

$$Z_{GDL-ch_{ac}}(y,s) = -\frac{b_{O_2}}{\bar{c}_{GDL}(\delta_{GDL},y,s)} \frac{\partial\bar{c}_{GDL}(\delta_{GDL},y,s)}{\partial\bar{i}_F(y,s)} \tag{5.31}$$

Solving Eq. (5.9) in the Laplace domain s and considering the concentration of oxygen in the channel $c_{ch}(y)$ at the initial condition $t=0$ and considering the same boundary conditions for the solution of Eq. (5.11) yields:

$$\bar{c}_{GDL}(x,y,s) = \bar{c}_{ch}(y,s) + \bar{i}_F(y,s)\frac{\exp(-\lambda_1 x) - \exp(\lambda_1 x)}{4FD_{GDL}\lambda_1[\exp(\lambda_1\delta_{GDL}) + \exp(-\lambda_1\delta_{GDL})]} \tag{5.32}$$

Evaluating Eq. (5.32) with $\bar{c}_{GDL}(x=\delta_{GDL},y,s)$ and differentiating it with respect to $\bar{i}_F\ (y,s)$ and substituting it into Eq. (5.31) yields:

$$Z_{GDL-ch_{ac}}(y,s) = \frac{b_{O_2}}{\bar{c}_{GDL}(\delta_{GDL},y,s)} \frac{\tanh(\lambda_1\delta_{GDL})}{4FD_{GDL}\lambda_1} \tag{5.33}$$

The transfer function of a linear, time-invariant system represented in the Laplace space s can be translated into the frequency domain by considering $s=j\omega$ [21]. The impedance of the GDL-channel can be represented in the frequency domain by considering $s=j\omega$ and $\lambda_1 = \sqrt{j\omega/D_{GDL}}$ in Eq. (5.33) as such:

$$\tilde{Z}_{GDL-ch_{ac}}(y,j\omega) = R_{GDL-ch_{ac}} \frac{\tanh\left(\sqrt{j\omega\tau_{O_2\_GDL}}\right)}{\sqrt{j\omega\tau_{O_2\_GDL}}} \tag{5.34}$$

where $\tau_{O_2\_GDL}$ is the oxygen diffusion time constant defined in Eq. (5.2). Eq. (5.34) represents the impedance response of the GDL-channel considering oxygen transport through the channel and oxygen diffusion from the channel to the GDL/CCL interface as a function of the frequency with

$$R_{GDL-ch_{ac}} = \frac{b_{O_2}\delta_{GDL}}{4FD_{GDL}\widetilde{c}_{GDL}(\delta_{GDL}, y, j\omega)} \tag{5.35}$$

defined as the local oxygen transport resistance in the GDL-channel considering the oscillating oxygen concentration in the GDL/CCL interface $\widetilde{c}_{GDL}(\delta_{GDL}, y, j\omega)$ with

$$\widetilde{c}_{GDL}(\delta_{GDL}, y, j\omega) = \widetilde{c}_{ch}(y, j\omega) - \frac{\widetilde{i_F}(y, j\omega)}{4F\sqrt{D_{GDL}j\omega}}\tanh\left(\sqrt{j\omega\tau_{O_2\_GDL}}\right) \tag{5.36}$$

Eq. (5.32) in the frequency domain with $s=j\omega$ and evaluated at $x=\delta_{GDL}$ corresponds to Eq. (5.36). The term $\widetilde{i_F}(y, j\omega)$ on the right-hand side of Eq. (5.36) represents the Fourier transform of the sinusoidal current density $i_F(y,t)=i_{F,SS}(y)+i_{AC}\exp(j[\omega t+\phi_I])$ [22] and considers the steady-state current density and oscillating contribution of the sinusoidal perturbation.

MATLAB exercise. Demonstrate that the Fourier analysis of the sinusoidal signal $I(t)=I_{SS}+I_{AC}\exp(j[\omega t+\phi_I])$, where I_{SS} is the steady-state contribution, I_{AC} is the amplitude of the superimposed oscillating perturbation and neglecting the phase shift $\phi_I=0$, yields the signal in the frequency domain as $\widetilde{I} = I_{SS} + I_{AC}$.

Solution. Considering the Fourier analysis presented in Chapter 1 Section 1.2.2, the script shown in Fig. 5.7 is written in MATLAB software.

Fourier_Transform_Signal_Chapter5.m

```
% Definition of symbolic parameters
syms t f T Iac Iss
% Period of the signal
T=1/f;
% Neglecting phase shift
phi=0;
% Coefficients of the Fourier transform
n=[-1 0 1];
% Angular frequency
w=2*pi*f;
% I(t) signal
It=Iss+Iac*exp(1i*(w*t+phi));
% Sum of Fourier coefficients  yields the signal in frequency domain
Iw=sum((1/T)*int((It.*exp(-n*1i*w*t)),t,[0 T]))
```

Command Window

```
>> Fourier_Transform_Signal_Chapter5

Iw =

Iac + Iss

>>
```

FIG. 5.7 Script in MATLAB to mathematically represent the frequency domain of the sinusoidal signal. Note, 1i in MATLAB means the imaginary component j.

Eq. (5.18) in the frequency domain with $s=j\omega$ corresponds to the oscillating oxygen concentration in the channel, $\widetilde{c}_{\mathrm{ch}}(y,j\omega)$ represented in Eq. (5.36), as such:

$$\widetilde{c}_{\mathrm{ch}}(y,j\omega)=\frac{p_{\mathrm{a}}}{5RT_{\mathrm{a}}}\frac{\widetilde{\varphi}_{\mathrm{ch}}(y,j\omega)}{\left[\frac{\varphi_{\mathrm{ch}}(y,j\omega)}{5}+0.8+H\right]} \tag{5.37}$$

with

$$\widetilde{\varphi}_{\mathrm{ch}}(y,j\omega)=1+\frac{u_{\mathrm{ch}}}{\cosh\left(\sqrt{j\omega\tau_{\mathrm{O_2\_GDL}}}\right)}\frac{1}{\beta_{\omega}\ell_{\mathrm{c}}\Lambda-u_{\mathrm{ch}}}\left[\frac{1}{\gamma_{\omega}}-\exp\left(-\frac{y}{\ell_{\mathrm{c}}\Lambda}\right)\right] \tag{5.38}$$

and

$$\gamma_{\omega}=\exp\left(y\frac{\beta_{\omega}}{u_{\mathrm{ch}}}\right) \tag{5.39}$$

and

$$\beta_{\omega}=j\omega+\sqrt{j\omega D_{\mathrm{GDL}}}\tanh\left(\sqrt{j\omega\tau_{\mathrm{O_2\_GDL}}}\right)h_{\mathrm{c}}^{-1} \tag{5.40}$$

5.5.2 Impedance model considering steady-state oxygen depletion along the air channel

The equation that represents the local impedance response of the GDL-channel considering steady-state oxygen depletion along the air channel can be defined as:

$$Z_{\mathrm{GDL-ch_{ss}}}=R_{\mathrm{GDL-ch_{ss}}}\frac{\tanh\left(\sqrt{j\omega\tau_{\mathrm{O_2\_GDL}}}\right)}{\sqrt{j\omega\tau_{\mathrm{O_2\_GDL}}}} \tag{5.41}$$

with

$$R_{\mathrm{GDL-ch_{ss}}}=\frac{b_{\mathrm{O_2}}\delta_{\mathrm{GDL}}}{4FD_{\mathrm{GDL}}c_{\mathrm{GDL}}(\delta_{\mathrm{GDL}},y)} \tag{5.42}$$

defined as the local oxygen transport resistance of the GDL-channel. $c_{\mathrm{GDL}}(\delta_{\mathrm{GDL}},y)$ defined in Eq. (5.42) represents the oxygen concentration at the GDL/CCL interface considering steady-state oxygen concentration in the air channel. $\tau_{\mathrm{O_2\_GDL}}$ is the oxygen diffusion time constant defined in Eq. (5.2).

The oxygen concentration at the GDL/CCL interface at steady-state conditions $c_{\mathrm{GDL}}(\delta_{\mathrm{GDL}},y)$ represented in Eq. (5.42) can be deduced from an analysis of Fick's first law:

$$c_{\mathrm{GDL}}(\delta_{\mathrm{GDL}},y)=c_{\mathrm{ch}}(y)-\frac{i_{\mathrm{F}}(y)\delta_{\mathrm{GDL}}}{4FD_{\mathrm{GDL}}} \tag{5.43}$$

where the first term on the right-hand side of Eq. (5.43) represents the oxygen concentration in the air channel at steady-state conditions (Eq. 5.27) and the current density distribution $i_F(y)$ is defined in Eq. (5.26).

Eq. (5.41) represents the local impedance response of the GDL-channel considering steady-state oxygen depletion along the air channel. Mainka et al. [3] reported the local impedance of the GDL-channel represented in Eq. (5.5) and the authors assumed that half of the water produced by the ORR in the CCL is transported towards the channel.

5.5.3 Current distribution along the channel

Different experimental techniques to measure the current distribution in the PEMFC have been developed such as the printed circuit developed by Cleghorn et al. [23], the subcell approach, the partial membrane electrode assembly (MEA) method and current mapping technique by Stumper et al. [24], the segmented subcells in conjunction with a multichannel potentiostatic system by Yang et al. [25], the segmented flow-field plate approach by Noponen et al. [26], and current mapping technique based on segmented flow field by Mench et al. [27]. The combination of locally resolved current density measurements and local EIS allows the identification of local in-homogeneities in the performance of a PEMFC [28].

Eq. (5.44) represents a simplification of the current density distribution represented in Eq. (5.26) where $\dot{y} = y/\ell_c$ is the nondimensional length of the channel, i_{F_in} is the inlet cell current density, i_{cell} is the total cell current density and S_{O_2} is the total oxygen stoichiometry.

$$i_F(\dot{y}) = i_{F\_in} \exp\left(-\dot{y}\frac{i_{F\_in}}{i_{cell}S_{O_2}}\right) \tag{5.44}$$

Eq. (5.44) has been considered in the study by Chevalier et al. [12] for the development of the impedance model of the PEMFC considering oxygen transport in the GDL and air channel. The integral of the current density distribution expressed in Eq. (5.44) represents the total current density of the PEMFC:

$$i_{cell} = \int_0^1 i_F(\dot{y})d\dot{y} \tag{5.45}$$

Fig. 5.8 shows the current density distribution generated by Eq. (5.44) at three different oxygen stoichiometries $S_{O_2}=2.78$, $S_{O_2}=2.3$, and $S_{O_2}=1.46$ and total cell current density of $i_{cell}=0.125$ A/cm^2.

MATLAB exercise. Calculation of the integral of the current density distribution shown in Fig. 5.8.

Solution. The integral of the current density distribution represents the area below the curve shown in Fig. 5.8 and can be calculated through the solution of Eqs (5.44) and (5.45). A script in MATLAB software can solve Eqs (5.44) and (5.45). The inlet current density i_{F_in} is calculated using nonlinear curve-fitting. The script shown in Fig. 5.9 calculates the inlet current density expressed in

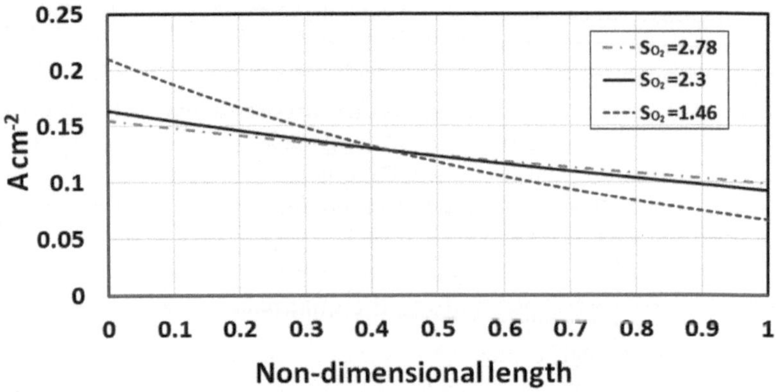

FIG. 5.8 Current density distribution along the nondimensional length of the channel at different oxygen stoichiometry.

Eq. (5.44) which is represented in Fig. 5.8 where the nondimensional length of the channel is equal to zero $\dot{y} = 0$. The script shown in Fig. 5.9 also calculates the integral of the current density distribution expressed in Eq. (5.45).

The mathematical treatment for the impedance model of the GDL-channel presented in this chapter considered the same mathematical expression for current density distribution reported by Chevalier et al. [12]. Kulikovsky [11] considered a different mathematical expression for the current density distribution [15] along the channel for the derivation of the local impedance model of a PEMFC. The mathematical expression for current density distribution reported by Kulikovsky [15] is:

$$i_{\mathrm{F}}(\dot{y}) = -S_{\mathrm{O}_2} i_{\mathrm{cell}} \ln\left(1 - \frac{1}{S_{\mathrm{O}_2}}\right)\left[1 - \frac{1}{S_{\mathrm{O}_2}}\right]^{\dot{y}} \tag{5.46}$$

Fig. 5.10 shows a comparison between the current density distribution along the nondimensional length of the channel generated by the analytical mathematical expression (Eq. 5.44) reported by Chevalier et al. [14] and reported (Eq. 5.46) by Kulikovsky [15]. Fig. 5.10 demonstrates that both analytical equations predict the same current density distribution at different oxygen stoichiometry S_{O_2}. Eq. (5.44) is a function of the inlet current density, total cell current density, and total oxygen stoichiometry. The application of the equation of the current density distribution represented in Eq. (5.46) is more straightforward than Eq. (5.44) because Eq. (5.46) is only a function of the total current density of the PEMFC and total oxygen stoichiometry. The mathematical framework for the derivation of the local impedance of the GDL-channel presented in this chapter considers the analytical expression for current density distribution Eq. (5.44) reported by Chevalier et al. [14]. The model of the local impedance of the GDL-channel could also be derived by considering Eq. (5.46) instead. This has been demonstrated in the analytical model of the local impedance of the PEMFC reported by Kulikovsky [6,11].

Current_distribution_Chapter_5.m

```
% Non-dimensional channel length
nLc=0:0.001:1;
% Oxygen stoichiometry
so2=2.3;
% Change of variable
ydata=nLc;
% Cell current density
I=0.125;

%%%Non-linear regression method
% Initial condition
y0=0.1;
% Function defining current density distribution
IY_S=@(y,ydata)(y(1).*exp(-ydata.*y(1)./(I.*so2)));
% Function defining average current density
Iy=@(y,ydata)(mean(IY_S(y,ydata)));
% Curve fitting
YY=lsqcurvefit(Iy,y0,ydata,I);
% Inlet current density
I_in=YY

% Definition of symbolic parameter
syms nLc
% Current density distribution, Eq. 5.44
If=(I_in.*exp(-nLc.*I_in./(I.*so2)));
% Integral of current density distribution, Eq. 5.45
Icell=int(If,nLc,[0 1]);
% Conversion from symbolic to real
Ic=double(Icell)
```

Command Window

```
<stopping criteria details>

I_in =

    0.1640


Ic =

    0.1250

fx >>
```

FIG. 5.9 Script in MATLAB for the calculation of the inlet current density and integral of the current density distribution.

5.5.4 Total impedance model of the GDL-channel

The total impedance of a segmented cell can be represented as the total impedance of individual local impedances connected in parallel as shown in Fig. 5.5.

Eq. (5.47) represents the total impedance response of the GDL-channel considering steady-state oxygen depletion along the air channel.

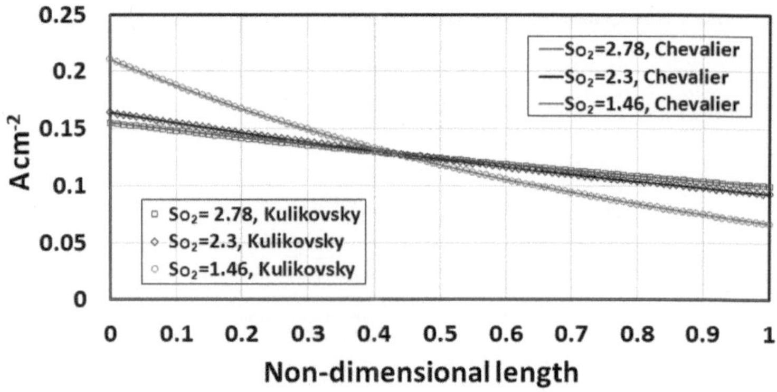

FIG. 5.10 Current density distribution generated by the analytical equation (Eq. 5.44) reported by Chevalier et al. [14] and the analytical equation (Eq. 5.46) reported by Kulikovsky [15].

$$\frac{1}{Z_{\text{T\_GDL}-\text{ch}_{\text{ss}}}} = \frac{1}{\ell_c}\int_0^{\ell_c} \frac{1}{Z_{\text{GDL}-\text{ch}_{\text{ss}}}(y)}dy \tag{5.47}$$

where $Z_{\text{GDL}-\text{ch}_{\text{ss}}}(y)$ is the local impedance of the GDL-channel expressed in Eq. (5.41).

Eq. (5.48) represents the total impedance response of the GDL-channel considering nonstationary oxygen depletion along the air channel during AC conditions.

$$\frac{1}{Z_{\text{T\_GDL}-\text{ch}_{\text{ac}}}} = \frac{1}{\ell_c}\int_0^{\ell_c} \frac{1}{Z_{\text{GDL}-\text{ch}_{\text{ac}}}(y)}dy \tag{5.48}$$

where $Z_{\text{GDL}-\text{ch}_{\text{ac}}}(y)$ is the local impedance of the GDL-channel expressed in Eq. (5.34).

Eqs (5.47) and (5.48) represent the total impedance of the GDL-channel considering local impedances connected in parallel. Eqs (5.47) and (5.48) are analogous to the equation of the total impedance of a segmented cell represented in Eq. (5.8). Eqs (5.47) and (5.48) can be solved in MATLAB software by defining the variable y (channel length coordinate) of the local impedance of the GDL-channel as a symbolic function and considering the function *int(expr, var,[a b])* defined in MATLAB [29] to solve the integral functions expressed in Eqs (5.47) and (5.48). The application of the function *int(expr,var,[a b])* defined in MATLAB has already been demonstrated in the script shown in Fig. 5.9 previously in Section 5.5.3, to calculate the integral of the current density distribution. The following commands in MATLAB allow the solution of the integral function expressed in Eqs (5.47) and (5.48):

```
Zt_gdl_ch=int(1./Z_gdl_ch,y,[0 Lc])
ZT_gdl_ch=double(Lc./ Zt_gdl_ch)
```

where Z_gdl_ch is the mathematical expression for $Z_{GDL-ch_{ss}}(y)$ Eq. (5.41) or $Z_{GDL-ch_{ac}}(y)$ Eq. (5.34) which can also be defined in MATLAB and Lc is the total length of the channel.

5.6 Simulation of total impedance response of the GDL-channel

The simulation of the total and local impedance response of the GDL-channel will be demonstrated by using parameters reported in the literature. Rubio et al. [30] reported EIS measurements carried out in a H_2/air PEMFC with an active area of 100 cm^2 at different oxygen stoichiometries as shown in Fig. 5.11. The EIS measurements were carried in the PEMFC operated at 0.125 A/cm^2 with oxygen stoichiometries of $S_{O_2}=2.78$, $S_{O_2}=2.3$, and $S_{O_2}=1.46$. The PEMFC was operated at ambient and dry conditions. A Randles circuit comprising a finite-length Warburg impedance was fitted to the EIS measurements. The configuration of the Randles circuit considering a CPE and finite-length Warburg impedance has been presented and discussed in Section 1.4.6 from Chapter 1. The Randles circuit considered in the study by Rubio et al. [30] considered a capacitor instead of a CPE. The difference in the impedance responses generated by a capacitor and a CPE has been depicted in Fig. 1.15 from Chapter 1. Fig. 5.11 shows the fitting results reported by Rubio et al. [30]. The dashed lines shown in Fig. 5.11 represent the impedance response of the Randles circuit fitted to the EIS measurements of the PEMFC.

Table 5.1 shows the estimated parameters of the Randles circuit resulting from fitting of the Randles circuit with the EIS measurements of the PEMFC shown in Fig. 5.11. The mathematical expression of the finite-length Warburg

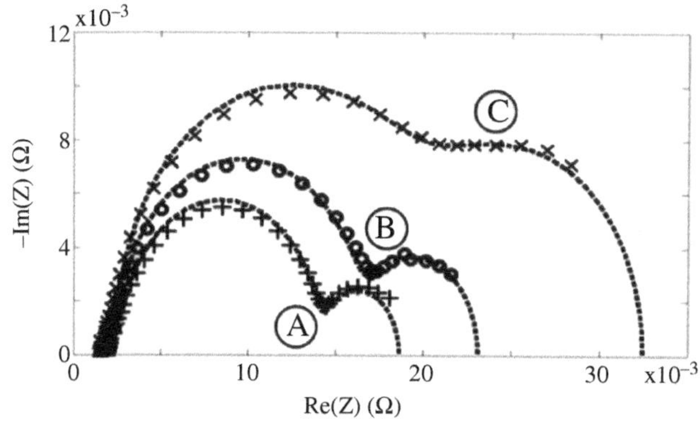

FIG. 5.11 Effect of oxygen stoichiometry on the impedance response of the PEMFC. The oxygen stoichiometry decreases from A to C. A is the highest stoichiometry (2.78) and C is the lowest stoichiometry (1.46). *(Reproduced from M.A. Rubio, A. Urquia, R. Kuhn, S. Dormido, Electrochemical parameter estimation in operating proton exchange membrane fuel cells, J. Power Sources 183 (2008), 123, with permission of Elsevier. Copyright Elsevier.)*

TABLE 5.1 Parameters reported in the study by Rubio et al. [30] and estimated by fitting the Randles circuit with EIS measurements carried out in a PEMFC operated at different oxygen stoichiometry.

Case	S_{O_2}	$R_{W\_O_2}$ mΩ	$\tau_{O_2\_GDL}$ s	R_C mΩ	C_{dl} F	R_{ohm} mΩ
A	2.78	5.769	0.751	10.6	0.253	2.562
B	2.3	7.969	0.698	13.2	0.372	2.343
C	1.46	12.69	0.409	18.1	0.587	2.304

impedance to represent the oxygen diffusion impedance in the GDL is defined in Eq. (5.1). The oxygen transport resistance $R_{W\_O_2}$ and oxygen diffusion time constant $\tau_{O_2\_GDL}$ are parameters of the finite-length Warburg impedance represented in Eq. (5.1). The oxygen transport resistance $R_{W\_O_2}$ increases with decreasing oxygen stoichiometry S_{O_2} as shown in Table 5.1. The increase in oxygen transport resistance $R_{W\_O_2}$ is also related to the increase of the second semicircle of the impedance spectrum at low frequencies as shown in Fig. 5.11. The increase of oxygen transport resistance $R_{W\_O_2}$ with decreasing oxygen stoichiometry S_{O_2} can be attributed to water accumulation in the GDL which hinders the oxygen transport towards the active sites in the CCL. The charge transfer resistance R_C also increases with decreasing oxygen stoichiometry S_{O_2}. The charge transfer resistance R_C is directly proportional to the Tafel slope b_{O_2} and inversely proportional to the cell current as such $R_C = b_{O_2}/I_{cell}$, an increase of the Tafel slope b_{O_2} has been attributed to oxygen transport limitations [31,32]. The double-layer capacitance C_{dl} increases with decreasing oxygen stoichiometry S_{O_2}, a decreasing oxygen stoichiometry S_{O_2} can result in water accumulation in the CCL where water accumulation can assist the expansion of the interface between the solid and solution phases in the CCL thus increasing the capacitance effect for accumulation of charge in the CCL. The ohmic resistance R_{ohm} of the PEMFC decreases with decreasing oxygen stoichiometry S_{O_2}, also water accumulation in the PEMFC with decreasing oxygen stoichiometry can enhance proton conduction in the PEM.

The total impedance of the GDL-channel can be simulated using Eqs (5.47) or (5.48). The difference between Eqs (5.47) and (5.48) is that Eq. (5.48) considers the local impedance of the GDL-channel with nonstationary or oscillating oxygen concentration in the channel during AC conditions. Eq. (5.47) considers the local impedance of the GDL-channel with steady-state oxygen concentration in the channel. Eq. (5.47) comprises the local impedance of the GDL-channel represented in Eq. (5.41). The current density distribution previously shown in Fig. 5.8 and the parameters represented in Table 5.1 were estimated from the same PEMFC. The total current density of the PEMFC is $i_{cell}=0.125$ A/cm^2. The parameters of the finite-length Warburg impedance shown in Table 5.1 are considered a baseline to assist in the simulation of the total impedance response of the GDL-channel generated by Eqs (5.47) and (5.48) because the parameters shown in Table 5.1 were estimated from the impedance spectra of the PEMFC shown in Fig. 5.11. Further information about the simulation of the total impedance response of the GDL-channel presented in this section can be found in the study by Cruz-Manzo and Greenwood [19].

5.6.1 Estimation of diffusion parameters in the gas diffusion layer

The Tafel slope b_{O_2}, oxygen diffusion coefficient D_{GDL} and oxygen diffusion distance δ_{GDL} represented in the local impedance of the GDL-channel are considered to be constant along the length of the channel. The local impedance models of the GDL-channel represented in Eqs (5.5) and (5.7) and developed by Mainka et al. [3] and Maranzana et al. [4] also considered a constant Tafel slope, oxygen diffusion coefficient and oxygen diffusion distance in the GDL along the length of the channel. The Tafel slope b_{O_2} can be calculated from the charge transfer resistance R_C shown in Table 5.1 and the total current of the cell 12.5 A as such $b_{O_2}=R_C I_{cell}$.

It is required to estimate the oxygen diffusion coefficient D_{GDL} and oxygen diffusion distance δ_{GDL} in the GDL to simulate the total and local impedance of the GDL-channel. The estimation of the diffusion parameters D_{GDL} and δ_{GDL} to simulate the total and local impedance of the GDL-channel can be carried out through a curve-fitting process in MATLAB.

Two approaches based on a curve-fitting process can be considered for the estimation of the diffusion parameters D_{GDL} and δ_{GDL} required for the simulation of the total impedance response of the GDL-channel.

1. Estimation of D_{GDL} and δ_{GDL} by fitting the impedance response generated by Eq. (5.47) that represents the total impedance of the GDL-channel with the impedance response generated by the finite-length Warburg impedance (Eq. 5.1). The finite-length Warburg impedance represented in Eq. (5.1) considering the parameters shown in Table 5.1 can simulate the total impedance of the GDL-channel at different oxygen stoichiometry. Nevertheless, this approach could result in a very high computational cost during the curve-fitting process. During the fitting process, it is required to solve the integral of the function represented in Eq. (5.47) at different frequencies. Therefore, the execution of the programmable routine during the curve-fitting process could result in a high number of iterations to solve the curve-fitting (data-fitting) problem.
2. Estimation of D_{GDL} and δ_{GDL} by fitting the local impedance response generated by Eq. (5.41) and considering the coordinate y of the channel length where the current density distribution shown in Fig. 5.8 is equal to

0.125 A/cm^2 with the impedance response generated by the finite-length Warburg impedance (Eq. 5.1). The local impedance of the GDL-channel represented in Eq. (5.41) evaluated at the position of the channel length y where the current density distribution shown in Fig. 5.8 is equal to 0.125 A/cm^2 could be considered a close representation of the total GDL-channel impedance. The fitting process considering this approach can be executed at low computational cost as it is only required to tune the diffusion parameters of the analytical Eq. (5.41) to predict the impedance response generated by Eq. (5.1).

The second aforementioned approach is the chosen approach in this book. The impedance response generated by Eq. (5.41) at the position of channel length y where the current density distribution shown in Fig. 5.8 is equal to 0.125 A/cm^2 is fitted to the simulated impedance response generated by the finite-length Warburg impedance represented in Eq. (5.1) considering the parameters shown in Table 5.1. The fitting process is carried out in MATLAB software using the function *lsqcurvefit* [33]. This function solves nonlinear curve-fitting (data-fitting) problems in least-squares sense using the Levenberg–Marquardt algorithm. The script for the fitting process written in MATLAB can be found in Appendix A. The goodness of fit is evaluated by the average of the real and imaginary relative residuals between the simulated and measured impedance data [19]. Fig. 5.12 shows a flowchart for the simulation of the total impedance of the GDL-channel.

The process for the simulation of the total GDL-channel impedance response shown in Fig. 5.12 can be summarised in the following steps:

1. The modelling architecture requires as inputs the oxygen stoichiometry S_{O_2}, Tafel slope b_{O_2}, pressure, temperature and RH of air at the inlet of the channel, cell current density and channel geometry such as channel length ℓ_c and depth h_c. Although the dimension of the channel length ℓ_c is not relevant because the channel length ℓ_c becomes nondimensional in the mathematical treatment.
2. Definition of the fraction of the channel length where the current density distribution is equal to the total current density of the cell. The current density distribution is a function of the oxygen stoichiometry as shown in Fig. 5.8.
3. Fitting the local impedance of the GDL-channel (Eq. 5.41) considering the position of the channel length y where the current density distribution is equal to the total current density with the impedance response generated by the finite-length Warburg impedance (Eq. 5.1) and considering parameters from Table 5.1.
4. Simulation of the total GDL-channel impedance generated by Eq. (5.47).

The aforementioned steps calculate the oxygen diffusion coefficient D_{GDL} and oxygen diffusion distance δ_{GDL} in the GDL. A hypothetical value of $h_c = 0.001$

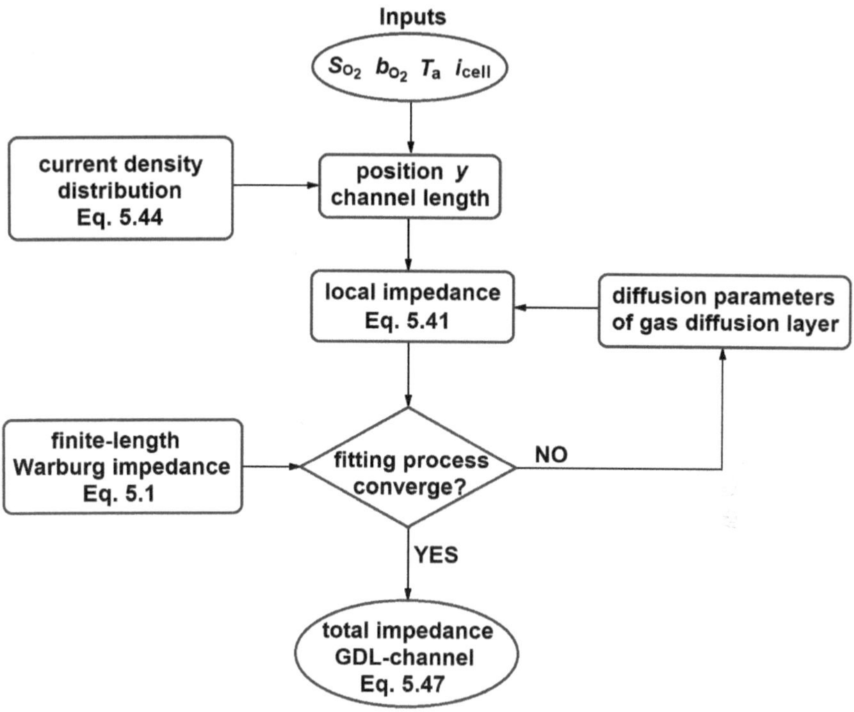

FIG. 5.12 Flowchart for simulation of the total impedance of the GDL-channel.

m has been considered in the analysis. Ambient conditions and dry conditions $H=0$ for air entering the channel are also considered in the analysis. Table 5.2 shows the resulting oxygen diffusion coefficient D_{GDL} and oxygen diffusion distance δ_{GDL}. In addition, Table 5.2 shows the Tafel slope b_{O_2} calculated from the charge transfer resistance R_C defined in Table 5.1 and considering the total

TABLE 5.2 Oxygen diffusion coefficient D_{GDL} and oxygen diffusion distance δ_{GDL} estimated through a nonlinear regression method in MATLAB software.

Case	S_{O_2}	b_{O_2} mV	D_{GDL} m^2/s	δ_{GDL} m
A	2.78	132.5	1.3024×10^{-6}	9.8899×10^{-4}
B	2.3	165	1.1673×10^{-6}	9.0265×10^{-4}
C	1.46	226.3	8.7633×10^{-7}	5.9868×10^{-4}

The parameters have also been reported in the study by Cruz-Manzo and Greenwood [19].

current of the cell 12.5 A as such $b_{O_2}=R_C I_{cell}$. The Tafel slope b_{O_2} increases with decreasing oxygen stoichiometry. An increase in the Tafel slope b_{O_2} with decreasing oxygen stoichiometry can be attributed to an increase in mass transport limitations [31,32]. A factor of 2.3 multiplied in the Tafel slope b_{O_2} shown in Table 5.2 yields the Tafel slope with units of V/dec [22]. The Tafel slope b_{O_2} shown in Table 5.2 seems to be overestimated. Perhaps the charge transfer resistance R_C should have been estimated through a different circuit or impedance model which considers other parameters that are not considered in the Randles circuit (e.g. oxygen transport limitations in the CCL). The Tafel slope calculated from the charge transfer resistance and EIS measurements carried out in a PEMFC with oxygen stoichiometry of 3 has been reported to increase from 23 mV/dec at a current density of 0.063 A/cm^2 to 384 mV/dec at a current density of 1.08 A/cm^2 [34]. In Chapter 4, the charge transfer resistance which is a function of the Tafel slope was derived, see Eqs (4.52) and (4.53) from Chapter 4. If a factor of 2.3 is considered in the Tafel slope expressed in Eq. (4.53) from Chapter 4, therefore the charge transfer resistance expressed in Eq. (4.52) from Chapter 4 would resemble the charge transfer resistance expressed in the study by Malevich et al. [2]. The considered factor of 2.3 in the charge transfer resistance from the study by Malevich et al. [2] can be related to the fact that the Butler–Volmer equation can be approximated to the Tafel equation when $|\eta| > 2.3RT/\alpha zF$ [22]. A value of the Tafel slope with units V/dec was considered in the charge transfer resistance R_C expressed in the impedance model reported by Mainka et al. [3]. If the charge transfer resistance considers the Tafel slope with units of V/dec, it makes an easy agreement of the empirical Tafel slope with EIS measurements. High values of the Tafel slope have been related to a double or a quadruple Tafel slope and have been attributed to electrodes of PEMFCs limited by migration and diffusion of electrochemical species [35,36].

The parameters shown in Table 5.1 were estimated from a PEMFC with SGL 10 BB material as the GDL. The thickness of the GDL (SGL 10 BB) is 420 μm [37]. The oxygen diffusion distance δ_{GDL} in the GDL represents the effective diffusion distance for oxygen to diffuse from the GDL/channel interface to the GDL/CCL interface. The oxygen diffusion distance δ_{GDL} shown in Table 5.2 resulted in a greater value than the thickness (420 μm) of the GDL. It has been demonstrated that the effective tortuous path length for gas diffusion in the GDL is about 2.6 times the GDL thickness [38]. The oxygen diffusion distance δ_{GDL} will depend on properties of the GDL such as permeability, porosity, tortuosity and hydrophobicity as well as operating conditions. The local impedance model developed by Mainka et al. [3] and represented in Eq. (5.5) was validated with the same parameters shown in Table 5.1. Mainka et al. [3] also considered an oxygen diffusion distance δ_{GDL} greater than the thickness of the GDL and concluded that the increase in oxygen diffusion distance could have been attributed to oxygen diffusion below the channel rib. Fig. 5.13 shows the diffusion of oxygen from the channel to the CCL considering that water

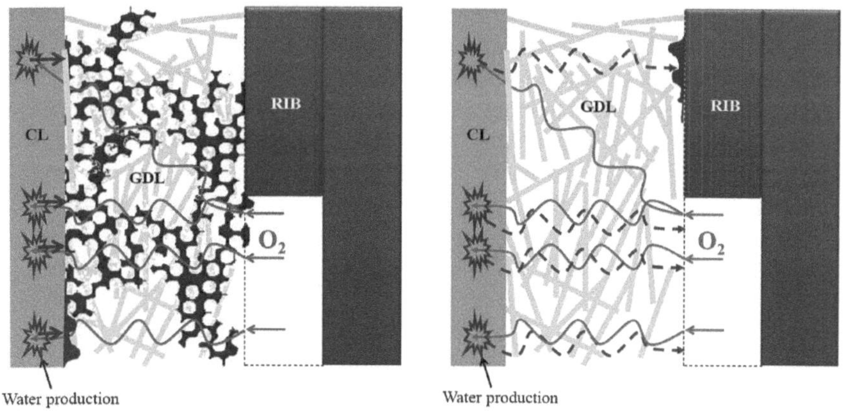

(a) (b)

FIG. 5.13 Oxygen diffusion from the channel to the cathode catalyst later, (a) water in the liquid phase is transported from the cathode catalyst layer to the gas diffusion layer, (b) water in vapour phase is transported from the cathode catalyst layer to the gas diffusion layer. *(Reproduced from Straubhaar, et al., Pore network modelling of condensation in gas diffusion layers of proton exchange membrane fuel cells, Int. J. Heat Mass Transf. 102 (2016) 892, with permission of Elsevier. Copyright Elsevier.)*

produced by the ORR is transported towards the GDL in liquid phase as shown in Fig. 5.13a and vapour phase as shown in Fig. 5.13b. It is clearly observed in Fig. 5.13 that the oxygen diffusion distance is greater than the GDL thickness. In addition, the diffusion of oxygen below the channel rib increases the oxygen diffusion distance as reported by Mainka et al. [3].

The oxygen diffusion distance δ_{GDL} and oxygen diffusion coefficient D_{GDL} from Table 5.2 calculates the oxygen diffusion time constant $\tau_{O_2\_GDL} = {\delta_{GDL}}^2/D_{GDL}$ presented in Table 5.1. The oxygen diffusion distance δ_{GDL} in the GDL decreases with decreasing oxygen stoichiometry as the oxygen diffusion distance δ_{GDL} is directly proportional to the oxygen diffusion time constant $\tau_{O_2\_GDL}$. Under certain conditions one of the diffusion parameters will have a stronger effect on the oxygen diffusion time constant and oxygen transport resistance [10]. For instance, high oxygen concentration decreases the oxygen transport resistance, but it increases the rate of water production during the ORR. An increase of water accumulation may yield a decrease in the oxygen diffusion coefficient D_{GDL}. The oxygen diffusion coefficient D_{GDL} decreased with decreasing oxygen stoichiometry S_{O_2} as shown in Table 5.2. The oxygen diffusion coefficient D_{GDL} is inversely proportional to the oxygen transport resistance as demonstrated in Eq. (5.42). The oxygen diffusion coefficient in the GDL has been reported with an order of magnitude of 10^{-6} m^2/s [11]. Although Kulikovsky [39] estimated the oxygen diffusion coefficient in the GDL with an order of magnitude of 10^{-7} m^2/s from EIS measurements of a PEMFC operated at low oxygen stoichiometry. An oxygen diffusion coefficient with an order of magnitude of 10^{-7} m^2/s was also considered in the study by Chevalier et al. [12] to calculate the oxygen diffusion time constant in the GDL and was represented in the low frequency impedance of the PEMFC operated at low oxygen stoichiometry. The reduction of the oxygen diffusion coefficient D_{GDL} shown in Table 5.2 with decreasing oxygen stoichiometry could be attributed to an increase of water concentration. Iranzo et al. [40] reported an equation to represent the effective diffusion coefficient in the GDL as a function of porosity of the GDL, and volume fraction of liquid water. The effective diffusion coefficient in the GDL decreases with increasing amounts of liquid water.

5.6.2 Simulation of the total GDL-channel impedance at different oxygen stoichiometry

A script in MATLAB software was considered for the simulation of the total GDL-channel impedance. The script considered the function *int(expr,var, [a b])* to solve the integral function defined in the equation (Eq. 5.47) of the total GDL-impedance response. The script written in MATLAB to simulate the total impedance of the GDL-channel can be found in Appendix A. Fig. 5.14 shows the impedance response generated by the finite-length Warburg impedance (Eq. 5.1) and parameters from Table 5.1 at different oxygen stoichiometry. Fig. 5.14 also shows the impedance response generated by Eq. (5.47) and considered that the parameters D_{GDL}, δ_{GDL}, and b_{O_2} shown in Table 5.2 are constant along the length of the channel. The simulation results generated by Eq. (5.47) represent the final step of the flowchart shown in Fig. 5.12. The range of frequencies in the simulation results was considered from 1 kHz to 0.001 Hz. A discrepancy between the impedance response generated by the finite-length Warburg impedance (Eq. 5.1) and the impedance response generated by the model of

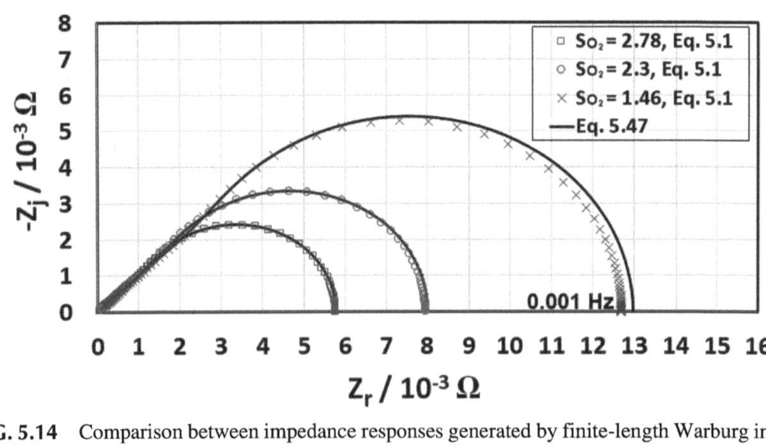

FIG. 5.14 Comparison between impedance responses generated by finite-length Warburg impedance (Eq. 5.1) and total GDL-channel impedance (Eq. 5.47) at different oxygen stoichiometry.

the total impedance response of the GDL-channel (Eq. 5.47) at low oxygen stoichiometry S_{O_2} is shown in Fig. 5.14. The discrepancy can be attributed to the fact that values of the Tafel slope b_{O_2} and diffusion parameters D_{GDL} and δ_{GDL} of the GDL have been considered to be constant in the local impedance of the GDL-channel along the length y of the channel. A better prediction of the total impedance of the GDL-channel generated by Eq. (5.47) at low oxygen stoichiometry could be achieved by considering variable values of the Tafel slope and diffusion parameters along the length of the channel. A variation of the Tafel slope and oxygen diffusion coefficient in the GDL along a segmented channel of a PEMFC has been reported by Reshetenko and Kulikovsky [41,42].

Fig. 5.15 shows a comparison between the total impedance of the GDL-channel generated by Eqs (5.47) and (5.48) considering parameters from Table 5.2 at oxygen stoichiometries of $S_{O_2}=2.78$ (case A) and $S_{O_2}=2.3$ (case B). The range of frequencies in the simulation results was considered from 1 kHz to 0.001 Hz. The low frequency limit (LFL) of the impedance spectra as the frequency is reduced $\omega \rightarrow 0$ is shown in Fig. 5.15. A discrepancy between the LFL at 0.001 Hz of the impedance spectra generated by Eqs (5.47) and (5.48) is observed in Fig. 5.15. This discrepancy in the LFL increases with decreasing oxygen stoichiometry and is attributed to the amplitude i_{AC} of the sinusoidal perturbation considered in the oscillating current density $\widetilde{i_F}(y, j\omega)$ from Eq. (5.36) (Eq. 5.48 comprises Eq. 5.36). The considered value of the amplitude i_{AC} in the oscillating contribution of the sinusoidal current density $\widetilde{i_F}(y, j\omega)$ from Eq. (5.36) was 10% of the DC current density. The value of the amplitude i_{AC} in the oscillating contribution of the sinusoidal current density was decreased to 2% of the DC current density in Eq. (5.36) and the resulting impedance spectrum (labelled as i_{AC} 2% DC) converged close to the LFL of the

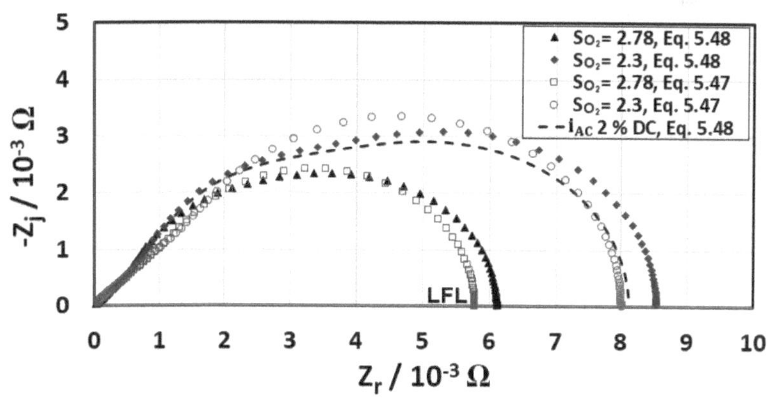

FIG. 5.15 Simulation of the total impedance of the GDL-channel considering steady-state oxygen depletion along the channel (Eq. 5.47) and nonstationary oxygen depletion along the channel (Eq. 5.48) during AC conditions. The impedance response generated by (Eq. 5.48) considered an amplitude i_{AC} of the oscillating contribution of the sinusoidal current density of 10% of the DC current density. An amplitude of 2% of the DC current density is also simulated.

impedance spectrum generated by Eq. (5.47) considering steady-state oxygen depletion along the air channel. The total impedance response of the GDL-channel considering oscillating oxygen concentration in the channel (Eq. 5.48) can become similar to the total impedance response of the GDL-channel considering steady-state oxygen concentration in the channel (Eq. 5.47) by increasing oxygen stoichiometry and decreasing the amplitude in the oscillating contribution of the sinusoidal current density.

5.7 Simulation of local impedance response of the GDL-channel

It is assumed that the Tafel slope b_{O_2}, oxygen diffusion coefficient D_{GDL}, and oxygen diffusion distance δ_{GDL} in the GDL are constant along the length of the channel. The parameters shown in Table 5.2 are considered in Eq. (5.34) to simulate the local impedance response of the GDL-channel at different positions of the length of the channel $0.1\ell_c$, $0.5\ell_c$, and ℓ_c and at different oxygen stoichiometry as shown in Figs 5.16 and 5.17. Also in Figs 5.16 and 5.17 the range of frequencies in the simulation results was considered from 1 kHz to 0.001 Hz where the impedance spectra had increased from the inlet to the outlet of the channel. The value of the LFL shown in Figs 5.16 and 5.17 was calculated with Eq. (5.41) and the parameters shown in Table 5.2 at the lowest frequency of 0.001 Hz. The LFL at 0.001 Hz in the figures can represent steady state conditions because the LFL is calculated with the local impedance of the GDL-channel (Eq. 5.41) that considers steady-state oxygen concentration along the channel. Similar to the results shown in Fig. 5.15, the discrepancy between the LFL at 0.001 Hz calculated by Eqs (5.34) and (5.41) is attributed to a

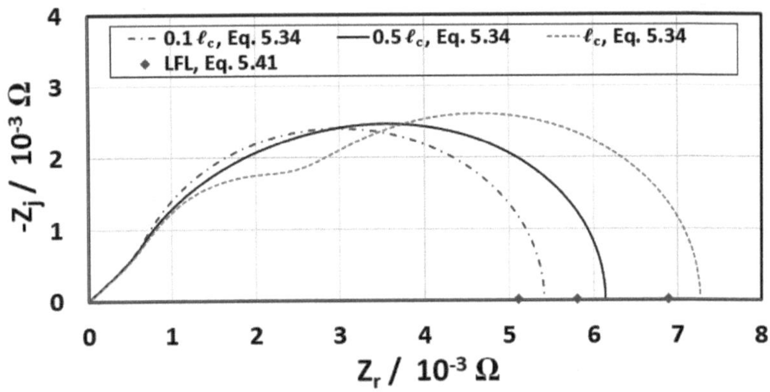

FIG. 5.16 Local impedance of the GDL-channel at different fraction of the channel length ℓ_c at a stoichiometry of $S_{O_2}=2.78$. The impedance response generated by Eq. (5.34) considers oscillating oxygen concentration in the channel during AC conditions. The low frequency limit (LFL) is calculated with Eq. (5.41) and represents the impedance of the GDL-channel considering steady-state oxygen depletion along the channel at the lowest frequency.

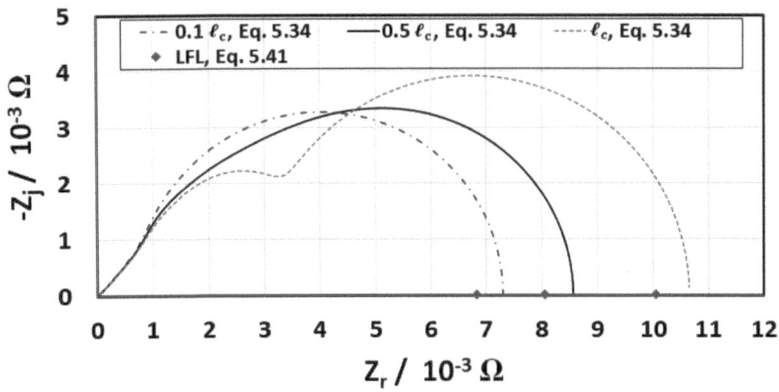

FIG. 5.17 Local impedance of the GDL-channel at different fraction of channel length ℓ_c at a stoichiometry of $S_{O_2}=2.3$. The impedance response generated by Eq. (5.34) considers oscillating oxygen concentration in the channel during AC conditions. The low frequency limit (LFL) is calculated with Eq. (5.41) and represents the impedance of the GDL-channel considering steady-state oxygen depletion along the channel at the lowest frequency.

nonlinear effect caused by the amplitude of the oscillating current density (10% of the DC current density). A linear EIS response can be measured with a sinusoidal perturbation with low amplitude superimposed onto the nonlinear and steady-state response of the PEMFC. The size of the local impedance spectra at stoichiometry $S_{O_2}=2.78$ was smaller than the size of the local impedance spectra at stoichiometry $S_{O_2}=2.3$ as demonstrated in Figs 5.16 and 5.17. The separation between the impedance response of the GDL and impedance response of the channel is observed in the last segment of the channel $y=\ell_c$ and is more noticeable at an oxygen stoichiometry of $S_{O_2}=2.3$ as shown in Fig. 5.17, where the contribution of the channel on the impedance spectrum is represented through the second loop at low frequencies. Schneider et al. [28] reported local impedance spectra from a segmented PEMFC where an increase in size of the low frequency loop on the local impedance spectra along the gas channels from the air inlet to the outlet was observed. The increase of impedance spectrum along the channel can be related to the cumulative flooding in the GDL and a decreasing molar fraction of oxygen in the cathode gas stream along the channel.

5.8 Effect of relative humidity on GDL-channel impedance spectrum

The impedance response of the GDL-channel will change at different RH of the air entering the channel. Fig. 5.18 shows the effect of increasing RH and temperature of air entering the channel on the total impedance spectrum of the GDL-channel. The simulated impedance spectra were generated by the equation of the total impedance of the GDL-channel considering steady-state

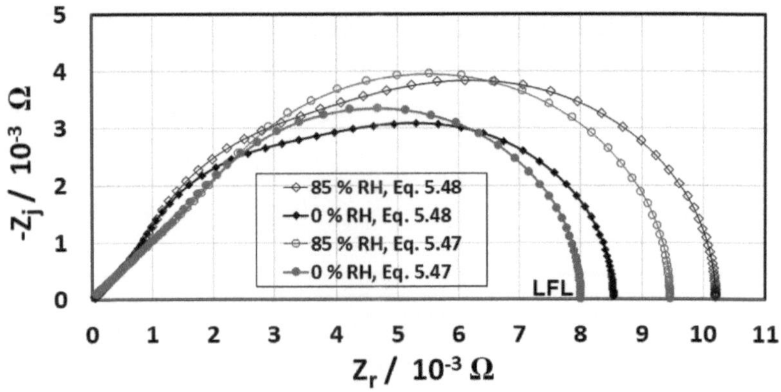

FIG. 5.18 Simulated effect of relative humidity (RH) on the total impedance spectrum of the GDL-channel.

oxygen concentration in the channel (Eq. 5.47) and oscillating oxygen concentration in the channel during AC conditions (Eq. 5.48). The parameters shown in Table 5.2 for case B at a stoichiometry of $S_{O_2}=2.3$ are considered in Eqs (5.47) and (5.48). The range of frequencies in the simulation results was considered from 1 kHz to 0.001 Hz. The simulated results shown in Fig. 5.18 considered an increase of RH from dry conditions 0% RH to wet conditions at 85% RH, in addition, the temperature of the air was increased from 20°C to 35°C. These ambient conditions of 85% RH and 35°C can be encountered in a hot and humid region, e.g. Veracruz, Mexico. The simulated impedance spectra of the GDL-channel increased with increasing RH and temperature of the air entering the channel. Fig. 5.18 also shows that the difference between the LFL at 0.001 Hz, generated by the impedance of the GDL-channel considering steady-state oxygen concentration in the channel (Eq. 5.47) and oscillating oxygen concentration in the channel (Eq. 5.48), increased with increasing RH and temperature. Yan et al. [7] reported that optimal performance in a PEMFC can occur at low air RH supplied at the cathode and high hydrogen RH supplied at the anode. Janicka et al. [43] reported that when the PEMFC is operated with gases with a high humidity, the oxygen diffusion resistance estimated from EIS measurements results in a higher value than the rest of the resistances of the PEMFC related to charge transfer resistance of the redox reaction and membrane resistance.

The increase of the impedance spectrum of the GDL-channel with increasing RH can be attributed to a reduction of the oxygen concentration in the channel. Under wet conditions, the oxygen concentration in the channel calculated through Eqs (5.27) or (5.37) is reduced when $H>0$, where H is the humidity molar ratio calculated through Eq. (5.22). The simulated impedance spectra shown in Fig. 5.18 considered a constant effective diffusion coefficient D_{GDL} with changing RH in the impedance model of the GDL-channel. It has been

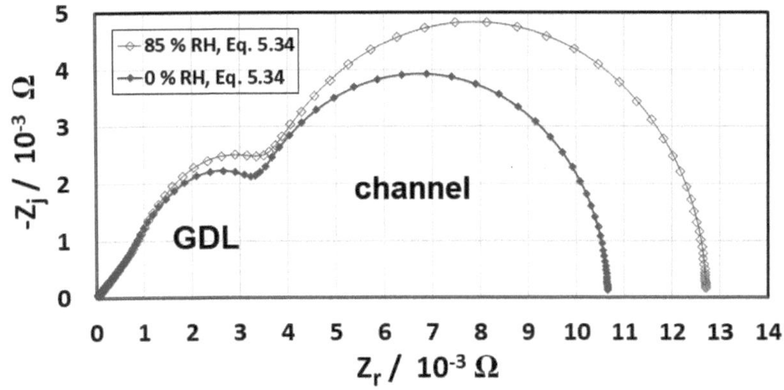

FIG. 5.19 Simulated effect of relative humidity (RH) on the local impedance spectrum of the GDL-channel at the position of the channel length $y=\ell_c$.

reported that there is a decrease in the effective diffusion coefficient in the GDL with increasing RH because the porosity of the GDL is reduced due to pores being partially saturated with water [2]. A greater difference between the simulated impedance spectra considering dry and wet conditions is expected if a reduction of the oxygen diffusion coefficient D_{GDL} with increasing RH of the air entering the channel is considered in the impedance model of the GDL-channel.

The effect of RH can also be analysed on the local impedance of the GDL-channel where the impedance response of the GDL can be separated from the impedance response of the channel. Fig. 5.19 shows the effect of RH and temperature of air entering the channel on the simulated local impedance of the GDL-channel at the position of the channel length $y=\ell_c$. The simulated impedance spectra were generated by the equation of the local impedance of the GDL-channel considering oscillating oxygen concentration in the channel during AC conditions (Eq. 5.34). The parameters shown in Table 5.2 for case B at a stoichiometry of $S_{O_2}=2.3$ have been considered in Eq. (5.34). An increase of the simulated local impedance of the GDL-channel with increasing conditions, 85% RH and air temperature of 35°C, is observed in Fig. 5.19. The second semicircle at low frequencies attributed to the oxygen transport in the channel increased with increasing air temperature and RH. An increase in the humidity molar ratio H and temperature yields a decrease in the oxygen concentration in the channel as expressed in Eq. (5.37) and hence there is an increase of the second semicircle of the GDL-channel impedance spectrum shown in Fig. 5.19 which is attributed to oxygen transport in the channel.

5.9 Oscillating oxygen concentration in the channel

It has been demonstrated [5] through an experimental analysis that the amplitude of the oscillating oxygen concentration increases along the air channel.

This can be reflected in the increase of a second loop at low frequencies in local EIS measurements carried out in a segmented PEMFC. The second loop at low frequencies attributed to the oscillating oxygen concentration along the channel could bias the relation between the low-frequency resistance of the impedance spectrum and the polarisation resistance estimated from the polarisation curve as demonstrated in the results presented by Maranzana et al. [4] and shown Fig. 5.4.

The impedance spectra shown in Fig. 5.20 are generated by Eq. (5.34) and considering different values of the amplitude in the oscillating contribution of the sinusoidal current density. Eq. (5.34) represents the local impedance of the GDL-channel considering oscillating oxygen concentration along the channel. The value of the LFL shown in Fig. 5.20 is calculated through Eq. (5.41) at the lowest frequency. Eq. (5.41) represents the local impedance of the GDL-channel considering steady-state oxygen concentration in the channel. The simulated local impedance response of the GDL-channel shown in Fig. 5.20 considers the coordinate position of the channel length $y=\ell_c$. The simulation results consider the parameters shown in Table 5.2 for case B at an oxygen stoichiometry of $S_{O_2}=2.3$ as a base-line and the range of frequencies was considered from 1 kHz to 0.001 Hz. Two semicircles are observed in the impedance spectra considering oscillating oxygen concentration in the channel shown in Fig. 5.20 and generated by Eq. (5.34). The first semicircle at high frequencies is attributed to the impedance response of the GDL and the second at low frequencies is attributed to the oxygen depletion and the oscillating oxygen concentration along the channel.

The simulated impedance response of the GDL-channel generated by Eq. (5.34) and shown in Fig. 5.20 considers different values of the amplitude i_{AC} in the oscillating contribution of the sinusoidal current density. The

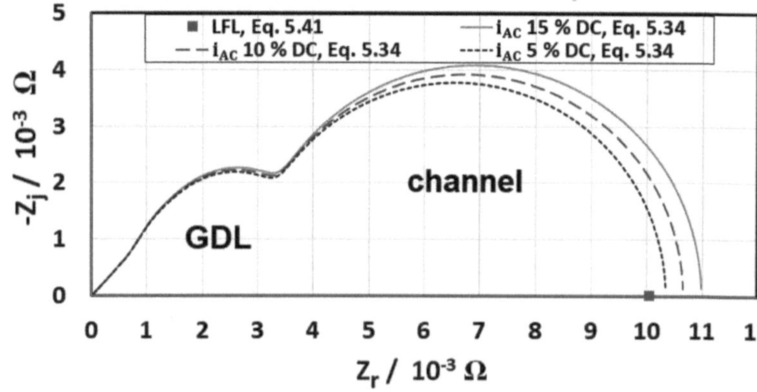

FIG. 5.20 Local GDL-channel impedance spectrum at coordinate $y=\ell_c$. Effect of the amplitude i_{AC} of the oscillating contribution of the sinusoidal current density on the local impedance spectrum of the GDL-channel. The amplitude in the oscillating contribution of the sinusoidal current density is considered to be 5%, 10%, and 15% of the DC current density.

considered amplitude i_{AC} in the oscillating contribution of the sinusoidal current density is 5%, 10%, and 15% of the DC current density. The LFL generated by Eq. (5.41) can represent the steady-state value because no oscillating effect of the oxygen concentration along the channel is considered in Eq. (5.41). The simulation of the impedance response of the GDL-channel considering oscillating oxygen concentration along the channel and generated by Eq. (5.34) demonstrates that increasing the amplitude i_{AC} in the oscillating contribution of the sinusoidal current density increases the second loop of the GDL-channel impedance spectrum. A drift of the GDL-channel impedance spectrum away from its steady-state value is observed which is related to the real component Z_r of the impedance response as the frequency is reduced to a value near zero. Fig. 5.20 also demonstrates that the LFL at 0.001 Hz of the impedance spectra considering oscillating oxygen concentration in the channel converges closer to the LFL generated by Eq. (5.41) as the amplitude i_{AC} in the oscillating contribution of the sinusoidal current density is decreased. Yuan et al. [44] reported an expansion of the EIS measurements in a 500 W PEMFC stack with increasing amplitude of the oscillating contribution of the sinusoidal current from 5% to 15% of the DC current. A nonlinear impedance response with increasing amplitude of the sinusoidal perturbation during EIS measurements of a PEMFC would not be consistent with Kramers–Kronig relations [45]. Maranzana et al. [4] proposed that the value of the amplitude of the oscillating contribution of the sinusoidal current should be 2% of the DC current to minimise the nondesired effect of the sinusoidal current amplitude which drifts the low frequency value of EIS measurements away from the value of the polarisation resistance. Nevertheless, a reduction in the amplitude of the sinusoidal perturbation could compromise the signal-to-noise ratio and affect the quality of the EIS measurements.

The bias between the low-frequency resistance of the impedance spectrum of the PEMFC and the polarisation resistance from the polarisation curve has been attributed to the oscillating oxygen concentration along the channel of the PEMFC. It has been proposed that the nondesired effect of the oscillating oxygen concentration along the channel in the impedance spectrum of the PEMFC could be reduced by oscillating the velocity of oxygen through the channel in phase with the oscillating current during EIS tests [4]. The effect of air flow oscillations in the air channel on EIS measurements of PEMFCs has been theoretically demonstrated through the impedance model developed by Kulikovsky [13]. The oscillating air flow velocity considered in the impedance model of Kulikovsky [13] was represented as:

$$\widetilde{v} = 1 + k_v \widetilde{\eta} \exp(j\omega t) \tag{5.49}$$

where $0 \leq k_v \leq 1$ is the modulation amplitude parameter. Fig. 5.21 shows the PEMFC impedance results reported by Kulikovsky [13]. The effect of the amplitude of the oscillating air flow on the PEMFC impedance spectrum is observed in Fig. 5.21. If the modulation amplitude is zero $k_v = 0$, the air flow velocity through the channel is static and constant and a second semicircle at

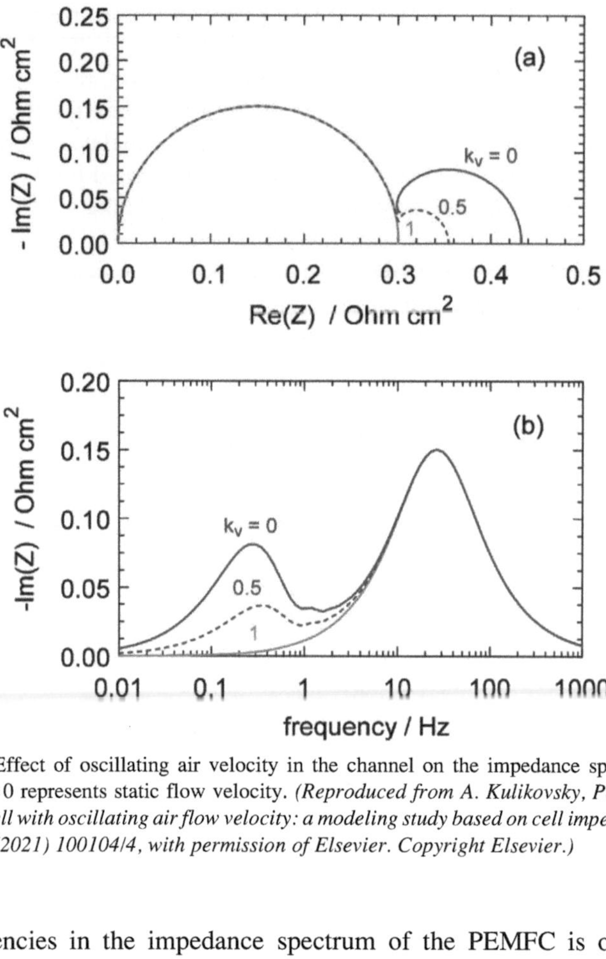

FIG. 5.21 Effect of oscillating air velocity in the channel on the impedance spectrum of the PEMFC. $k_v = 0$ represents static flow velocity. *(Reproduced from A. Kulikovsky, Performance of a PEM fuel cell with oscillating air flow velocity: a modeling study based on cell impedance, eTransportation 7 (2021) 100104/4, with permission of Elsevier. Copyright Elsevier.)*

low frequencies in the impedance spectrum of the PEMFC is observed in Fig. 5.21. The second semicircle at low frequencies decreases with increasing air flow velocity oscillations $k_v > 0$ and disappears if the amplitude of the air flow velocity oscillations is equal to the amplitude of the potential oscillations $k_v = 1$.

The mathematical expression of the oscillating oxygen flow velocity could be considered in the mathematical framework of the GDL-channel impedance presented in Section 5.4. The diffusion parameters of the impedance model of the GDL-channel considering oscillating oxygen velocity in the channel have to be estimated from EIS measurements carried out in a PEMFC with variable air flow. This could be achieved through the use of modern air flow controllers and the use of electronic hardware to ensure that the oscillating air flow is in phase with the oscillating current under galvanostatic experimental conditions.

Meyer and Zhao [46] carried out EIS measurements in a PEMFC by oscillating the inlet air flow in phase with the oscillating current during galvanostatic

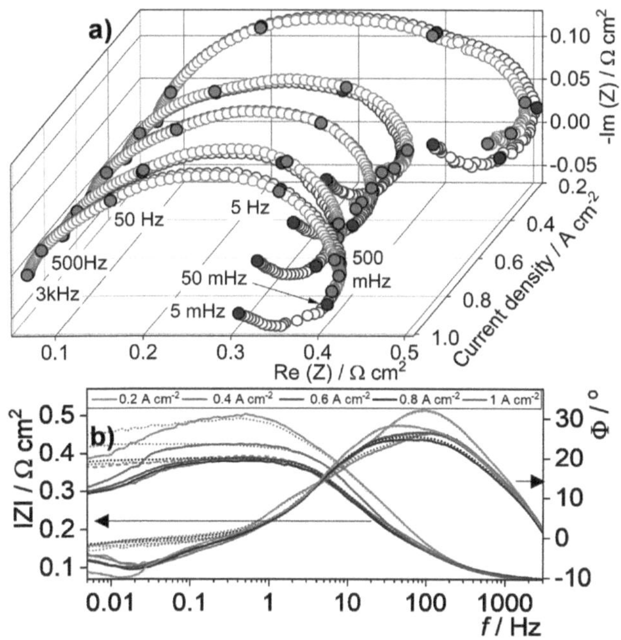

FIG. 5.22 EIS measurements carried out in a PEMFC at different current density and considering oscillating air flow (*blue circles*; *dark grey* in print version) and neglecting oscillating air flow (*red circles*; *grey* in print version) during the EIS measurements. *(Reproduced from Q. Meyer, C. Zhao, Air perturbation-induced low-frequency inductive electrochemical impedance arc in proton exchange membrane fuel cells, J. Power Sources 488 (2021), 229245/7, with permission of Elsevier. Copyright Elsevier. For interpretation of the references to colour in this figure legend, the reader is referred to the web version of this article.)*

conditions. The experimental results demonstrated that the low-frequency resistance of the EIS measurements converges with the polarisation resistance from polarisation curves. The EIS measurements resulting from oscillating the inlet air flow during PEMFC operation featured inductive loops (EIS measurements with positive imaginary components) at low frequencies as shown in Fig. 5.22. Chapter 7 discusses the formation of inductive loops at low frequencies and considers that the oscillating water vapour concentration in the CCL can yield the formation of inductive loops at low frequencies. Further experimental studies and mathematical modelling are required to interpret the EIS response of the PEMFC under the oscillating air flow rate approach. The impedance model reported by Kulikovsky [13] considering oscillating air flow velocity does not simulate inductive loops at low frequencies in the impedance spectrum of the PEMFC.

5.10 Randles circuit with analytical Warburg impedance model

The Randles circuit is commonly considered to estimate charge transfer resistance and oxygen transport resistance represented in the impedance spectrum of the PEMFC. The Randles circuit considers a resistor representing the ohmic resistance in the PEMFC connected in series with a resistor-capacitor parallel configuration representing the charge transfer process during the ORR and double-layer capacitance at the electrode/electrolyte interface. The electrochemical mechanisms during the hydrogen oxidation reaction (HOR) are not commonly considered in the Randles circuit because the HOR is a facile and less complicated sequential reaction than the ORR. The finite-length Warburg impedance represented in Eq. (5.1) has been commonly considered in the Randles circuit to estimate the oxygen transport resistance present in the GDL from EIS measurements of PEMFCs. Eq. (5.50) represents the impedance response of the Randles circuit considering the finite-length Warburg impedance.

$$Z_{\text{Randles}} = R_{\text{ohm}} + \frac{R_C + Z_{\text{Warburg}}}{1 + j\omega C_{\text{dl}}\left[R_C + Z_{\text{Warburg}}\right]} \tag{5.50}$$

where R_{ohm} represents the total ohmic resistance of the PEMFC, R_C represents the charge transfer resistance during the ORR, C_{dl} represents the double-layer capacitance in the CCL, and Z_{Warburg} is the Warburg impedance representing diffusion impedance in the GDL and can be represented through Eq. (5.1). Note that the impedance of the Randles circuit expressed in Eq. (5.50) can also be expressed with a CPE instead of the capacitor C_{dl} to represent nonhomogeneous charge distribution in the CCL as demonstrated in Chapter 1 Section 1.4.6.

The analytical mathematical expression to represent the impedance of the GDL-channel developed in this chapter is analogous to the mathematical expression of the finite-length Warburg impedance. Therefore, the impedance model of the GDL-channel developed in this study can be implemented with the Randles circuit to represent the contribution of the impedance response of the GDL-channel on EIS measurements of PEMFCs.

Rubio et al. [30] fitted Eq. (5.50), considering the finite-length Warburg impedance represented in Eq. (5.1), with EIS measurements carried out in a PEMFC at different oxygen stoichiometry as shown in Fig. 5.11. The fitting process resulted in the parameters shown in Table 5.1. Fig. 5.23 shows the impedance spectrum labelled as 'Randles with Eq. (5.1)' generated by Eq. (5.50) whilst considering Z_{Warburg} = Eq. (5.1) and the parameters reported in the study by Rubio et al. [30] as shown in Table 5.1 at an oxygen stoichiometry $S_{O_2} = 2.78$. Fig. 5.23 also shows the impedance spectrum labelled as 'Randles with Eq. (5.47)' generated by Eq. (5.50) whilst considering Z_{Warburg} = Eq. (5.47) and parameters shown in Tables 5.1 and 5.2 at an oxygen stoichiometry $S_{O_2} = 2.78$. The considered range of frequencies in the simulation results was from 10 kHz to 0.001 Hz.

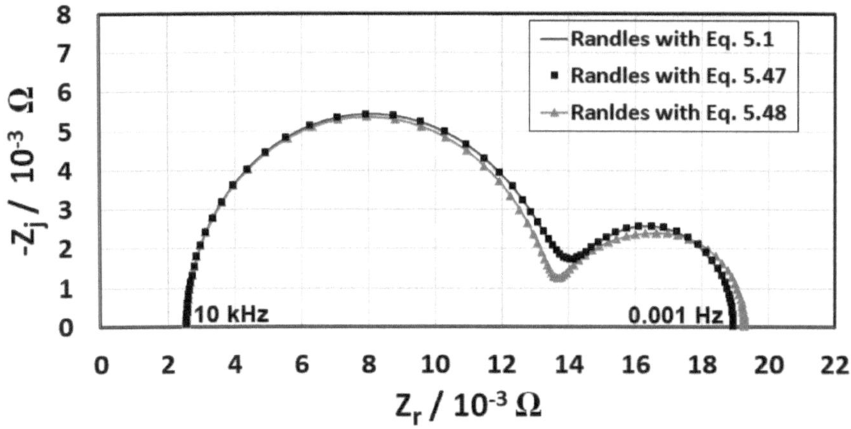

FIG. 5.23 Impedance response labelled as 'Randles with Eq. (5.1)' is generated by Eq. (5.50) considering Z_{Warburg} = Eq. (5.1) with parameters from Table 5.1 at $S_{O_2}=2.78$. Impedance response labelled as 'Randles with Eq. (5.47)' is generated by Eq. (5.50) considering Z_{Warburg} = Eq. (5.47) with parameters from Tables 5.1 and 5.2 at $S_{O_2}=2.78$. Impedance response labelled as 'Randles with Eq. (5.48)' is generated by Eq. (5.50) considering Z_{Warburg} = Eq. (5.48) with parameters from Tables 5.1 and 5.2 at $S_{O_2}=2.78$.

The impedance spectrum labelled as 'Randles with Eq. (5.1)' is similar to the impedance spectrum labelled as 'Randles with Eq. (5.47)' because the impedance response generated by Eq. (5.47) is similar to the impedance response generated by Eq. (5.1) at an oxygen stoichiometry of $S_{O_2}=2.78$ as demonstrated in Fig. 5.14. In addition, Fig. 5.23 shows the impedance spectrum labelled as 'Randles with Eq. (5.48)' generated by Eq. (5.50) whilst considering Z_{Warburg} = Eq. (5.48) and parameters shown in Tables 5.1 and 5.2 at an oxygen stoichiometry $S_{O_2}=2.78$. The impedance spectrum labelled as 'Randles with Eq. (5.48)' does not match the second semicircle at low frequencies represented in the impedance spectra labelled as 'Randles with Eq. (5.47)' and 'Randles with Eq. (5.1)'. The second semicircle at low frequencies of the impedance spectrum generated by Eq. (5.47) considers that the time constant of the GDL-channel impedance is only attributed to oxygen diffusion in the GDL. Eq. (5.48) simulates the second semicircle at low frequencies considering that the time constant of the GDL-channel impedance is attributed to not only oxygen diffusion in the GDL, but also oxygen transport through the channel. Eq. (5.48) represents the impedance of the GDL-channel and considers the oscillating oxygen concentration along the channel and the effect of the amplitude of the oscillating contribution of the sinusoidal current density on the low-frequency impedance spectrum. Eq. (5.47) represents the impedance of the GDL-channel considering steady-state oxygen concentration in the channel.

Bolillat et al. [47] discussed that the effect of oxygen oscillations along the channel and represented in EIS measurements of PEMFCs at low frequencies is commonly ignored. The Randles circuit considering the finite-length Warburg impedance can be applied to EIS measurements of a PEMFC to qualitatively estimate the oxygen transport resistance in the GDL and not to quantitively estimate diffusion parameters of solely the GDL because the impedance response of the GDL can be overlapped with the impedance response for oxygen transport in the channel.

The electrochemical mechanisms represented in the impedance spectrum of the PEMFC at low frequencies are not only governed by a single time constant, but also multitime constants related to the different transport mechanisms during PEMFC operation e.g. convective-diffusion transport of oxygen through thc GDL and channel. A different approach to estimate diffusion parameters in the GDL can be obtained by fitting the Randles circuit expressed in Eq. (5.50) and considering $Z_{Warburg}=$Eq. (5.34) with the impedance spectrum of the PEMFC shown in Fig. 5.11 at high oxygen stoichiometry. The impedance spectrum labelled as 'Randles with Eq. (5.1)' previously shown in Fig. 5.23 is considered a baseline because it resembles the experimental impedance spectrum of the PEMFC shown in Fig. 5.11 for case A at $S_{O_2}=2.78$.

Eq. (5.34) represents the local impedance of the GDL-channel and considers the oscillating oxygen concentration along the channel. Eq. (5.34) evaluated at the coordinate of the channel $y=0.481\ell_c$ and at $S_{O_2}=2.78$ can represent the total impedance of the GDL-channel because at $y=0.481\ell_c$ the current density distribution is equal to the total current density of the cell (0.125 A/cm^2) at $S_{O_2}=2.78$ as shown in Fig. 5.8. The fitting results between the impedance spectrum generated by the Randles circuit considering $Z_{Warburg}=$Eq. (5.34) and the impedance spectrum generated by the Randles circuit considering $Z_{Warburg}=$Eq. (5.1) are shown in Fig. 5.24. The script for the fitting process written in MATLAB and using the function *lsqcurvefit* [33] can be found in

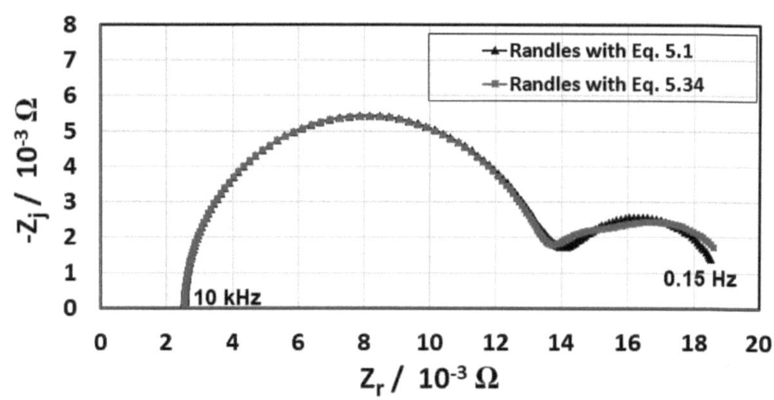

FIG. 5.24 Comparison between impedance spectrum generated by Eq. (5.50) considering $Z_{Warburg}=$Eq. (5.34) evaluated at $y=0.481\ell_c$ and the impedance spectrum generated by Eq. (5.50) considering $Z_{Warburg}=$Eq. (5.1) at $S_{O_2}=2.78$.

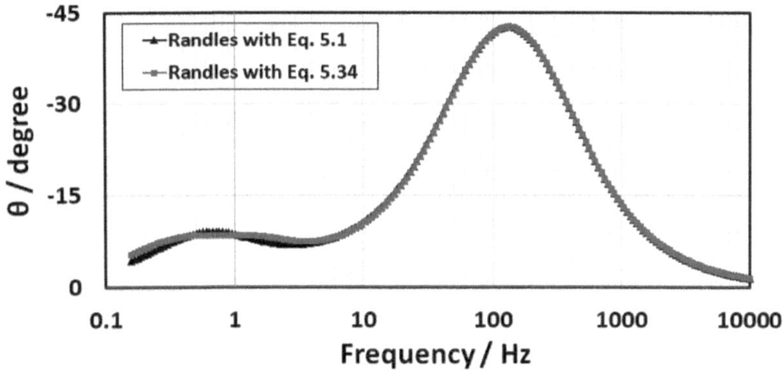

FIG. 5.25 Comparison between phase angle of the impedance response generated by Eq. (5.50) considering $Z_{Warburg}$ = Eq. (5.34) evaluated at $y = 0.481\ell_c$ and phase angle of the impedance response generated by Eq. (5.50) considering $Z_{Warburg}$ = Eq. (5.1) at $S_{O_2} = 2.78$.

Appendix A. The fitting process considered the local impedance of the GDL-channel (Eq. 5.34) at $y = 0.481\ell_c$ instead of the total impedance of the GDL-channel (Eq. 5.48) because this approach can be executed at low computational cost as previously discussed in Section 5.6.1. Fig. 5.25 shows the fitting results of the phase angle of the impedance response generated by the Randles circuit considering $Z_{Warburg}$ = Eq. (5.1) and Randles circuit considering $Z_{Warburg}$ = Eq. (5.34). A better agreement in the impedance results would be achieved at higher oxygen stoichiometry.

The impedance spectrum generated by Eq. (5.50) considering $Z_{Warburg}$ = Eq. (5.34) evaluated at $y = 0.481\ell_c$ also considers an oxygen diffusion coefficient of $D_{GDL} = 5.2974 \times 10^{-7} m^2/s$ and oxygen diffusion distance of $\delta_{GDL} = 4.0551 \times 10^{-4} m$. These values predicted a lower oxygen diffusion time constant $\tau_{O_2\_GDL} = \delta_{GDL}{}^2/D_{GDL} = 0.3104$ s than the oxygen diffusion time constant shown in Table 5.1 $\tau_{O_2\_GDL} = 0.751$ s at $S_{O_2} = 2.78$. The difference in the oxygen diffusion time constant $\tau_{O_2\_GDL}$ is attributed to the fact that the impedance spectrum labelled as 'Randles with Eq. (5.34)' considers oxygen transport mechanisms in the GDL and channel and sinusoidal current amplitude effect which are not considered in the impedance spectra labelled as 'Randles with Eq. (5.1)'. A different impedance model considering impedance of the CCL, and GDL-channel may be required to estimate electro-diffusion mechanisms represented in EIS measurements of PEMFCs.

5.11 Effect of double-layer capacitance of catalyst layer on the GDL-channel impedance

The simulation results shown in previous sections demonstrated the effect of oxygen transport in the channel and do not consider the effect of the charge double-layer of the CCL on the impedance spectrum of the GDL-channel.

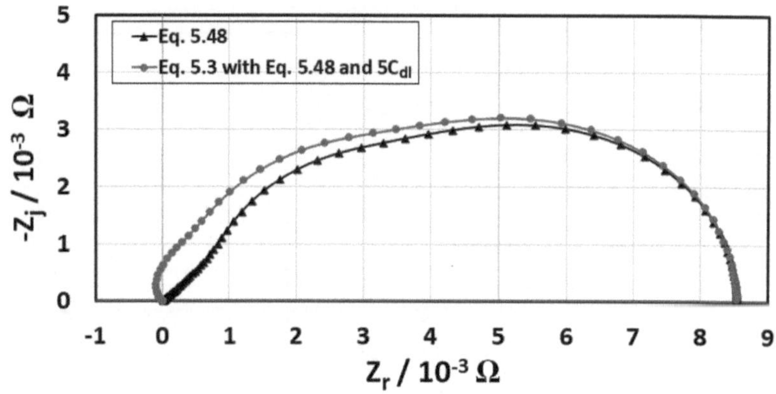

FIG. 5.26 Effect of double-layer capacitance of the CCL on the GDL-channel impedance spectrum. Impedance spectrum labelled 'Eq. (5.3) with Eq. (5.48) and $5C_{dl}$' is generated by Eq. (5.3) considering Eq. (5.48) in the numerator of Eq. (5.3). The spectrum labelled 'Eq. (5.48)' was generated by Eq. (5.48) and does not consider the effect of the double-layer capacitance on the GDL-channel impedance. The values of the parameters shown in Tables 5.1 and 5.2 at an oxygen stoichiometry of $S_{O_2}=2.3$ were considered in the simulation results. A value of capacitance 5 times higher, $5C_{dl}$, than the value represented in Table 5.1 was also considered.

Eq. (5.3) considers the effect of the double-layer capacitance of the CCL on the GDL impedance. The contribution of the double-layer capacitance of the CCL on the GDL impedance is reduced with reducing frequency $\omega \rightarrow 0$ as demonstrated in the denominator of Eq. (5.3). The impedance model of the GDL-channel represented through either Eqs (5.47) or (5.48) can be considered in the numerator of Eq. (5.3) to simulate the effect of the charge double-layer of the CCL on the impedance spectrum of the GDL-channel. The impedance spectrum shown in Fig. 5.26 labelled 'Eq. (5.3) with Eq. (5.48) and $5C_{dl}$' was generated by Eq. (5.3) considering Eq. (5.48) in the numerator of Eq. (5.3) and considering the parameters shown in Tables 5.1 and 5.2 at an oxygen stoichiometry of $S_{O_2}=2.3$ and a capacitance value 5 times higher, $5C_{dl}$, than the value represented in Table 5.1. The impedance spectrum labelled 'Eq. (5.48)' was generated by Eq. (5.48) considering the parameters shown in Table 5.2 and does not consider the effect of double-layer capacitance of the CCL. The impedance spectrum labelled 'Eq. (5.48)' shown in Fig. 5.26 is similar to the impedance spectrum shown in Fig. 5.15 labelled '$S_{O_2}=2.3$, Eq. (5.48)'. The simulations were carried out in MATLAB software considering a range of frequencies from 10 kHz to 0.001 Hz. Fig. 5.26 shows the effect of increasing double-layer capacitance of the CCL on the GDL-channel impedance spectrum. The double-layer capacitance of the CCL mainly distorts the high frequency region of the impedance spectrum of the GDL-channel as shown in the impedance spectrum labelled 'Eq. (5.3) with Eq. (5.48) and $5C_{dl}$' compared with the impedance spectrum labelled 'Eq. (5.48)'. The impedance spectrum of the GDL-channel labelled 'Eq. (5.3) with Eq. (5.48) and $5C_{dl}$' considers negative real components $-Z_r$ at high frequencies. These become positive when considering the impedance of the GDL-channel with the impedance of the rest of the layers of the PEMFC as the total ohmic resistance of the PEMFC shifts the overall impedance spectrum of the PEMFC towards the upper end of the real axis Z_r of the complex-impedance plot.

Next in Chapter 6, a cathodic impedance model of the PEMFC considering the impedance model of the CCL presented previously in Chapter 4 will be combined with the analytical impedance model of the GDL-channel presented in this chapter and will be applied to real-world EIS-PEMFC measurements. The cathodic impedance model of Chapter 6 will implement the effect of the charge-double layer in the CCL on the impedance response of the GDL-channel.

References

[1] N. Fouquet, C. Doulet, C. Nouillant, D. Dauphin-Tanguy, B. Ould-Bouamama, Model based PEM fuel cell state-of-health monitoring via AC impedance measurements, J. Power Sources 159 (2006) 905–913.

[2] D. Malevich, E. Halliop, B.A. Peppley, J.G. Pharoah, K. Karan, Investigation of charge-transfer and mass-transport resistances in PEMFCs with microporous layer using electrochemical impedance spectroscopy, J. Electrochem. Soc. 156 (2009) B216–B224.

[3] J. Mainka, G. Maranzana, J. Dillet, S. Didierjean, O. Lottin, Effect of oxygen depletion along the air channel of a PEMFC on the Warburg diffusion impedance, J. Electrochem. Soc. 157 (2010) B1561–B1568.

[4] G. Maranzana, J. Mainka, O. Lottin, J. Dillet, A. Lamibrac, A. Thomas, S. Didierjean, A proton exchange membrane fuel cell impedance model taking into account convection along the air channel: on the bias between the low frequency limit of the impedance and the slope of the polarization curve, Electrochim. Acta 83 (2012) 13–27.

[5] I.A. Schneider, S.A. Freunberger, D. Kramer, A. Wokaun, G.G. Scherer, Oscillations in gas channels: part I. the forgotten player in impedance spectroscopy in PEFCs, J. Electrochem. Soc. 154 (2007) B383–B388.

[6] A. Kulikovsky, O. Shamardina, A model for PEM fuel cell impedance: oxygen flow in the channel triggers spatial and frequency oscillations of the local impedance, J. Electrochem. Soc. 162 (2015) F1068–F1077.

[7] Q. Yan, H. Toghiani, H. Causey, Steady state and dynamic performance of proton exchange membrane fuel cells (PEMFCs) under various operating conditions and load changes, J. Power Sources 161 (2006) 492–502.

[8] T.V. Reshetenko, G. Bender, K. Bethune, R. Rocheleau, Effects of local variations of the gas diffusion layer properties on PEMFC performance using a segmented cell system, Electrochim. Acta 80 (2012) 368–376.

[9] A. Kulikovsky, Why impedance of the gas diffusion layer in a PEM fuel cell differs from the Warburg finite-length impedance? Electrochem. Commun. 84 (2017) 28–31.

[10] S. Cruz-Manzo, P. Greenwood, Study of oxygen diffusion in the cathode catalyst layer and gas diffusion layer for polymer electrolyte fuel cells with EIS, J. Electroanal. Chem. 892 (2021) 115270.

[11] A. Kulikovsky, Analytical impedance of oxygen transport in a PEM fuel cell channel, J. Electrochem. Soc. 166 (2019) F306–F311.

[12] S. Chevalier, C. Josset, B. Auvity, Analytical solution for the low frequency polymer electrolyte membrane fuel cell impedance, J. Power Sources 407 (2018) 123–131.

[13] A. Kulikovsky, Performance of a PEM fuel cell with oscillating air flow velocity: a modeling study based on cell impedance, eTransportation 7 (2021) 100104.

[14] S. Chevalier, C. Josset, B. Auvity, Analytical solutions and dimensional analysis of pseudo 2D current density distribution model in PEM fuel cells, Renew. Energy 125 (2018) 738–746.

[15] A. Kulikovsky, The effect of stoichiometric ratio λ on the performance of a polymer electrolyte fuel cell, Electrochim. Acta 49 (2004) 617–625.

[16] S. Chevalier, C. Josset, A. Bazylak, B. Auvity, Measurements of air velocities in polymer electrolyte membrane fuel cell channels using electrochemical impedance spectroscopy, J. Electrochem. Soc. 163 (2016) F816–F823.

[17] A. Kulikovsky, Semi-analytical 1D + 1D model of a polymer electrolyte fuel cell, Electrochem. Commun. 6 (2004) 969–977.

[18] A. Bard, L. Faulkner, Electrochemical Methods Fundamentals and Applications, John Wiley, 2001.

[19] S. Cruz-Manzo, P. Greenwood, Analytical Warburg impedance model for EIS analysis of the gas diffusion layer with oxygen depletion in the air channel of a PEMFC, J. Electrochem. Soc. 168 (2021), 074502.

[20] F. Barbir, PEM Fuel Cells, Academic Press, Cambridge, MA, 2005.

[21] K. Ogata, Modern Control Engineering, fourth ed., Pearson, 2005.

[22] M.E. Orazem, B. Tribollet, Electrochemical Impedance Spectroscopy, second ed., Wiley, Hoboken, NJ, 2017.

[23] S.J.C. Cleghorn, C.R. Derouin, M.S. Wilson, S. Gottesfeld, A printed circuit board approach to measuring current distribution in a fuel cell, J. Appl. Electrochem. 28 (1998) 663–672.

[24] J. Stumper, S.A. Campbell, D.P. Wilkinson, M.C. Johnson, M. Davis, In-situ methods for the determination of current distributions in PEM fuel cells, Electrochim. Acta 43 (1998) 3773–3783.

[25] X.G. Yang, N. Burke, C.Y. Wang, K. Tajiri, K. Shinohara, Simultaneous measurements of species and current distributions in a PEFC under low-humidity operation, J. Electrochem. Soc. 152 (2005) A759–A766.

[26] M. Noponen, T. Mennola, M. Mikkola, T. Hottinen, P. Lund, Measurement of current distribution in a free-breathing PEMFC, J. Power Sources 106 (2002) 304–312.

[27] M.M. Mench, C.Y. Wang, M. Ishikawa, In situ current distribution measurements in polymer electrolyte fuel cells, J. Electrochem. Soc. 150 (2003) A1052–A1059.

[28] I.A. Schneider, H. Kuhn, A. Wokaun, G.G. Scherer, Fast locally resolved electrochemical impedance spectroscopy in polymer electrolyte fuel cells, J. Electrochem. Soc. 152 (2005) A2092–A2103.

[29] MathWorks®. https://uk.mathworks.com/help/symbolic/sym.int.html. last access, June 2022.

[30] M.A. Rubio, A. Urquia, R. Kuhn, S. Dormido, Electrochemical parameter estimation in operating proton exchange membrane fuel cells, J. Power Sources 183 (2008) 118–125.

[31] M.L. Perry, J. Newman, E.J. Cairns, Mass transport in gas-diffusion electrodes: a diagnostic tool for fuel-cell cathodes, J. Electrochem. Soc. 145 (1998) 5–15.

[32] M. Chatenet, J. Benziger, M. Inaba, S. Kjelstrup, T. Zawodzinski, R. Raccichini, Good practice guide for papers on fuel cells and electrolysis cells for the journal of power sources, J. Power Sources 451 (2020) 227635.

[33] MathWorks®. https://uk.mathworks.com/help/optim/ug/lsqcurvefit.html. last access, June 2022.

[34] J. Mainka, G. Maranzana, J. Dillet, S. Didierjean, O. Lottin, On the estimation of high frequency parameters of proton exchange membrane fuel cells via electrochemical impedance spectroscopy, J. Power Sources 253 (2014) 381–391.

[35] F. Jaouen, G. Lindbergh, Transient techniques for investigating mass-transport limitations in gas diffusion electrodes I. Modeling the PEFC cathode, J. Electrochem. Soc. 150 (2003) A1699–A1710.

[36] F. Jaouen, G. Lindbergh, K. Wiezell, Transient techniques for investigating mass-transport limitations in gas diffusion electrodes II. Experimental characterization of the PEFC cathode, J. Electrochem. Soc. 150 (2003) A1711–A1717.

[37] J.T. Gostick, M.W. Fowler, M.A. Ioannidis, M.D. Pritzker, Y.M. Volfkovich, A. Sakars, Capillary pressure and hydrophilic porosity in gas diffusion layers for polymer electrolyte fuel cells, J. Power Sources 156 (2006) 375–387.

[38] T.E. Springer, T.A. Zawodzinski, M.S. Wilson, S. Golfesfeld, Characterization of polymer electrolyte fuel cells using AC impedance spectroscopy, J. Electrochem. Soc. 143 (1996) 587–599.

[39] A. Kulikovsky, A fast low-current model for impedance of a PEM fuel cell cathode at low air stoichiometry, J. Electrochem. Soc. 164 (2017) F911–F915.

[40] A. Iranzo, S.J. Navas, F. Rosa, M.R. Berber, Determination of time constants of diffusion and electrochemical processes in polymer electrolyte membrane fuel cells, Energy 221 (2021) 119833.

[41] T. Reshetenko, A. Kulikovsky, A model for local impedance: validation of the model for local parameters recovery from a single Spectrum of PEM fuel cell, J. Electrochem. Soc. 166 (2019) F431–F439.

[42] T. Reshetenko, A. Kulikovsky, Comparison of two physical models for fitting PEM fuel cell impedance spectra measured at a low air flow stoichiometry, J. Electrochem. Soc. 163 (2016) F238–F246.

[43] E. Janicka, M. Mielniczek, L. Gawel, K. Darowicki, Optimization of the relative humidity of reactant gases in hydrogen fuel cells using dynamic impedance measurements, Energies 14 (2021) 3038.

[44] X. Yuan, J.C. Sun, H. Wang, J. Zhang, AC impedance diagnosis of a 500 W PEM fuel cell stack part II: individual cell impedance, J. Power Sources 161 (2006) 929–937.

[45] F. Fasmin, R. Srinivasan, Detection of nonlinearities in electrochemical impedance spectra by Kramers–Kronig transforms, J. Solid State Electrochem. 19 (2015) 1833–1847.

[46] Q. Meyer, C. Zhao, Air perturbation-induced low-frequency inductive electrochemical impedance arc in proton exchange membrane fuel cells, J. Power Sources 488 (2021) 229245.

[47] P. Boillat, F.N. Büchi, L. Gubler, T.J. Schmidt, Use and misuse of electrochemical impedance spectroscopy (EIS) in fuel cell research, in: ECS Meeting Abstracts, 2019, p. 1435. MA2019-02.

Part III

Modelling application

Chapter 6

Electrochemical phenomena in the cathodic impedance spectrum

6.1 Cathodic impedance model

Losses in the anode of a polymer electrolyte membrane fuel cell (PEMFC) are expected during long-term operation (e.g. after 250 h) [1] and during start-up and shut-down events [2]. Moreover, losses related to the hydrogen oxidation reaction (HOR) are expected to be minimum if pure hydrogen is supplied to the anode. Some studies [3–6] have neglected the contribution of the anode on the impedance response of the PEMFC. A cathodic impedance model can be applied to electrochemical impedance spectroscopy (EIS) measurements carried out in a PEMFC if the anode contribution is neglected. However, the best practise to study the electrochemical phenomena in the cathode through EIS measurements is through the experimental separation of the cathodic impedance response from the overall impedance response of the PEMFC. The separation of the cathodic impedance response from the PEMFC impedance response can be achieved through the use of a reference electrode or a three-electrode configuration in the measurement system. In Chapter 3 Section 3.3, the experimental architecture of a three-electrode configuration in the EIS measurement system for the measurement of the cathodic impedance was discussed.

In Chapter 4, the impedance model of the cathode catalyst layer (CCL) was presented. In Chapter 5 the impedance model of the gas diffusion layer (GDL) considering oxygen transport in the channels of the flow field plate (FFP) was developed. In this chapter, it is possible to define a cathodic impedance model by combining the CCL impedance model from Chapter 4 with the GDL-channel impedance model presented in Chapter 5. The main equations describing the impedance models of the CCL and GDL-channel are summarised in the following sections. The impedance model of the GDL-channel represented in Chapter 5 does not consider the effect of the charge double-layer of the CCL on the GDL-channel impedance spectrum. The effect of the charge double-layer capacitance in the CCL on the GDL-channel impedance will be considered in

Electrochemical Phenomena in the Cathode Impedance Spectrum of PEM Fuel Cells.
https://doi.org/10.1016/B978-0-323-90607-4.00002-1
Copyright © 2022 Elsevier Inc. All rights reserved.

the cathodic impedance model to be defined later in this chapter. The analysis presented in this chapter estimates electro-diffusion parameters such as Tafel slope and oxygen diffusion coefficients from EIS measurements. The estimated electro-diffusion parameters are considered a baseline in the developed cathodic impedance model to qualitatively simulate the change of the impedance responses of the CCL and GDL-channel at different PEMFC operating conditions.

6.1.1 Impedance model of the gas diffusion layer and air channel

In Chapter 5, two modelling approaches for the local impedance of the GDL-channel were developed; an impedance model for the GDL-channel considering oxygen depletion along the channel during steady conditions and an impedance model for the GDL-channel considering nonstationary oxygen depletion along the channel caused by the sinusoidal perturbation (AC conditions).

6.1.1.1 *Local impedance of the GDL-channel with steady-state oxygen depletion along the channel*

The equation for the local impedance of the GDL-channel defined in Chapter 5 considering steady-state oxygen depletion along the channel was:

$$Z_{\mathrm{GDL-ch_{ss}}} = R_{\mathrm{GDL-ch_{ss}}}(y)\frac{\tanh\left(\sqrt{j\omega\tau_{\mathrm{O_2\_GDL}}}\right)}{\sqrt{j\omega\tau_{\mathrm{O_2\_GDL}}}} \tag{6.1}$$

where $\tau_{\mathrm{O_2\_GDL}} = \delta_{\mathrm{GDL}}{}^2/D_{\mathrm{GDL}}$ is the time constant for oxygen diffusion in the GDL defined in Eq. (5.2) from Chapter 5, δ_{GDL} is the oxygen diffusion distance in the GDL, D_{GDL} is the oxygen diffusion coefficient in the GDL, j is the imaginary component of a complex number, and ω is the angular frequency and

$$R_{\mathrm{GDL-ch_{ss}}}(y) = \frac{b_{\mathrm{O_2}}\delta_{\mathrm{GDL}}}{4FD_{\mathrm{GDL}}\xi c_{\mathrm{GDL}}(\delta_{\mathrm{GDL}}, y)} \tag{6.2}$$

where $b_{\mathrm{O_2}}$ is the absolute value of the Tafel slope, F is the Faraday constant, and $c_{\mathrm{GDL}}(\delta_{\mathrm{GDL}}, y)$ is the oxygen concentration in the GDL/CCL interface defined in Eq. (5.43) from Chapter 5. The impedance model of the GDL-channel developed in Chapter 5 ignored the water produced by the oxygen reduction reaction (ORR) in the CCL. A nondimensional parameter $0 < \xi \leq 1$ is considered in Eq. (6.2) to account for the reduction of oxygen concentration in the GDL/CCL interface by water produced by the ORR. An excess of water concentration hinders the transport of oxygen in the GDL to reach the active sites in the CCL. The parameter ξ to reduce the oxygen concentration at the GDL/CCL interface $\xi c_{\mathrm{GDL}}(\delta_{\mathrm{GDL}}, y)$ could also account for the pressure drop during the liquid–gas two-phase flow in the channel [7] and the clamping pressure on the GDL performance [8]. The value of the parameter ξ could also depend on the material properties comprising the cathode and operating conditions of the PEMFC.

6.1.1.2 Local impedance of the GDL-channel considering nonstationary oxygen depletion along the channel during AC conditions

The equation for the local impedance of the GDL-channel defined in Chapter 5 considering oscillating oxygen concentration in the channel during AC conditions was:

$$Z_{\mathrm{GDL-ch_{ac}}} = R_{\mathrm{GDL-ch_{ac}}}(y, j\omega)\frac{\tanh\left(\sqrt{j\omega\tau_{\mathrm{O_2\_GDL}}}\right)}{\sqrt{j\omega\tau_{\mathrm{O_2\_GDL}}}} \tag{6.3}$$

with

$$R_{\mathrm{GDL-ch_{ac}}}(y, j\omega) = \frac{b_{\mathrm{O_2}}\delta_{\mathrm{GDL}}}{4FD_{\mathrm{GDL}}\xi\widetilde{c}_{\mathrm{GDL}}(\delta_{\mathrm{GDL}}, y, j\omega)} \tag{6.4}$$

where $\widetilde{c}_{\mathrm{GDL}}(\delta_{\mathrm{GDL}}, y, j\omega)$ is the oxygen concentration at the GDL/CCL interface defined in Eq. (5.36) from Chapter 5.

6.1.1.3 Total impedance of the GDL-channel

The equation for the total impedance of the GDL-channel defined in Chapter 5 was represented as the total impedance of individual local impedances connected in parallel as such:

$$\frac{1}{Z_{\mathrm{T\_GDL-ch_{ss,ac}}}} = \frac{1}{\ell_c}\int_0^{\ell_c}\frac{1}{Z_{\mathrm{GDL-ch_{ss,ac}}}(y)}dy \tag{6.5}$$

where ℓ_c is the total length of the channel. The local impedance of the GDL-channel $Z_{\mathrm{GDL-ch_{ss,ac}}}(y)$ can be represented through Eqs (6.1) or (6.3). Eq. (6.5) can simulate the total impedance of the GDL-channel considering either local impedances with steady-state oxygen concentration (Eq. 6.1) or oscillating oxygen concentration in the channel (Eq. 6.3). Eq. (6.5) requires operating condition inputs such as current density, relative humidity (RH), temperature, pressure and stoichiometry of air and diffusion parameters of the GDL such as oxygen diffusion coefficient and oxygen diffusion distance in the GDL. The simulation of the impedance response generated by Eq. (6.5) results in a very high computational cost (e.g. execution of several minutes in MATLAB software) because it is required to compute the integral of the inverse of the function of the local impedance along the entire length of the channel and at different frequencies.

The total impedance of the GDL-channel can also be represented through the local impedance of the GDL-channel evaluated at the position y^* of the channel length where the current density distribution is equal to the total current density of the PEMFC $i_{\mathrm{F}}(y^*) = i_{\mathrm{cell}}$. The current density distribution along the channel was defined in Eq. (5.44) from Chapter 5.

$$Z_{\mathrm{GDL-ch_{ss,ac}}}(y^*) \approx Z_{\mathrm{T\_GDL-ch_{ss,ac}}} \tag{6.6}$$

The simulation of the impedance response generated by Eq. (6.6) requires low computational cost and it can be considered a close representation of the total impedance of the GDL-channel. Eq. (6.6) can be considered in the impedance model of the cathode but is not satisfied at low oxygen stoichiometry $S_{O_2} < 2$ because the current density distribution is an exponential function with dependency on oxygen stoichiometry (as shown in Fig. 5.8 from Chapter 5), the Tafel slope b_{O_2} and oxygen diffusion coefficient D_{GDL} are considered constant along the length of the channel in the mathematical treatment.

6.1.2 Impedance model of the cathode catalyst layer

The impedance model of the CCL has been developed in Chapter 4 and considers that the resistance for proton conduction in the electrolyte is higher than the resistance for electron conduction in the carbon structure of the CCL by several orders of magnitude $R_P \gg R_e$. In addition, it is considered that the resistance for electron conduction is considered negligible $R_e = 0$. The impedance response of the CCL can be represented with the approximation that the total impedance of a segmented cell can be represented with its local impedance evaluated at the coordinate y^* of the channel length where the current density distribution is equal to the total current density of the cell $i_F(y^*) = i_{cell}$ whilst considering an oxygen stoichiometry $S_{O_2} > 2$. Oxygen is transported from the air channel towards the active sites in the CCL; therefore, it is possible to relate the oxygen concentration expressed in the equation of the impedance of the CCL with the oxygen concentration expressed in the equation of the impedance of the GDL-channel. The impedance equation of the CCL neglecting the electron resistance in the carbon structure and considering the oxygen transport limitation in the CCL is represented as:

$$Z_{CCL}(y^*) = \frac{R_P}{\gamma}[\coth(\gamma)] \tag{6.7}$$

with

$$\gamma = \sqrt{R_P\left[\frac{1}{R_C(y^*) + Z_{O_2\_CCL}(y^*)} + Y[j\omega]^P\right]} \tag{6.8}$$

where R_P is the proton resistance and Y and P are parameters of a constant phase element (CPE) and

$$Z_{O_2\_CCL}(y^*) = R_{O_2\_CCL}(y^*)\frac{\tanh\left(\sqrt{j\omega\tau_{O_2\_CCL}}\right)}{\sqrt{j\omega\tau_{O_2\_CCL}}} \tag{6.9}$$

Eq. (6.9) is defined as the diffusion impedance for oxygen diffusion across the CCL with

$$R_{O_2\_CCL}(y^*) = \frac{RT_{cell}\delta_{CCL}}{z^2F^2D_{CCL}\xi c_{GDL}(\delta_{GDL}, y^*)} \tag{6.10}$$

defined as the resistance for oxygen diffusion in the CCL where R is the gas constant, T_{cell} is the PEMFC temperature, δ_{CCL} is the oxygen diffusion distance in the CCL, z is the number of electrons consumed during the ORR, F is the Faraday constant, D_{CCL} is the oxygen diffusion coefficient in the CCL. The terms $\xi c_{GDL}(\delta_{GDL}, y^*)$ represent the oxygen concentration at the GDL/CCL interface defined in Eq. (6.2). $\tau_{O_2\_CCL} = \delta_{CCL}{}^2/D_{CCL}$ is the time constant for oxygen diffusion in the CCL defined in Eq. (4.99) from Chapter 4.

Eq. (6.11) is the charge transfer resistance during the ORR. Note that the denominator of the Eq. (6.11) is the current density distribution at the position y^* of the channel length where it is equal to the total current density of the cell $i_F(y^*) = i_{cell}$.

$$R_C(y^*) = \frac{b_{O_2}}{i_F(y^*)} \tag{6.11}$$

The impedance of the CCL could consider electron resistance in the carbon structure or considering an external capacitance attributed to the electric double layer at the carbon/Nafion and carbon/water interface in the catalyst layer. These effects on the impedance response of the CCL have been discussed in Chapter 4 Sections 4.5.3 and 4.5.4.

6.1.3 Impedance model of the cathodic side of the PEMFC

The impedance of the layers comprising the cathode can be defined as the sum of the impedance responses of the CCL and GDL-channel as such:

$$Z_{CCL+GDL-ch}(y^*) = Z_{CCL}(y^*) + Z_{K\_GDL-ch}(y^*) \tag{6.12}$$

with

$$Z_{K\_GDL-ch}(y^*) = \frac{Z_{GDL-ch_{ss,ac}}(y^*)}{1 + Y[j\omega]^P R_C(y^*)} \tag{6.13}$$

where $Z_{CCL}(y^*)$ corresponds to the impedance of the CCL represented in Eq. (6.7) and $Z_{GDL-ch_{ss,ac}}(y^*)$ represents an approximation of the total impedance of the GDL-channel as expressed in Eq. (6.6). Eq. (6.13) can consider the impedance of the GDL-channel with either steady-state oxygen concentration in the channel $Z_{GDL-ch_{ss}}(y^*)$ (Eq. 6.1) or with oscillating oxygen concentration in the channel $Z_{GDL-ch_{ac}}(y^*)$ (Eq. 6.3) during AC conditions. Eq. (6.13) also considers the effect of the nonhomogeneous accumulation of charge in the CCL on the impedance response of the GDL-channel. Eq. (6.13) has been reported by Kulikovsky [9], but considering a capacitor instead of a CPE and the finite-length Warburg impedance.

The impedance of the PEMFC ignoring the contribution of the anode can be expressed as:

$$Z_{PEMFC-cathode} = Lj\omega + R_{ohm} + Z_{CCL+GDL-ch}(y^*) \tag{6.14}$$

where L represents the inductance of the cables of the EIS measurement system, R_{ohm} represents the resistance to flow of protons in the polymer electrolyte membrane (PEM) and electrons in the GDL and FFP from the cathode. The third parameter on the right-hand side of Eq. (6.14) represents the sum of the impedance responses of the CCL and GDL-channel as expressed in Eq. (6.12). In this chapter, Eq. (6.14) will be coined 'cathodic impedance model' or 'impedance model of the cathode' for simplicity, because Eq. (6.14) simulates the experimental cathodic impedance response that can be extracted from the overall impedance response of the PEMFC through the use of a reference electrode in the EIS measurement system.

The cathodic impedance model (Eq. 6.14) considers electrochemical and diffusion parameters such as Tafel slope, oxygen diffusion coefficient, oxygen diffusion distance and operating conditions such as current density, stoichiometry and RH of air. The model can be applied to real-world EIS measurements to reveal the contribution of the impedance response of the CCL, GDL, and air channel which are overlapped in the cathodic impedance spectrum of the PEMFC. In addition, it will be possible to gain an insight into the contribution of the impedance response of the individual layers comprising the cathode on the cathodic impedance spectrum of PEMFCs operated at different operating conditions. The impedance model of the cathode expressed in Eq. (6.14) can be applied to simulate the cathodic impedance response of a PEMFC operated at an oxygen stoichiometry $S_{O_2}>2$. At an oxygen stoichiometry of $S_{O_2}<2$ and considering a variation of the Tafel slope along the channel length in a segmented cell the third term on the right-hand side of Eq. (6.14) should be integrated along the channel length as such:

$$\frac{1}{Z_{T\_CCL+GDL-ch}} = \frac{1}{\ell_c}\int_0^{\ell_c} \frac{1}{Z_{CCL}(y) + Z_{K\_GDL-ch}(y)}dy \tag{6.15}$$

Eq. (6.15) represents the total impedance of local impedances connected in parallel in a segmented cell. The graphical representation of a segmented cell has been depicted in Fig. 5.5 from Chapter 5.

6.2 Modelling architecture of the cathodic impedance model

Commercial software such as ZView, ZMAN, ZSimpWin, etc. allow fitting of electrical circuits with EIS measurements. The impedance model of the cathode represented in Eq. (6.14) comprises analytical expressions that cannot be represented through electrical components. The construction of the impedance model of the cathode represented in Eq. (6.14) and its application with real-world EIS measurements can be achieved through the use of commercial software that allows the implementation of analytical mathematical expressions and algorithms such as MATLAB.

6.2.1 Model execution

MATLAB software is considered to fit the cathodic impedance model with the EIS measurements using a nonlinear regression strategy. The fitting process is carried out through the function *lsqcurvefit* defined in MATLAB software [10]. This function solves nonlinear curve-fitting (data-fitting) problems in a least-squares sense using the Levenberg–Marquardt algorithm. The quality of the fit can be assessed through the real and imaginary relative residuals between the simulated and measured impedance data. Eqs (6.16) and (6.17) represent the average of the real and imaginary relative residuals of the simulated and measured impedance data and have been adapted from the study of Song and Bazant [11].

$$\Delta Z_{\mathrm{r}} = \frac{1}{K}\sum_{\mathrm{k}} \frac{Z_{\mathrm{r\,data}}(\omega_{\mathrm{k}}) - Z_{\mathrm{r\,model}}(\omega_{\mathrm{k}})}{\left|Z_{\mathrm{data}(\omega_{\mathrm{k}})}\right|} \tag{6.16}$$

$$\Delta Z_{\mathrm{j}} = \frac{1}{K}\sum_{\mathrm{k}} \frac{Z_{\mathrm{j\,data}}(\omega_{\mathrm{k}}) - Z_{\mathrm{j\,model}}(\omega_{\mathrm{k}})}{\left|Z_{\mathrm{data}(\omega_{\mathrm{k}})}\right|} \tag{6.17}$$

where K is the number of experimental frequencies ω, Z_{data} is the measured EIS data, $Z_{\mathrm{r_{data}}}$ and $Z_{\mathrm{j_{data}}}$ are the real and imaginary components of the measured EIS data and correspond to the frequency ω_{k} and $Z_{\mathrm{r_{model}}}$ and $Z_{\mathrm{j_{model}}}$ are the real and the imaginary components from the simulated impedance response and correspond to the frequency ω_{k}. In the book of Orazem and Tribollet [12], different methods to assess the quality of a regression of a model with EIS data are discussed.

Fig. 6.1 shows a flowchart describing the functionality of the modelling architecture. The modelling architecture can be applied to simulate the cathodic impedance response of PEMFCs operated at oxygen stoichiometry $S_{\mathrm{O_2}} > 2$ because the relation $i_{\mathrm{F}}(y^*) = i_{\mathrm{cell}}$ previously discussed in Section 6.1.1.3 is considered. Four main steps are carried out during the execution of the modelling architecture. Each step is represented in the flowchart diagram with a representative colour as shown in Fig. 6.1.

1. As a first step, a version of the cathodic impedance model considering lumped parameters is fitted to cathodic EIS measurements carried out in a PEMFC. The lumped parameters consist of electro-diffusion resistances and time constants. For instance, the parameters comprising the oxygen transport resistance in the CCL (Eq. 6.10) can be represented as a lumped parameter $R_{\mathrm{W\_CCL}} = R_{\mathrm{O_2\_CCL}}(y^*)$.
2. The lumped parameters estimated from the first step allow the estimation of diffusion parameters of the GDL through a nonlinear curve-fitting. The calculation of the diffusion parameters of the GDL requires the definition of operating conditions of the PEMFC at which the EIS measurements were carried out.

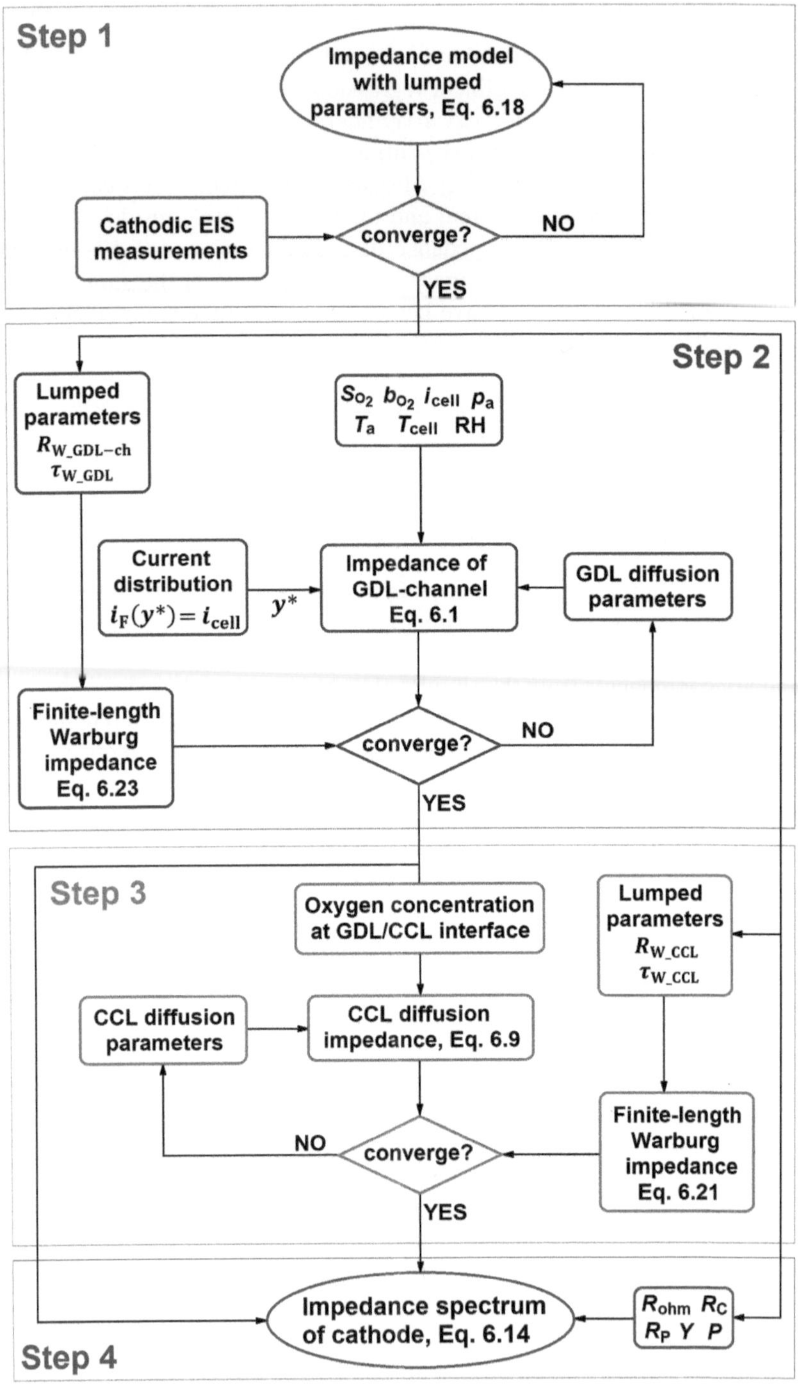

FIG. 6.1 Flowchart for simulation of impedance response of the cathode considering diffusion parameters and PEMFC operating conditions.

3. The oxygen concentration in the GDL/CCL interface estimated by the diffusion impedance of the GDL-channel is linked to the impedance diffusion of the CCL. The lumped parameters estimated in the first step and oxygen concentration at the GDL/CCL interface allow the estimation of diffusion parameters of the CCL through a nonlinear curve-fitting.
4. Once the diffusion parameters of the CCL and GDL are defined, it is possible to simulate the effect of operating conditions (stoichiometry, RH, pressure, air temperature and current density) on the impedance response of the cathode.

6.2.2 Lumped impedance model of the cathode

The cathodic impedance model represented in Eq. (6.14) ignoring the inductor element L can be represented with lumped parameters (LPs) as such:

$$Z_{\text{PEMFC-cathode\_LP}} = R_{\text{ohm}} + Z_{\text{CCL\_LP}} + Z_{\text{GDL-ch\_LP}} \tag{6.18}$$

with

$$Z_{\text{CCL\_LP}} = \frac{R_{\text{P}}}{\beta}[\coth(\beta)] \tag{6.19}$$

$$\beta = \sqrt{R_{\text{P}}\left[\frac{1}{R_{\text{C\_LP}} + Z_{\text{W\_CCL}}} + Y[j\omega]^{P}\right]} \tag{6.20}$$

$$Z_{\text{W\_CCL}} = R_{\text{W\_CCL}}\frac{\tanh\left(\sqrt{j\omega\tau_{\text{W\_CCL}}}\right)}{\sqrt{j\omega\tau_{\text{W\_CCL}}}} \tag{6.21}$$

$$Z_{\text{GDL-ch\_LP}} = \frac{Z_{\text{W\_GDL-ch}}}{1 + Y[j\omega]^{P}R_{\text{C\_LP}}} \tag{6.22}$$

$$Z_{\text{W\_GDL-ch}} = R_{\text{W\_GDL-ch}}\frac{\tanh\left(\sqrt{j\omega\tau_{\text{W\_GDL}}}\right)}{\sqrt{j\omega\tau_{\text{W\_GDL}}}} \tag{6.23}$$

Eq. (6.21) is the finite-length Warburg impedance with lumped parameters to represent the diffusion impedance of the CCL. The numerator of Eq. (6.22) also comprises a finite-length Warburg impedance with LPs and represents the impedance of the GDL-channel. Fig. 6.2 shows an electrical circuit to represent the cathodic impedance model expressed in Eq. (6.18) and considering lumped parameters. The ohmic resistance R_{ohm} can be represented as a set of resistances connected in series to represent the resistance to flow of protons in the PEM and the resistances to flow of electrons in the GDL and FFP as such $R_{\text{ohm}} = R_{\text{PEM}_{H^+}} + R_{\text{GDL}_{e^-}} + R_{\text{FFP}_{e^-}}$. The fitting process of Eq. (6.18) with real-world PEMFC-EIS measurements can be carried out with commercial software such as MATLAB. Eq. (6.18) represents LPs and cannot simulate the effect of operating conditions of the PEMFC on the cathodic impedance response.

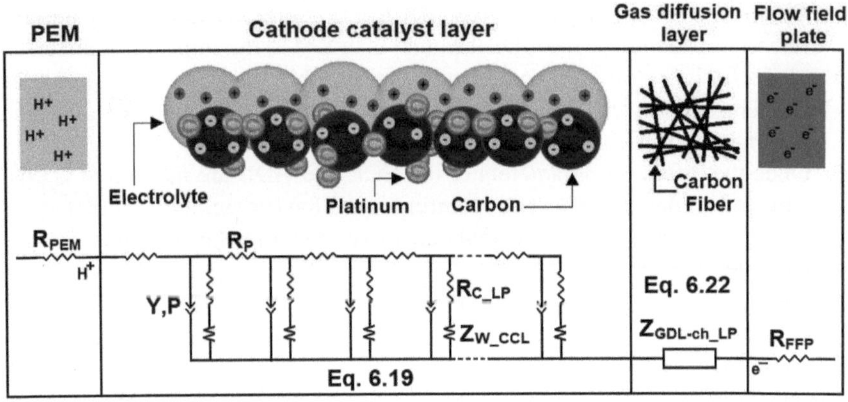

FIG. 6.2 Equivalent electrical circuit to represent the cathodic impedance model with lumped parameters, Eq. (6.18).

The lumped impedance model of the CCL represented in Eq. (6.19) could only be represented as the impedance of a transmission-line model defined in commercial software if the finite-length Warburg impedance for oxygen diffusion in the CCL is neglected, $Z_{W\_CCL}=0$. For instance, Eq. (6.19) considering $Z_{W\_CCL}=0$ would be analogous to the impedance of the transmission-line model defined in ZView software (Transmission-Line Open Circuit Terminus, DX-Type 6, Scribner Associates, Inc.). The third term on the right-hand side of Eq. (6.18) representing the impedance of the GDL-channel is an analytical equation and cannot be represented as an electrical component but could be represented with the finite-length Warburg impedance (Eq. 6.23). Under the aforementioned considerations, it would be possible to fit the electrical circuit representing the cathode impedance with PEMFC-EIS measurements using commercial software, e.g. ZView software. Nevertheless, these considerations would ignore the oxygen diffusion impedance in the CCL and effect of charge double-layer of the CCL on the GDL-channel impedance.

6.2.3 Estimation of lumped parameters

The fitting process of an impedance model containing a high number of parameters with EIS measurements can be a challenge because an impedance model with different values of the estimated parameters could represent the same impedance response. The estimation of the parameters when fitting an impedance model with PEMFC-EIS measurements featuring a single semicircle in the impedance spectrum becomes a challenging task because the time constants related to the electro-diffusion processes taking place on the different layers comprising the PEMFC are overlapped on the impedance spectrum. This could also mislead the calculation of the electro-diffusion parameters in the PEMFC from an impedance model considering a high number of parameters. It seems that the best practise would be to reduce the number of fitted parameters by establishing the value of the electro-diffusion parameters in the model for instance Tafel slope, diffusion coefficients, etc. Kulikovsky and Eikerling [13] have applied a physics-based impedance model of the CCL with EIS measurements without fitting the parameters. The results demonstrate that electrochemical parameters of the CCL can be calculated from the impedance spectrum without a fitting process.

If it is not possible to establish the values of the electro-diffusion parameters in the impedance model prior to the fitting process with EIS measurements, it is advised to reduce the number of fitting parameters in the impedance model. For instance, ZView software (Scribner Associates, Inc.) recommends to fit the transmission-line model (Transmission-Line Open Circuit Terminus, DX-Type 6) with EIS measurements by considering a fixed and not variable value of the resistor connected between the upper and lower rails of the transmission-line model during the fitting process. A two-step fitting process is considered to fit the cathodic impedance model considering LPs (Eq. 6.18) with EIS measurements as shown in Fig. 6.3. First Eq. (6.18) is fitted with the EIS measurements but neglecting the expression $Y[j\omega]^P R_{C\_LP}=0$ in Eq. (6.22). Under this condition, the impedance of the GDL-channel represented in the third term on the right-hand side of Eq. (6.18) is only represented with the finite-length Warburg impedance Eq. (6.23). Secondly, the full expression of Eq. (6.18) is considered in the fitting process, but the parameters R_{C_LP} and R_{ohm}, are kept fixed and the value of the rest of the parameters calculated from the first-step fitting process are considered as initial values during the second-step fitting process. The fixed value of R_{C_LP} during the second-step fitting process may not be considered the same value as calculated from the first-step fitting process. Noting that the second-step fitting process is analogous to the fitting process of the transmission-line model recommended in ZView software, the charge transfer resistance R_{C_LP} connected between the upper and lower rails shown in Fig. 6.2 is considered fixed during the fitting process.

If Eq. (6.18) is applied to PEMFC-EIS measurements using the two-step fitting process depicted in Fig. 6.3, the parameters shown in Table 6.1 are calculated. Table 6.1 also shows the analogy between the parameters of the lumped model (Eq. 6.18) and the cathodic model (Eq. 6.14). The finite-length Warburg impedance from the lumped model represented in Eq. (6.23) considers lumped parameters, e.g. R_{W_GDL-ch}, τ_{W_GDL}. The Tafel slope b_{O_2}, oxygen diffusion distance δ_{GDL} in the GDL, and oxygen diffusion coefficient D_{GDL} in the GDL, are enclosed within the oxygen transport resistance of the GDL-channel (Eq. 6.2) from the cathodic model (Eq. 6.14). The lumped model represented in Eq. (6.18) has been applied with EIS measurements of PEMFCs to estimate oxygen diffusion resistance in the CCL and GDL [14].

A graphical user interface (GUI) developed in MATLAB as shown in Fig. 6.4 can be used to fit the impedance model with LPs represented in Eq. (6.18) with experimental EIS measurements. The two-step fitting process

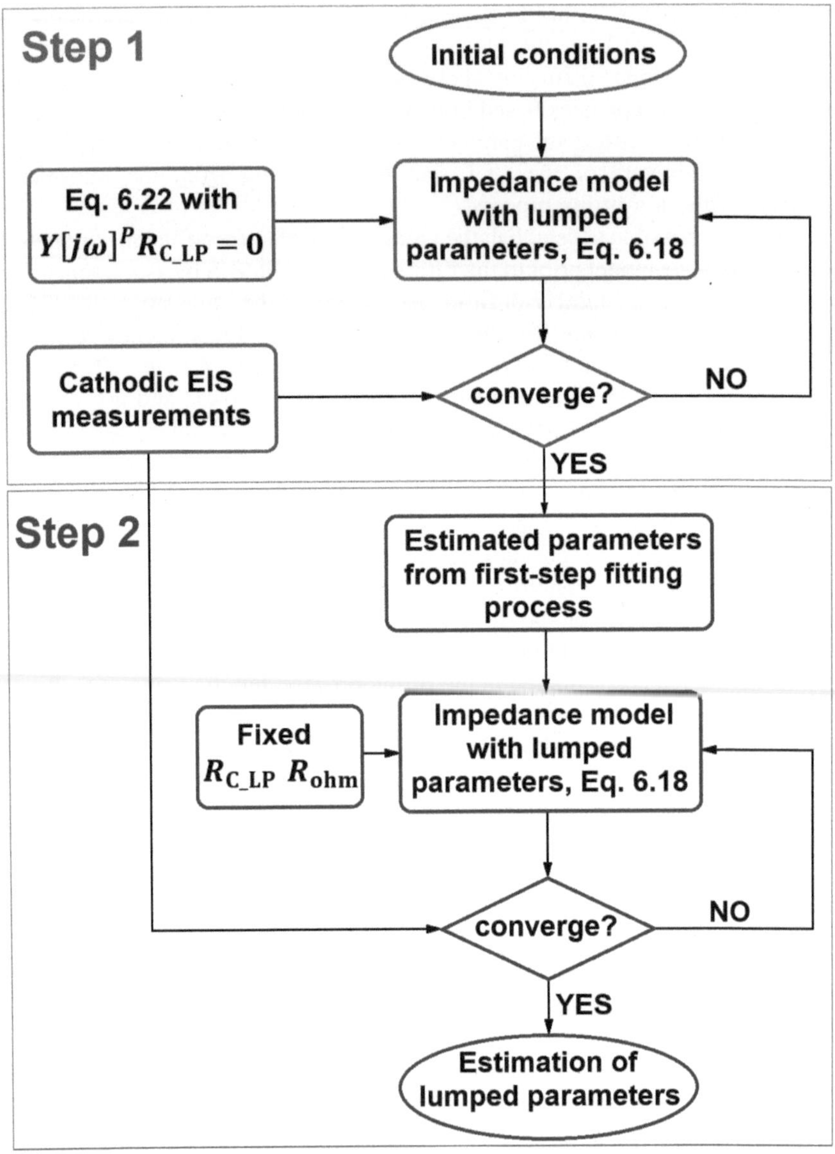

FIG. 6.3 Two-step fitting process for the estimation of the lumped parameters represented in Eq. (6.18).

shown in Fig. 6.3 is executed in the GUI. The GUI is not only able to compare measured and simulated data in the Nyquist plot, but also compare measured and simulated data in the Bode plot (modulus and phase). Sensitivity to discrepancies between model and experimental data can be assessed by comparing the

TABLE 6.1 Analogy between parameters from lumped model (Eq. 6.18) and cathodic model (Eq. 6.14).

Lumped model	R_{C_LP} Ωcm^2	R_{W_CCL} Ωcm^2	τ_{W_CCL} s	R_{W_GDL-ch} Ωcm^2	τ_{W_GDL} s
Cathodic model	$R_C(y^*)$ Ωcm^2	$R_{O_2\_CCL}(y^*)$ Ωcm^2	$\tau_{O_2\_CCL}$ s	$R_{GDL-ch_{ss}}(y^*)$ Ωcm^2	$\tau_{O_2\_GDL}$ s

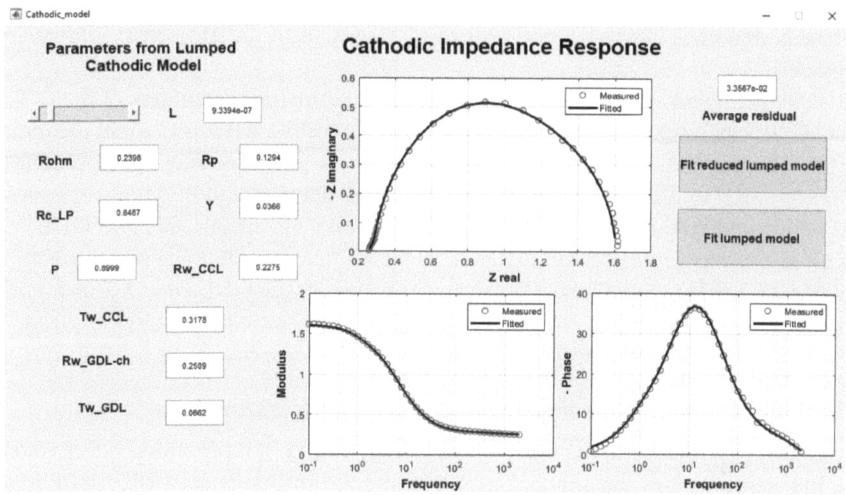

FIG. 6.4 Graphical user interface in MATLAB for estimation of lumped parameters in Eq. (6.18).

experimental data with the simulated data in a Bode format (modulus and phase). Orazem and Tribollet [12] reported in their study that the real component and Bode modulus plots are relatively insensitive to the quality of the fit of a model to impedance data. The imaginary component of the impedance and Bode phase angle plots are modestly sensitive to fit quality. The use of the GUI to fit the impedance response of the CCL with EIS measurements has been considered in the study of Cruz-Manzo and Chen [15].

6.2.4 Estimation of diffusion parameters

The simulation of the impedance response generated by the cathodic impedance model represented in Eq. (6.14) requires the definition of operating conditions such as stoichiometry, pressure, temperature and RH of air, current density and PEMFC temperature. Diffusion parameters in the CCL and GDL are also required for the simulation of the cathodic impedance response generated by

Eq. (6.14). The calculation of the diffusion parameters in the GDL and CCL can be achieved through a fitting process between the equations of the lumped impedance model previously described in Section 6.2.2 and equations of the cathodic impedance model described in Section 6.1. Table 6.2 shows the diffusion parameters of the CCL and GDL estimated by considering values of lumped parameters from Eq. (6.18) and operating conditions of the PEMFC as inputs. The modelling architecture considers an international system of units (e.g. Kelvin, Pascal, metre, etc.), therefore the units of the input parameters of the modelling architecture are converted to the international system of units (e.g. bar to Pascal, Celsius degrees to Kelvin, etc.) during the calculation of the impedance response of the cathode. The estimation of the diffusion parameters is depicted in the second and third stages from the flowchart shown in Fig. 6.1.

The fitting process for the estimation of the diffusion parameters of the CCL and GDL can be executed in a GUI developed in MATLAB as shown in Fig. 6.5. The GUI requires operating conditions such as stoichiometry, temperature, pressure, RH of air entering the channel, current density, and PEMFC core temperature and calculates the oxygen diffusion coefficient in CCL and GDL as well as the oxygen diffusion distance in the CCL and GDL. The script written in MATLAB to estimate the diffusion parameters of the GDL and CCL is found in Appendix A.

6.2.4.1 Diffusion parameters of the gas diffusion layer

The local impedance of the GDL-channel $Z_{GDL-ch_{ss}}(y^*)$ represented in Eq. (6.1) at the position y^* of the channel length where the current density distribution is equal to the total current density of the PEMFC $i_F(y^*) = i_{cell}$ can be considered a close representation of the total impedance of the GDL-channel. The estimation of the diffusion parameters in the GDL considers the equation of the diffusion impedance of the GDL-channel considering oxygen depletion along the air channel at steady-state conditions $Z_{GDL-ch_{ss}}(y^*)$. This consideration is supported from the fact that the impedance response of the GDL-channel considering oxygen depletion along the air channel at steady-state conditions is similar to the impedance response of the finite-length Warburg impedance. This comparison has been demonstrated in Fig. 5.14 from Chapter 5.

The lumped parameters of the finite-length Warburg impedance R_{W_GDL-ch} and τ_{W_GDL} represented in Eq. (6.23) are estimated by fitting Eq. (6.18) with EIS measurements carried out in a PEMFC as previously discussed in Section 6.2.3. Thereafter, a fitting process between the impedance response generated by Eq. (6.1) $Z_{GDL-ch_{ss}}(y^*)$ and the impedance response generated by the finite-length Warburg impedance (lumped) in Eq. (6.23) is carried out to calculate diffusion parameters of the GDL. This process can be depicted in the section coloured in blue (grey in print version) (second stage) of the flowchart shown in Fig. 6.1.

TABLE 6.2 Diffusion parameters of the cathode catalyst layer (CCL) and gas diffusion layer (GDL) estimated with a nonlinear curve fitting in MATLAB software.

Inputs								Estimated			
LPs[a]	S_{O_2}	i_{cell} A/cm^2	T_a °C	p_a bar	T_{cell} °C	RH %	b_{O_2} V	δ_{GDL} m	δ_{CCL} m	D_{CCL} m^2/s	D_{GDL} m^2/s

Operating conditions of the PEMFC and values of lumped parameters from Eq. (6.18) are required as inputs.
[a]*LPs means lumped parameters estimated by Eq. (6.18).*

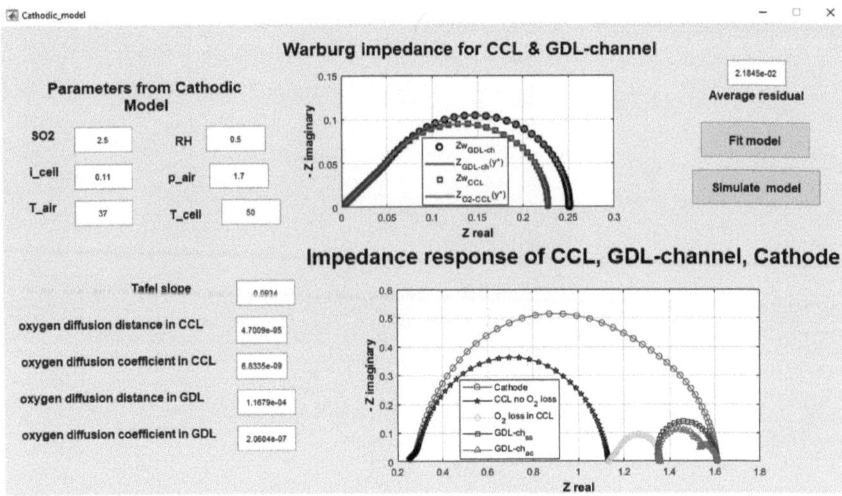

FIG. 6.5 Graphical User Interface developed in MATLAB for estimation of diffusion parameters in the cathode catalyst layer (CCL) and gas diffusion layer (GDL).

$Z_{GDL-ch_{ss}}(y^*)$ provides a close representation of the total impedance response of the GDL-channel. The fitting process to estimate diffusion parameters of the GDL could have also been carried out considering the total impedance of the GDL-channel $Z_{T\_GDL-ch_{ss}}$ expressed in Eq. (6.5) instead. However, this process results in a very high computational cost during the curve-fitting process because it is required to solve the integral function expressed in Eq. (6.5) at different frequencies. This could also result in a high number of iterations to solve the curve-fitting (data-fitting) problem. Hence the fitting process considering $Z_{GDL-ch_{ss}}(y^*)$ instead of $Z_{T\_GDL-ch_{ss}}$ is carried out in MATLAB. The fitting process between the impedance responses generated by Eqs (6.23) and (6.1) $Z_{GDL-ch_{ss}}(y^*)$ calculates the oxygen diffusion coefficient D_{GDL} and oxygen diffusion distance δ_{GDL} in the GDL. During the fitting process, it may be required to consider different values of the parameter $0<\xi\leq 1$ defined in Eq. (6.1) to estimate the correct order of magnitude of the diffusion parameters of the GDL. The estimated diffusion parameters of the GDL allow the calculation of the oxygen concentration at the GDL/CCL interface which is required in the impedance diffusion model of the CCL.

The Tafel slope b_{O_2} and operating conditions shown in Table 6.2 are required as inputs during the fitting process of Eq. (6.1) $Z_{GDL-ch_{ss}}(y^*)$ and Eq. (6.23) for the estimation of the diffusion parameters in the GDL. The Tafel slope b_{O_2} is calculated from the charge transfer resistance of the lumped model and current density as such $b_{O_2}=R_{C\_LP}i_{cell}$. Noting that the value of the charge transfer resistance R_{C_LP} can be calculated by fitting Eq. (6.18) with the PEMFC-EIS measurements as previously discussed in Section 6.2.3.

6.2.4.2 *Diffusion parameters of the cathode catalyst layer*

The lumped parameters R_{W_CCL} and τ_{W_CCL} related to the finite-length Warburg impedance and represented in Eq. (6.21) are estimated by fitting Eq. (6.18) with EIS measurements carried out in a PEMFC as previously discussed in Section 6.2.3. The fitting process between the impedance response generated by the finite-length Warburg impedance comprising the LPs (Eq. (6.21)) and the impedance response generated by the oxygen diffusion impedance of the CCL represented in Eq. (6.9) is carried out as well. The fitting process between the impedance response generated by Eqs (6.21) and (6.9) results in the calculation of the oxygen diffusion distance δ_{CCL} and oxygen diffusion coefficient D_{CCL} in the CCL, respectively. The fitting process requires as an input the oxygen concentration in the GDL/CCL interface represented in the diffusion impedance of the GDL-channel (Eq. 6.1). This process can be depicted in the section coloured in green (dark grey in print version) (third stage) of the flowchart shown in Fig. 6.1.

6.2.5 Simulation results from cathodic modelling architecture

The last stage of the application of the modelling architecture represented in the flowchart shown in Fig. 6.1 consists of the simulation of the impedance response of the CCL, GDL-channel, and cathode using the cathodic impedance model expressed in Eq. (6.14). The impedance model of the cathode requires operating conditions such as current density, RH, temperature, pressure, and stoichiometry of air. Also required are the diffusion parameters of the CCL and GDL which are calculated through the fitting process discussed in a previous Section 6.2.4.

The inductor L is not considered in the fitting process of the lumped model previously described in Section 6.2.2. The value of the inductor L is considered in Eq. (6.14) to make the simulated cathodic impedance spectrum intercept the real component Z_r (high frequency region) of the complex-impedance plot. It will be possible to simulate the diffusion impedance responses of the CCL and GDL-channel which are commonly overlapped in the impedance spectrum of the cathode at low frequencies. The impedance spectrum of the GDL-channel considering oscillating oxygen concentration in the channel during AC conditions can be simulated as well by considering Eq. (6.3) $Z_{GDL-ch_{ac}}(y^*)$ instead of Eq. (6.1) $Z_{GDL-ch_{ss}}(y^*)$ in the cathodic impedance model (Eq. 6.14). In addition, it will be possible to assess the effect of different operating conditions as well as the effect of diffusion parameters on the overall impedance spectrum of the cathode and on the individual impedance spectra of the CCL and GDL-channel.

During the simulation of the cathodic impedance response, it is also possible to simulate the change of the Tafel slope as a function of operating conditions. This is possible by incorporating a factor in the Tafel slope which considers a deviation of the charge transfer coefficient from nominal conditions. The nominal conditions represent the operating conditions of the PEMFC at which the experimental cathodic impedance response was measured. Eq. (6.24) is the Tafel slope represented in the cathodic impedance model (Eq. 6.14) considering the factor $\hat{\alpha}$ related to the deviation of the charge transfer coefficient from nominal conditions.

$$b_{O_2} = \frac{b^{n}_{O_2}}{\hat{\alpha}} \tag{6.24}$$

Where $b^{n}_{O_2}$ is the Tafel slope at nominal conditions and can be calculated from the charge transfer resistance represented in the lumped model expressed in Eq. (6.18) as such $b^{n}_{O_2} = R_{C\_LP} i_{cell}$ and

$$\hat{\alpha} = \frac{\alpha_s}{\alpha_n} \tag{6.25}$$

with

$$\alpha_{s,n} = \frac{RT_{cell_{s,n}}}{4F[R_{C\_LP} + R_{W\_GDL-ch}]i_F} - \frac{RT_{cell_{s,n}}}{4F[R_{C\_LP} + R_{W\_GDL-ch}]c_{GDL_{s,n}}(\delta_{GDL})}\left[\frac{\partial c_{GDL_{s,n}}(\delta_{GDL})}{\partial i_F}\right] \tag{6.26}$$

Eq. (6.26) represents the charge transfer coefficient and can be calculated from the Tafel equation [16]. Eq. (6.26) assumes that the oxygen diffusion in the CCL is fast and the charge transfer coefficient is influenced by oxygen diffusion in the GDL. This assumption is considered because the cathodic impedance model represented in Eq. (6.14) decouples the charge transfer resistance $R_C(y^*)$ from the oxygen diffusion impedance of the CCL $Z_{O_2\_CCL}(y^*)$. The subscript n represents a value of the operating condition of the PEMFC at which the cathodic impedance response was measured (nominal conditions). The subscript s represents a value of the operating condition different from the nominal condition. For the specific case, when $\hat{\alpha} = 1$ the value of the Tafel slope b_{O_2} expressed in Eq. (6.24) represents the Tafel slope of the measured cathodic impedance response. During the simulation of the cathodic impedance response at different operating conditions for $\hat{\alpha} \neq 1$, the magnitude of the Tafel slope b_{O_2} considered in the cathodic impedance model (Eq. 6.14) will change.

In the next sections, the modelling architecture previously presented will be applied to cathodic EIS measurements. The simulation of the impedance response of the CCL, GDL-channel, and cathode at different PEMFC operating conditions will be simulated as well.

6.3 Application of cathodic impedance model with EIS measurements

A diagnostic of the performance of the PEMFC through EIS could be assessed though the application of an impedance model with PEMFC-EIS measurements at different operating conditions. Studies in the literature have fitted electrical circuits with EIS measurements carried out in PEMFCs for the diagnosis and monitoring the PEMFC performance at critical conditions [17–19]. The aim of this chapter is not to provide a diagnostic of the processes causing a reduction in performance of the PEMFC. In this chapter, the cathodic impedance model is applied to the measured impedance spectrum of the cathodic side of a PEMFC to estimate base-line parameters at nominal conditions and simulate the impedance response of the CCL and GDL-channel with changing PEMFC operating conditions such as current density, pressure, temperature, and air stoichiometry. The impedance model could be applied to cathodic EIS measurements of PEMFCs operated at critical conditions (flooding, drying, reactant starvation, etc.) to assess the individual impedance responses of the CCL and GDL-channel causing a reduction in performance in the PEMFC. The cathodic modelling architecture will be applied to EIS measurements to simulate the impedance responses of the CCL and GDL-channel which are commonly overlapped in the overall cathodic impedance response of the PEMFC. A script written in MATLAB software for the simulation of the impedance spectra of the cathode, CCL, and GDL-channel at different operating conditions can be found in Appendix A.

6.3.1 Validity of EIS measurements

EIS measurements carried out in a 25 cm^2 PEMFC using a test rig acquired from Baltic Fuel Cells are considered for the analysis of the cathodic impedance. The membrane electrode assembly (MEA) consisted of a catalyst coated membrane, DuPont Nafion-115, with a platinum loading of 0.4 mg/cm^2 and carbon black for the electrodes. The thickness of the CCL is 12 μm and the GDL was a Toray TGP-H-030 carbon paper with thickness of 110 μm. The temperature of the PEMFC was controlled to 50°C by the test rig and the gas pressure was held to 1.7 bar for both anode and cathode. The stoichiometry of hydrogen was 2 and for air was 2.5. The RH of hydrogen entering the PEMFC was 100% and the RH of the air supplied to the cathode was 50%. The contact pressure on the active cell area was adjusted through a pneumatic air cylinder from the Baltic Fuel Cell compression unit. The contact pressure on the active area was fixed to 1.4 N/mm^2. The PEMFC was operated at steady-state with a current density of 0.11 A/cm^2 using a RBL488 Dynaload load bank. Thereafter EIS measurements in the PEMFC were carried out using a Solartron 1280C frequency response analyser (FRA) with 10 mV amplitude of the sinusoidal perturbation. The cathodic EIS measurements were obtained through a three-electrode configuration and the considered frequency range was from 20 kHz down to 0.1 Hz. The experimental set-up is depicted in Fig. 6.6 and further information about the configuration between the load and the FRA can be found in the technical bulletin from Solartron Analytical [20].

Fig. 6.7 shows the cathodic EIS measurements measured through a three-electrode configuration and the experimental set-up shown in Fig. 6.6. EIS

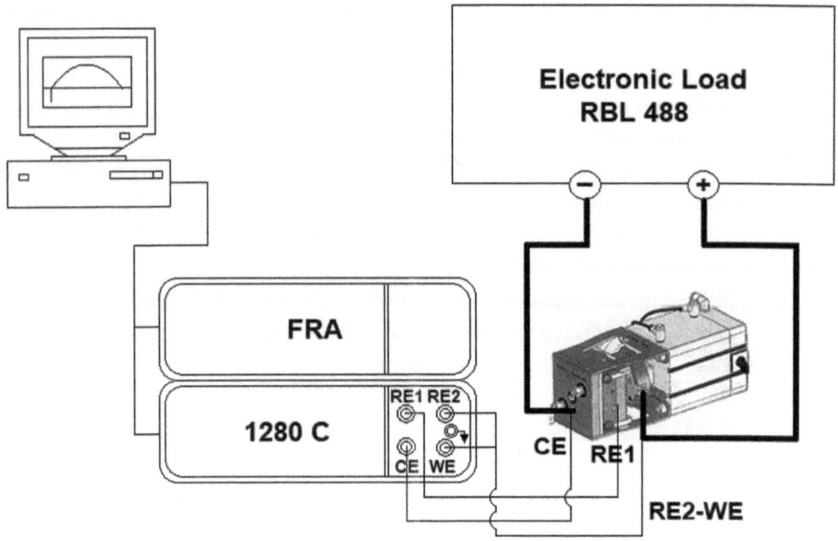

FIG. 6.6 Experimental set-up to measure the cathodic impedance response of the PEMFC.

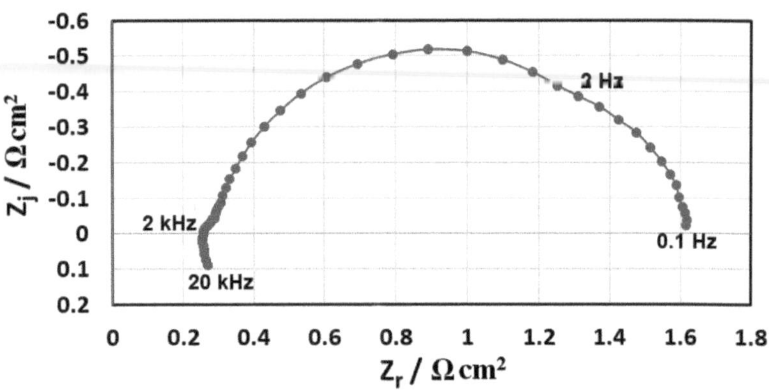

FIG. 6.7 Cathodic EIS measurements carried out in a 25 cm^2 PEMFC using a three-electrode configuration in the measurement system. The PEMFC was operated at 0.11 A/cm^2 with a core temperature of 50°C, pressure of the gas reactants at 1.7 bar, relative humidity (RH) in the anode of 100% and cathode of 50% RH, and a stoichiometry of 2 and 2.5 for hydrogen and air, respectively.

measurements with positive imaginary components Z_j were apparent at frequencies higher than 2 kHz and these measurements are attributed to the inductance of the electrical cables of the measurement system. The effect of the inductance of the electrical cables of the measurement system on EIS measurements has been discussed previously in Chapter 3 Section 3.4.1.

The cathodic impedance model will be applied to cathodic EIS measurements to estimate electro-diffusion parameters and analyse the contribution of the CCL, and GDL-channel impedance responses on the cathodic impedance spectrum. Before trying to estimate electro-diffusion parameters from the experimental cathodic impedance response, an evaluation of the cathodic EIS measurements is carried out to ensure that the EIS measurements comply with Kramers–Kronig (K–K) properties such as linearity and stability.

In Chapter 1 Section 1.5.4, it was demonstrated that the Hilbert transformation method using the fast Fourier transform (FFT) can be applied to EIS measurements to evaluate consistency with linearity and stability properties of K–K relations. A script written in MATLAB software for the application of the Hilbert transformation method described in Section 1.3.2 from Chapter 1 with EIS measurements is shown in Appendix A. Fig. 6.8 shows a comparison between the transformed data using the Hilbert transformation method and the cathodic EIS measurements. EIS measurements with positive imaginary components Z_j at high frequencies were not considered in the Hilbert transformation because these EIS measurements in the high frequency region are attributed to the external inductance of the measurement system. The Hilbert transformation was carried out around a reference frequency ω_o of the impedance data to be transformed which is related to the highest point of the experimental impedance spectrum. The EIS measurements were shifted along the real component Z_r of the complex-impedance plot for the comparison shown in Fig. 6.8. The scattered transformed data at the lowest frequency shown in Fig. 6.8 can be related to the low frequency limit considered in the EIS measurements (0.1 Hz). The Hilbert transformation requires a much lower frequency limit (e.g. 0.001 Hz) as FFT requires linearly spaced frequencies. Fig. 6.8 shows that it is possible to transform the EIS measurements using the Hilbert transformation.

It is also possible to evaluate consistency with K–K relations by fitting the Voigt circuit shown in Fig. 1.22 from Chapter 1 with the cathodic EIS measurements. The Voigt circuit satisfies K–K properties and consists of a resistor connected in series with a resistor-capacitor parallel configuration and then repeated a finite number of times. A script written in MATLAB software for

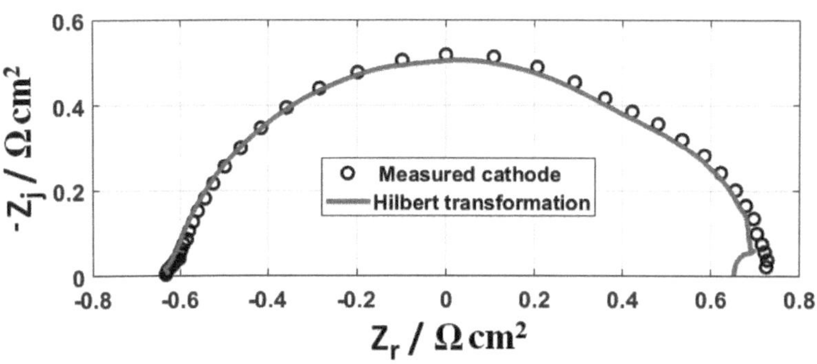

FIG. 6.8 Comparison between measured EIS data and transformed data through Hilbert transformation using MATLAB software.

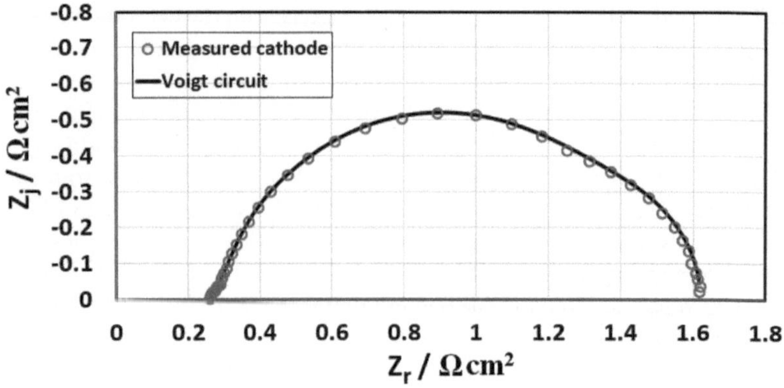

FIG. 6.9 Comparison between measured cathodic impedance response and generated impedance response by the Voigt circuit. The estimated parameters of the Voigt circuit using MATLAB software are $R_0=0.2548\ \Omega cm^2$, $R_1=0.8287\ \Omega cm^2$, $C_1=0.0373$ F/cm^2, $R_2=0.1787\ \Omega cm^2$, $C_2=0.9152$ F/cm^2, $R_3=0.1576\ \Omega cm^2$, $C_3=0.0610$ F/cm^2, $R_4=0.0305\ \Omega cm^2$, $C_4=0.0537$ F/cm^2, $R_5=0.03\ \Omega cm^2$, $C_5=0.0069$ F/cm^2, $R_6=0.1411\ \Omega cm^2$, $C_6=1.1591$ F/cm^2.

fitting the impedance response of the Voigt circuit with EIS data can be found in the Appendix A. A configuration of six parallel resistor-capacitors connected with a resistor in series was considered in the Voigt circuit. Fig. 6.9 shows that the Voigt circuit can reproduce the cathodic EIS measurements. Based on the results shown in Figs 6.8 and 6.9, it is possible to confidently continue with the analysis of the cathodic EIS measurements to estimate the electro-diffusion parameters of the cathode and simulate the impedance response of the CCL and GDL-channel at different operating conditions.

6.3.2 Parameter estimation from EIS measurements

The impedance model considering lumped parameters represented in Eq. (6.18) was fitted to the EIS measurements considering the fitting process discussed in Section 6.2.3. Fig. 6.10 shows a comparison in Bode plot between the EIS measurements and the fitted impedance response from Eq. (6.18). The EIS measurements with positive imaginary components Z_j at high frequencies attributed to the inductance of the electrical cables of the measurement system and shown in Fig. 6.7 were not considered during the fitting process. The impedance of the inductor $Lj\omega$ was considered in Eq. (6.18) ($Lj\omega+R_{ohm}+Z_{CCL\_LP}+Z_{GDL-ch\_LP}$) to compare the measured and simulated data at frequencies higher than 2kHz as shown in Fig. 6.10. The lumped impedance model represented in Eq. (6.18) is able to accurately reproduce the modulus of the EIS measurements as shown in Fig. 6.10a. The phase angle at frequencies higher than 2kHz shown in Fig. 6.10b resulted in positive values and is attributed to the inductance effect of the measurement system. A small discrepancy between the phase angle of the EIS

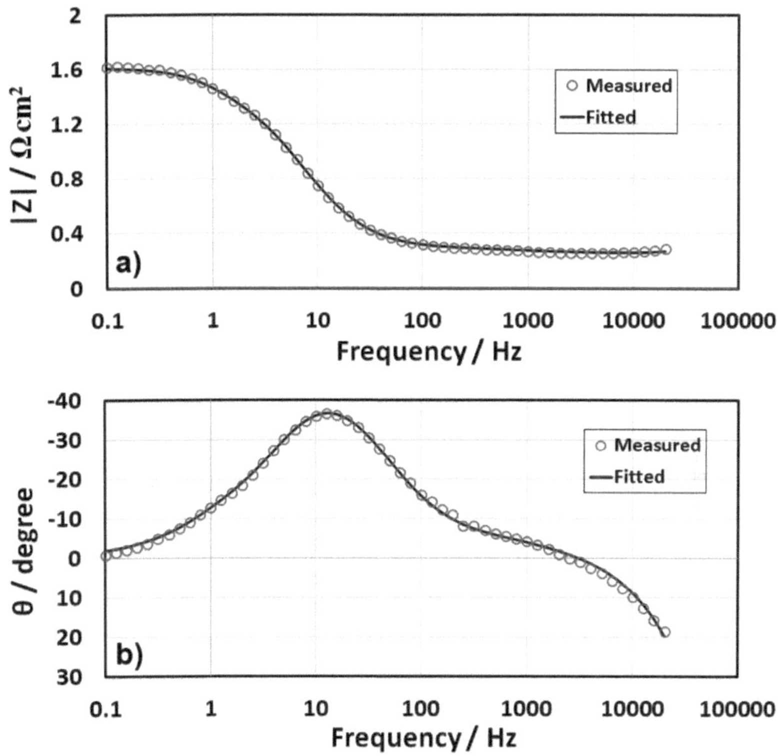

FIG. 6.10 Comparison in Bode plot between measured cathodic impedance response of the PEMFC and fitted impedance response using lumped model (Eq. 6.18), (a) modulus, (b) phase.

measurements and the phase angle of the simulated impedance response is encountered at frequencies higher than 2 kHz as shown in Fig. 6.10b.

Table 6.3 shows the resulting parameters from fitting the cathodic impedance model considering lumped parameters represented in Eq. (6.18) with the EIS measurements shown in Fig. 6.7. The parameters were estimated through the fitting process described in Section 6.2.3 and depicted in Fig. 6.3. The goodness of the fit is evaluated by the average of the real (Eq. 6.16) and imaginary (Eq. 6.17) relative residuals. The sum of Eqs (6.16) and (6.17) $\Delta Z = \Delta Z_r + \Delta Z_j$ resulted in 0.0336. A good quality fit is obtained when the real and imaginary relative residuals have a minimum value.

The lumped oxygen transport resistance in the GDL-channel resulted in a higher value than the lumped oxygen transport resistance in the CCL, $R_{W\_GDL-ch} > R_{W\_CCL}$ as shown in Table 6.3. The total oxygen transport resistance in the cathode considering LPs can be represented through the sum of the oxygen transport resistance in the CCL and the oxygen transport resistance in the GDL-channel as such $R_{W\_CCL} + R_{W\_GDL-ch}$. The total DC resistance of the

TABLE 6.3 Parameters estimated from cathodic EIS measurements carried out in a 25 cm^2 PEMFC operated at 0.11 A/cm^2 and the cathodic impedance model considering lumped parameters (Eq 6.18).

R_{ohm} Ωcm^2	R_P Ωcm^2	Y $s^P/\Omega cm^2$	P	R_{C_LP} Ωcm^2	R_{W_CCL} Ωcm^2	τ_{W_CCL} s	R_{W_GDL-ch} Ωcm^2	τ_{W_GDL} s
0.2398	0.1294	0.0366	0.8999	0.8487	0.2275	0.3178	0.2509	0.0662

cathode impedance spectrum shown in Fig. 6.7 represented at the low frequency limit $\omega \to 0$ can be calculated through the following mathematical relation:

$$R_{\mathrm{cathode_{DC}}} = R_{\mathrm{ohm}} + R_{\mathrm{P}}/3 + R_{\mathrm{C\_LP}} + R_{\mathrm{W\_CCL}} + R_{\mathrm{W\_GDL-ch}} \tag{6.27}$$

Eq. (6.27) calculates the real component of the cathodic impedance spectrum shown in Fig. 6.7 where the impedance spectrum intercepts or approaches the real component Z_r of the Nyquist plot at the low frequency limit $\omega \to 0$.

The estimation of the diffusion parameters of the GDL and CCL was carried out through the curve-fitting method discussed previously in Section 6.2.4. The LPs: $R_{\mathrm{C\_LP}}$, $R_{\mathrm{W\_GDL-ch}}$, $\tau_{\mathrm{W\_GDL}}$, $R_{\mathrm{W\_CCL}}$ and $\tau_{\mathrm{W\_CCL}}$ shown in Table 6.3, and the operating conditions of the PEMFC at which the cathodic impedance response was measured allowed the estimation of the diffusion parameters of the GDL and CCL. The Tafel slope b_{O_2} is also required for the estimation of the diffusion parameters of the GDL. The Tafel slope calculated from the charge transfer resistance shown in Table 6.3 $b_{O_2} = R_{\mathrm{C\_LP}} i_{\mathrm{cell}}$ resulted in $b_{O_2} = 0.0934$ V or expressed as $b_{O_2} = 0.2147$ V/dec if it is multiplied by a factor of 2.3. This value of the Tafel slope seems to be very high for PEMFCs with Pt as a catalyst. Bultel et al. [21] estimated a Tafel slope from EIS measurements carried out in a PEMFC which resulted in a value of $b_{O_2} = 0.22$ V/dec. The high value of the Tafel slope b_{O_2} was a result of doubling the value of the Tafel slope $2b_{O_2}$ related to kinetics of the ORR. Doubling the Tafel slope has been related to kinetics of the ORR, proton migration and oxygen diffusion in the agglomerates of the CCL [22]. Malevich et al. [16] proposed that doubling the value of the Tafel slope $2b_{O_2}$ is a result of oxygen transport limitations from the air channel to the active sites of the CCL as demonstrated in Fig. 2.10 from Chapter 2.

Table 6.4 shows the operating conditions considered for the estimation of the diffusion parameters of the CCL and GDL. In addition, Table 6.4 shows the resulting diffusion parameters of the GDL and CCL through the curve-fitting method discussed previously in Section 6.2.4. The factor ξ expressed in the oxygen transport resistance of the GDL-channel (Eq. 6.2) and in the oxygen transport resistance of the CCL (Eq. 6.10) was initially considered with a value of 1. The factor ξ reduces oxygen concentration in the GDL/CCL interface $\xi c_{\mathrm{GDL}}(\delta_{\mathrm{GDL}}, y^*)$ and is attributed to water produced by the ORR, pressure drop during the liquid–gas two-phase flow in the channel [7], and the clamping pressure on the GDL [8]. However, if $\xi = 1$ is considered in Eq. (6.2), the value of the oxygen diffusion distance in the GDL δ_{GDL} resulted in a lower value than the GDL thickness. Fig. 5.13 from Chapter 5 shows that the diffusion distance for oxygen transport through the GDL is higher than the thickness of the GDL. A value of $\xi = 0.6$ yielded an oxygen diffusion distance in the CCL and GDL greater than the thicknesses of the CCL $\ell_{\mathrm{CCL}} = 12$ μm and GDL $\ell_{\mathrm{GDL}} = 110$ μm as shown in Table 6.4. In addition, the value of $\xi = 0.6$ allowed for the estimation of diffusion coefficients in the CCL and GDL also shown in Table 6.4 within the order of magnitude reported in the literature [16,23]. The mathematical treatment for the derivation of the GDL-channel impedance

TABLE 6.4 Estimated diffusion parameters of the cathode catalyst layer (CCL) and gas diffusion layer (GDL) from cathodic EIS measurements.

Inputs							Estimated			
S_{O_2}	i_{cell} A/cm^2	T_a °C	p_a bar	T_{cell} °C	RH %	b_{C_2} V	δ_{GDL} μm	δ_{CCL} μm	D_{CCL} $\times 10^{-9}$ m^2/s	D_{GDL} $\times 10^{-7}$ m^2/s
2.5	0.11	37	1.7	50	50	0.0934	116.8	47	6.8335	2.0604

presented in Chapter 5 neglected the water produced by the ORR. The considered factor ξ for the estimation of diffusion parameters in the GDL and CCL implies the importance of considering water generated by the ORR and water transported through the cathode in an impedance model. This can be addressed in future improvements of the modelling framework.

The considered factor of $\xi=0.6$ to reduce the oxygen concentration in the GDL/CCL interface $\xi c_{\mathrm{GDL}}(\delta_{\mathrm{GDL}}, y^*)$ expressed in the oxygen transport resistance of the GDL-channel (Eq. 6.2) and of the CCL (Eq. 6.10) yielded the diffusion distances in the GDL and CCL shown in Table 6.4. The correct estimation of the diffusion parameters will also depend on the considered value of the Tafel slope b_{O_2}. The oxygen diffusion distance δ_{GDL} in the GDL was 6% higher than the GDL thickness, $\ell_{\mathrm{GDL}}=110\ \mu\mathrm{m}$. The oxygen diffusion distance δ_{CCL} in the CCL was 3.9 times higher than the CCL thickness, $\ell_{\mathrm{CCL}}=12\ \mu\mathrm{m}$. The oxygen diffusion distance in the GDL is attributed to properties such as tortuosity, permeability, etc. The oxygen diffusion distance in the CCL considers an effective distance for the diffusion of oxygen in the multiphase structure of the CCL (oxygen dissolved in water, oxygen dissolved in electrolyte (ionomer), and diffused through the porous media). The modelling assumptions for the development of the impedance model of the CCL considered that the diffusion distance of oxygen through the multiphase structure of the CCL is greater than the thickness of the CCL. Perhaps Fickian diffusion is limited for the estimation of diffusion parameters in the CCL, and a different modelling approach, e.g. statistical theory of random composite media [24] is required to accurately estimate diffusion parameters in the multiphase interface comprising the structure of the CCL. Nevertheless, in the following sections it will be possible to qualitatively simulate the impedance responses of the CCL and GDL-channel at different PEMFC operating conditions considering the parameters shown in Table 6.4 within the cathodic impedance model represented in Eq. (6.14). A hypothetical value for the depth of the air channel $h_c=0.001$ m is considered in the cathodic impedance model (see Section 5.4.2 from Chapter 5 corresponding to the oxygen transport in the air channel). The simulation results shown in the following sections are carried out from 10 kHz to the minimum frequency of 0.01 Hz to converge the impedance spectrum at low frequencies closer to the real component Z_r of the Nyquist plot.

6.3.3 Simulation of impedance responses of the CCL, GDL-channel, and cathode

The simulated impedance spectrum of the cathode generated by Eq. (6.14) and considering the parameters shown in Table 6.4 is shown in Fig. 6.11. The considered values of the parameters Y and P of the CPE, R_{ohm}, and R_{P} in the cathodic impedance model (Eq. 6.14) were similar to the values shown in Table 6.3. In addition, Fig. 6.11 shows the impedance responses of the CCL and GDL-channel which are masked on the overall cathodic impedance

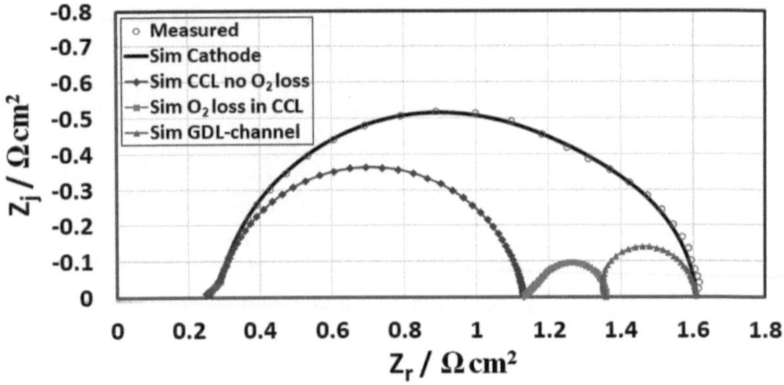

FIG. 6.11 Simulation of cathodic impedance spectrum (Sim cathode); cathode catalyst layer with no oxygen diffusion impedance (Sim CCL no O_2 loss); oxygen diffusion impedance in the cathode catalyst layer (Sim O_2 loss in CCL); impedance of the gas diffusion layer and channel (Sim GDL-channel).

spectrum. The individual impedance spectra of the CCL and GDL-channel are simulated through the local impedances represented in Eqs (6.7), (6.9) and (6.13), but considering the position of the channel length y^* where the current density distribution is equal to the cell current density $i_F(y^*) = i_{cell}$.

The impedance response of the CCL considering only charge transfer and proton transport and neglecting oxygen diffusion impedance, labelled as 'Sim CCL no O_2 loss', is simulated through Eq. (6.14) considering $Z_{O_2\_CCL}(y^*)=0$ and $Z_{K\_GDL-ch}(y^*)=0$. The oxygen diffusion impedance response of the CCL labelled as 'Sim O_2 loss in CCL' is simulated through Eq. (6.9) $Z_{O_2\_CCL}(y^*)$ and is shifted towards the right of the real component Z_r of the Nyquist plot by the amount $R_{ohm}+R_P/3+R_C(y^*)$. The impedance response of the GDL-channel labelled as 'Sim GDL-channel' is simulated through Eq. (6.13) $Z_{K\_GDL-ch}(y^*)$ considering Eq. (6.1) $Z_{GDL-ch_{ss}}(y^*)$ and is shifted towards the right of the real component Z_r of the Nyquist plot by the amount $R_{ohm}+R_P/3+R_C(y^*)+R_{O_2\_CCL}(y^*)$.

The simulated impedance response of the whole CCL is shown in Fig. 6.12. The diameter of the impedance spectrum of the CCL is calculated through the mathematical relation $R_P/3+R_C(y^*)+R_{O_2\_CCL}(y^*)$. The impedance response of the CCL considers the contribution of the charge transfer process during the ORR, proton transport and oxygen diffusion. The diameter of the impedance spectrum of the CCL considering only the charge transfer and proton transport is represented through $R_P/3+R_C(y^*)$. The simulated impedance response of the CCL shown in Fig. 6.12 features a 45-degree straight line at high frequencies attributed to the proton resistance in the electrolyte of the CCL. The electron resistance in the carbon structure has been neglected. An external capacitance attributed to the electric double-layer at the carbon/Nafion and carbon/water interface in the catalyst layer can be connected in parallel with the impedance

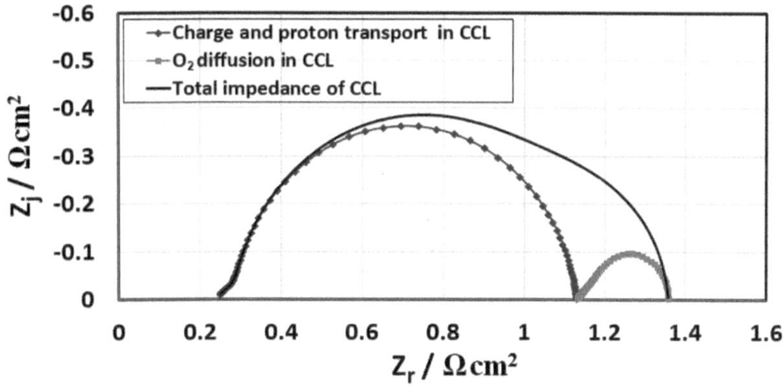

FIG. 6.12 Simulated impedance responses of the cathode catalyst layer.

of the CCL to change the slope of straight-line at high frequencies. The effect of electron resistance and external capacitance on the impedance response of the CCL has been discussed in Chapter 4 Sections 4.5.3 and 4.5.4.

The impedance spectra of the GDL-channel shown in Fig. 6.13 are shifted towards the right of the real component Z_r of the Nyquist plot by the amount $R_{ohm}+R_P/3+R_C(y^*)+R_{O_2\_CCL}(y^*)$ and consider steady-state and oscillating oxygen concentration in the channel. The contribution of the oxygen diffusion in GDL and contribution of the oxygen transport in the channel are masked on the impedance spectrum of the GDL-channel if oxygen concentration in the channel at steady-state conditions is considered (Eq. 6.13 considering Eq. 6.1 $Z_{GDL-ch_{ss}}(y^*)$). The impedance spectrum of the GDL-channel considering oscillating oxygen concentration during AC conditions and simulated through Eq. (6.13) with Eq. (6.3) $Z_{GDL-ch_{ac}}(y^*)$ reveals the contribution of the oxygen

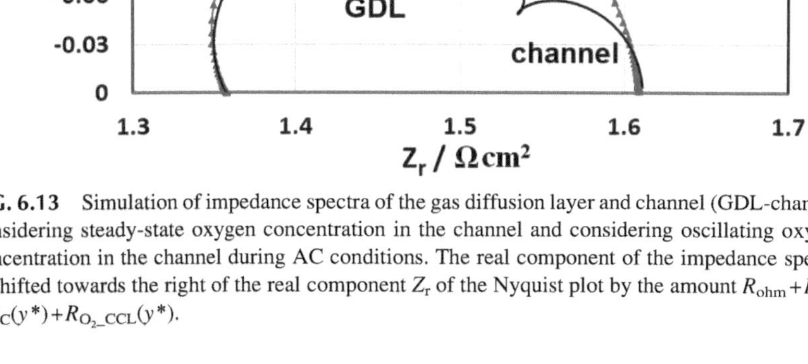

FIG. 6.13 Simulation of impedance spectra of the gas diffusion layer and channel (GDL-channel) considering steady-state oxygen concentration in the channel and considering oscillating oxygen concentration in the channel during AC conditions. The real component of the impedance spectra is shifted towards the right of the real component Z_r of the Nyquist plot by the amount $R_{ohm}+R_P/3+R_C(y^*)+R_{O_2\_CCL}(y^*)$.

transport impedance in the channel at the lowest frequencies as shown in Fig. 6.13. The impedance spectrum of the GDL-channel considering steady-state oxygen concentration in the channel shown in Fig. 6.13 does not feature a 45-degree straight line at high frequencies which is commonly represented in the impedance response of the finite-length Warburg impedance. The shape of the impedance spectrum of the GDL-channel with steady-state oxygen concentration in the channel shown in Fig. 6.13 is attributed to the parameters such as Y and P from the CPE and represented in the denominator of Eq. (6.13). In Chapter 5 Section 5.2.1 it was demonstrated that the impedance spectrum of the GDL is not similar to the impedance spectrum of the finite-length Warburg impedance when the double-layer capacitance in the CCL is high.

Kulikovsky [23] also demonstrated that it is possible to simulate the contribution of the impedance responses of the CCL, GDL and channel on the impedance spectrum of the PEMFC using a physics-based impedance model describing the charge, proton and oxygen transport through the PEMFC. Fig. 6.14 shows the simulated impedance responses of the CCL, GDL, and

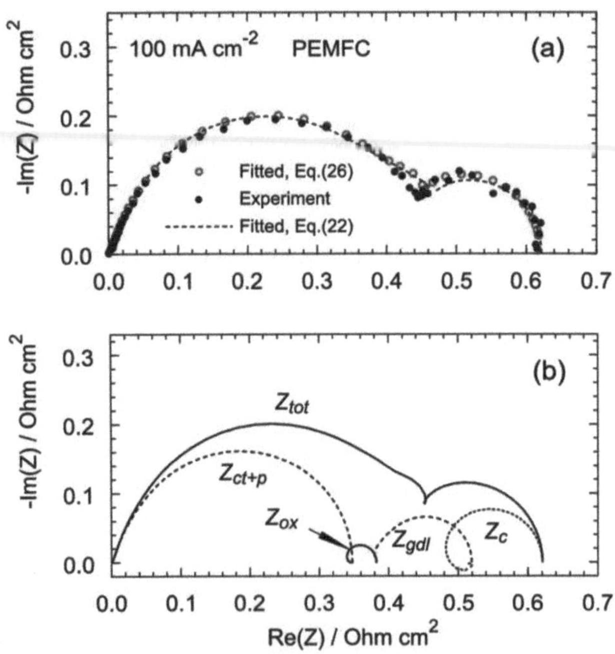

FIG. 6.14 Comparison between EIS measurements and simulated cathodic impedance spectrum reported by Kulikovsky [23], (a) EIS measurements and fitted simulated impedance response, (b) Simulated impedance response Z_{tot} with components: charge-transfer plus proton transport Z_{ct+p}, oxygen transport in the CCL Z_{ox}, oxygen transport in the GDL Z_{gdl} and in the channel Z_c. *(From A. Kulikovsky, A fast low-current model for impedance of a PEM fuel cell cathode at low air stoichiometry, J. Electrochem. Soc. 164 (2017), F911–F915. Figure Published under license of Creative Commons Attribution 4.0 License (CC BY).)*

channel reported in the study of Kulikovsky [23]. Similar to Fig. 6.11, the impedance responses of the individual layers comprising the PEMFC shown in Fig. 6.14 are masked on the overall impedance response of the PEMFC. The second semicircle at low frequencies of the experimental EIS spectrum shown in Fig. 6.14a is commonly attributed to oxygen transport limitations in the GDL. The simulation results from the impedance model reported by Kulikovsky [23] shown in Fig. 6.14b demonstrated that the semicircle at low frequencies is attributed to not only oxygen diffusion in the GDL but also oxygen transport in the channel.

6.4 Analysis of the cathodic impedance spectrum at low current density

EIS measurements at low current density are mainly attributed to the charge transfer process related to kinetics of the ORR [25]. At low current density the oxygen diffusion impedance in the cathodic impedance spectrum could be regarded as negligible because the production of water is very low to hinder the transport of oxygen towards the active sites in the CCL. Fig. 6.15 shows cathodic EIS measurements carried out in the same 25 cm^2 PEMFC described in Section 6.3.1 but at a current density of $i_{cell}=0.008$ A/cm^2. The operating conditions were the same as described in Section 6.3.1, but with the exception that the gas pressure was held at 1.9 bar for both anode and cathode with a RH in the anode of 98% and cathode of 55% RH.

The Voigt circuit shown in Fig. 1.22 from Chapter 1 was fitted to the cathodic EIS measurements at low current to ensure consistency with K–K relations. The fitting process was carried out using a script written in MATLAB software and shown in Appendix A. A configuration of six parallel resistor-capacitors connected with a resistor in series in the Voigt circuit was considered during the fitting process. Fig. 6.16 shows that the Voigt circuit can reproduce the cathodic EIS measurements at low current density. The EIS measurements with positive imaginary components Z_j at high frequencies attributed to the external inductance of the measurement system were not considered in the fitting results shown in Fig. 6.16.

It is possible to simulate the cathodic impedance spectrum shown in Fig. 6.15 using the cathodic impedance model expressed in Eq. (6.14) without fitting the impedance model with EIS measurements. The simulation of the cathodic impedance spectrum shown in Fig. 6.11 at 0.11 A/cm^2 considered a factor of $\xi=0.6$ expressed in Eqs (6.2) and (6.10) from the cathodic impedance model (Eq. 6.14) to reduce the oxygen concentration in the GDL/CCL interface. This factor ξ was related to a reduction of oxygen concentration by the water produced by the ORR and pressure drop during the liquid–gas two-phase flow in the channel. The simulation of the cathodic impedance response at a low current density of 0.008 A/cm^2 ignores the factor or is considered $\xi=1$ in the impedance model because at a low current density the water produced by the

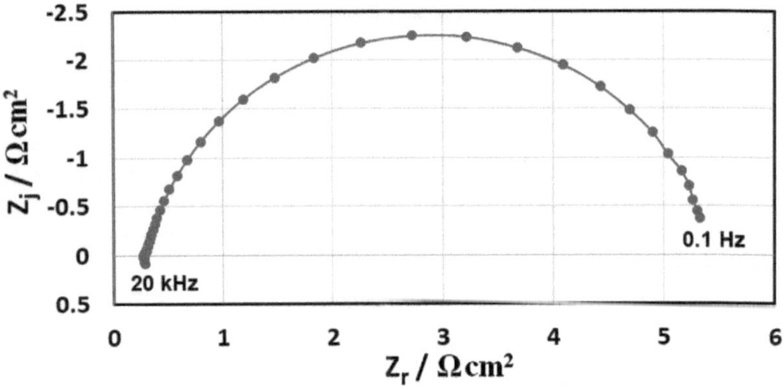

FIG. 6.15 Cathodic EIS measurements carried out in a 25 cm² PEMFC using a three-electrode configuration in the measurement system. The PEMFC was operated at 0.008 A/cm² with a core temperature of 50°C, pressure of the gas reactants at 1.9 bar, relative humidity (RH) in the anode of 98% and cathode of 55% RH, and stoichiometry of 2 and 2.5 for hydrogen and air, respectively.

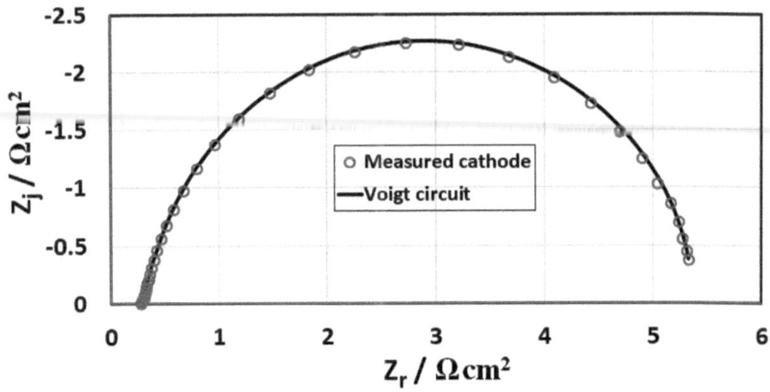

FIG. 6.16 Comparison between measured cathodic impedance response and generated impedance response by the Voigt circuit. The estimated parameters of the Voigt circuit using MATLAB software are $R_0=0.2792$ Ωcm², $R_1=3.7533$ Ωcm², $C_1=0.0285$ F/cm², $R_2=0.6437$ Ωcm², $C_2=0.4261$ F/cm², $R_3=363.1210$ Ωcm², $C_3=330.715$ F/cm², $R_4=0.5395$ Ωcm², $C_4=0.0668$ F/cm², $R_5=0.1078$ Ωcm², $C_5=0.0564$ F/cm², $R_6=0.0371$ Ωcm², $C_6=0.0112$ F/cm².

ORR has a negligible effect on the reduction of oxygen concentration in the GDL/CCL interface. The ohmic resistance R_{ohm} was considered to be the value of the real component where the EIS measurements shown in Fig. 6.15 intercept the real axis Z_r of the Nyquist plot. The diffusion parameters of the GDL and CCL are similar to the values shown in Table 6.4.

The Tafel slope shown in Table 6.4 $b_{O_2}=0.0934$ V was estimated from the cathodic EIS measurements carried out in a PEMFC operated with a gas pressure to 1.7 bar. The Tafel slope shown in Table 6.4 was considered to be twice the value of the Tafel slope attributed to kinetics of the ORR. The simulation of the cathodic impedance spectrum at a low current density of 0.008 A/cm^2 can be carried out by not only halving the Tafel slope b_{O_2} shown in Table 6.4 as $b_{O_2}=0.0467$ V, but also considering a slightly reduced value as $b_{O_2}=0.04$ V to account for the change in pressure from 1.7 to 1.9 bar. Wang et al. [26] reported a decrease in Tafel slope in the high potential region (low currents) in a PEMFC operated at 80°C with increasing oxygen partial pressure.

Based on the aforementioned considerations and parameters shown in Table 6.5, Fig. 6.17 demonstrates that it is possible to simulate the cathodic impedance spectrum at low current densities without fitting the impedance model with the EIS measurements. Although a better agreement between the measured and simulated data would result if the same fitting process discussed in Section 6.3.2 is carried out on the measured impedance spectrum shown in Fig. 6.15. The considered values of the parameters Y and P of the CPE in the cathodic impedance model (Eq. 6.14) were similar to the values shown in Table 6.3. The cathodic impedance spectrum labelled as 'Sim Cathode' and shown in Fig. 6.17 was simulated at frequencies from 10 kHz to 0.01 Hz. The impedance spectrum labelled as 'Sim CCL no O_2 loss' is simulated through Eq. (6.14) and considered $Z_{O_2\_CCL}(y^*)=0$ and $Z_{K\_GDL-ch}(y^*)=0$. The oxygen diffusion impedance response of the CCL labelled as 'Sim O_2 loss in CCL' is simulated through Eq. (6.9) $Z_{O_2\_CCL}(y^*)$ and is shifted towards the right of the real component Z_r of the Nyquist plot by the amount $R_{ohm}+R_P/3+R_C(y^*)$. The impedance response of the GDL-channel labelled as 'Sim GDL-channel' is simulated through Eq. (6.13) $Z_{K\_GDL-ch}(y^*)$ considering Eq. (6.1) and is shifted towards the right of the real component Z_r of the Nyquist plot by the amount $R_{ohm}+R_P/3+R_C(y^*)+R_{O_2\_CCL}(y^*)$. The charge transfer resistance and capacitance effect of charge accumulation in the CCL mainly represent the impedance spectrum of the cathode as demonstrated in the simulated impedance spectrum labelled as 'Sim CCL no O_2 loss'. Fig. 6.17 shows that the contribution of the oxygen diffusion impedance of the CCL and impedance of the GDL-channel can be considered negligible in the cathodic impedance spectrum at low current densities.

6.5 Simulation of cathodic impedance spectrum at different PEMFC temperatures

In this section and following sections, the change of the magnitude of the impedance spectra of the cathode, CCL and GDL-channel with changing PEMFC operating conditions such as PEMFC temperature, oxygen stoichiometry and current density will be simulated. The simulation of the impedance spectra will be carried out using Eq. (6.14) and considering as a baseline the parameters shown in Table 6.4. Some electro-diffusion parameters from Eq. (6.14) will be recalculated with varying PEMFC operating conditions whilst the rest of the parameters will be considered similar to the baseline. Although

TABLE 6.5 Considered parameters in cathodic impedance model for simulation of the cathodic impedance spectrum at a low current density of 0.008 A/cm^2.

S_{O_2}	i_{cell} A/cm^2	T_a °C	p_a bar	T_{cell} °C	RH %	b_{O_2} V	δ_{GDL} µm	δ_{CCL} µm	D_{CCL} $\times 10^{-9}$m^2/s	D_{GDL} $\times 10^{-7}$m^2/s
2.5	0.008	38.5	1.9	50	55	0.04	116.8	47	6.8335	2.0604

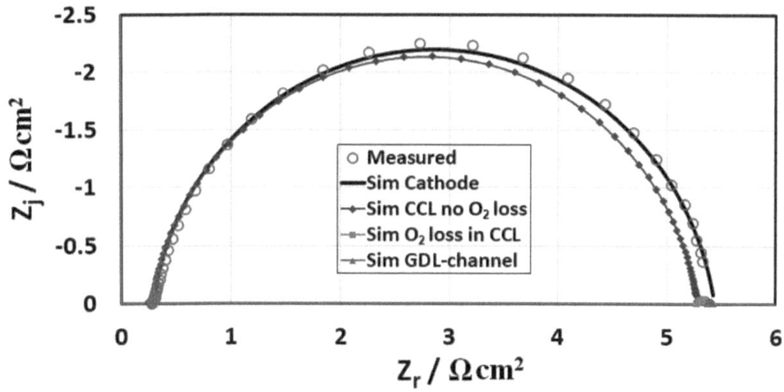

FIG. 6.17 Simulated cathodic impedance spectrum generated by Eq. (6.14) and considering parameters from Table 6.5. Simulation of cathodic impedance spectrum (Sim cathode); cathode catalyst layer with no oxygen diffusion impedance (Sim CCL no O_2 loss); oxygen diffusion impedance in the cathode catalyst layer (Sim O_2 loss in CCL); impedance of the gas diffusion layer and channel (Sim GDL-channel).

this approach represents an oversimplification for the simulation of the impedance spectra, it will be possible to qualitatively simulate the change (increase/decrease) of the magnitude of the impedance spectra of the cathode, CCL and GDL-channel with changing PEMFC operating conditions.

Increasing the temperature of the PEMFC improves the kinetics of the ORR, improves proton conduction in the PEM, improves the transport of oxygen in the GDL and CCL, and yields a reduction in the size of the impedance spectrum [27]. It is possible to qualitatively simulate the effect of increasing/decreasing PEMFC temperature on the cathodic impedance spectrum by considering as a baseline the parameters shown in Table 6.4 in the cathodic impedance model represented in Eq. (6.14). Iranzo et al. [28] estimated the time constants of electro-diffusion processes in a PEMFC using EIS measurements. Iranzo et al. [28] reported an empirical equation to estimate the effective diffusion coefficient of the GDL as a function of the GDL porosity, operating temperature, and operating pressure of the PEMFC. The empirical equation reported by Iranzo et al. [28] neglecting the effect of reactant pressure on the diffusion coefficient of the GDL is:

$$D_{\mathrm{GDL}} = D_{\mathrm{r}}\left[\frac{\varepsilon - 0.11}{1 - 0.11}\right]^{0.785}[1 - w_{\mathrm{s}}]^2\left[\frac{T_{\mathrm{cell}}}{T_{\mathrm{r}}}\right]^{1.5} \tag{6.28}$$

where ε is the porosity of the GDL, T_{r} is the reference temperature of air entering the cathode, D_{r} is the diffusion coefficient at the reference temperature, T_{cell} is the temperature of the PEMFC, and w_{s} is the water saturation (volume fraction of liquid water). It is possible to incorporate Eq. (6.28) in the cathodic impedance model expressed in Eq. (6.14) to simulate the effect of PEMFC temperature on the diffusion impedance in the cathode. As the parameters shown in Table 6.4 are considered a baseline, it is necessary to find the values of the parameters D_r and w_s in Eq. (6.28) to calculate the same diffusion coefficient of the GDL expressed in Table 6.4.

MATLAB exercise. Estimate parameters D_r and w_s to calculate $D_{GDL}=2.0604\times10^{-7}\mathrm{m}^2/\mathrm{s}$ using Eq. (6.28).

Solution. The porosity of the GDL TGP-H-030 is $\varepsilon=0.646$ [29]. The reference temperature of air entering the cathode and temperature of the PEMFC from Table 6.4 are $T_r=T_a=37\,^\circ\mathrm{C}$ and $T_{cell}=50\,^\circ\mathrm{C}$. The script in MATLAB software to estimate the oxygen diffusion coefficient as a function of temperature of the PEMFC is shown in Fig. 6.18.

The parameters $D_r=2.9443\times10^{-7}\mathrm{m}^2/\mathrm{s}$ and $w_s=0.0102$ from Eq. (6.28) have been estimated through a fitting process in MATLAB software by considering $T_r=37\,^\circ\mathrm{C}$ and $T_{cell}=50\,^\circ\mathrm{C}$ as shown in Fig. 6.18. The estimated parameters $D_r=2.9443\times10^{-7}\mathrm{m}^2/\mathrm{s}$ and $w_s=0.0102$ considering $T_r=37\,^\circ\mathrm{C}$ and $T_{cell}=50\,^\circ\mathrm{C}$ in Eq. (6.28) calculate $D_{GDL}=2.0604\times10^{-7}\mathrm{m}^2/\mathrm{s}$ which is the same value of the oxygen diffusion coefficient D_{GDL} expressed in Table 6.4. Once the parameters D_r and w_s in Eq. (6.28) are estimated, it is possible to calculate a new diffusion coefficient of the GDL by increasing/decreasing the PEMFC temperature T_{cell} expressed in Eq. (6.28). An increase in the operating temperature of the PEMFC yields an increase in the effective diffusion coefficient of the GDL. In addition, if the water saturation w_s is increased, a decrease in the oxygen diffusion coefficient results; a saturation value of $w_s=0.5$ reduces the oxygen diffusion coefficient by one order of magnitude [30]. Table 6.6 shows the considered parameters for the simulation of the cathodic impedance spectrum at different PEMFC temperatures. Eq. (6.24) calculates a change of the Tafel slope from 50 °C (base-line) to a different temperature and considers a factor $\widehat{\alpha}$ related to the deviation of the charge transfer coefficient from the base-line condition. The absolute value of the Tafel slope b_{O_2} decreases with increasing PEMFC temperature [31] as shown in Table 6.6.

Fig. 6.19 shows the effect of increasing the PEMFC temperature from 50°C (baseline) to 70°C and decreasing from 50°C (baseline) to 40°C on the cathodic impedance spectrum. The impedance spectra shown in Fig. 6.19 were simulated by considering negligible ohmic resistance $R_{ohm}=0\ \Omega\mathrm{cm}^2$, a very low CCL proton resistance $R_P=1\times10^{-6}\ \Omega\mathrm{cm}^2$ and the parameters shown in Table 6.6 in the cathodic impedance model (Eq. 6.14). The considered values of the parameters Y and P of the CPE in the cathodic impedance model (Eq. 6.14) were similar to the values shown in Table 6.3. The simulated cathodic impedance spectrum at a PEMFC temperature of 70°C considers air entering the channel at a temperature of 37°C (37 °C is the temperature setting of the cathode gas humidifier), under this condition a 20% RH in the air entering the channel is considered. The diffusion coefficient in the GDL calculated through Eq. (6.28) at a PEMFC temperature of 70°C resulted in $D_{GDL}=2.2546\times10^{-7}\mathrm{m}^2/\mathrm{s}$. A decrease in the impedance responses of the cathode and GDL-channel with

GDL_diffusion_coefficient.m

```
% Reference temperature at channel inlet
Tr=37+273.15;
% PEMFC temperature
Tc=50+273.15;
% Porosity of GDL
Ep=0.646;
% Diffusion coefficient of the GDL from EIS measurements
D_gdl=2.0604e-07;

% Initial conditions for estimation of parameters
x0=[0.01 0.01];
% Function defining empirical equation of GDL diffusion coefficient
f_D=@(x,Ep)(x(1)*(((Ep-0.11)/(1-0.11))^0.785)*((1-x(2))^2)*(Tc/Tr)^1.5);
% Fitting parameters
x = lsqcurvefit(f_D,x0,Ep,D_gdl);
% Estimated GDL diffusion coefficient at reference temperature
Dr=abs(x(1))
% Estimated parameter of water saturation
ws=abs(x(2))

%%% Estimation of diffusion coefficient in GDL at different PEMFC temperature
% PEMFC temperature
Tc=50+273.15;
% Empirical equation of diffusion coefficient in GDL, Eq. 6.28
D_GDL=(Dr*(((Ep-0.11)/(1-0.11))^0.785)*((1-ws)^2)*(Tc/Tr)^1.5)
```

Command Window

```
Dr =

   2.9443e-07

ws =

    0.0102

D_GDL =

   2.0604e-07

>>
```

FIG. 6.18 Script in MATLAB for the estimation of parameters comprising the semiempirical equation of the diffusion coefficient of the gas diffusion layer represented in Eq. (6.28).

increasing PEMFC temperature from 50°C (base-line) to 70°C is observed in Fig. 6.19. The decrease of the impedance spectra is not only attributed to an increase in PEMFC temperature from 50°C (baseline) to 70°C but also to a decrease in the RH of air entering the cathode from 50% (baseline) to 20%.

Fig. 6.19 also shows the effect of decreasing the PEMFC temperature from 50°C (baseline) to 40°C on the cathodic impedance spectrum. The simulated cathodic impedance spectrum at a PEMFC temperature of 40°C also considers air entering the channel at a temperature of 37°C and the RH of air entering the cathode is 85%. The diffusion coefficient in the GDL calculated through Eq. (6.28) at a PEMFC temperature of 40°C resulted in $D_{GDL} = 1.9655 \times 10^{-7} m^2/s$. An increase in the

TABLE 6.6 Parameters considered for the simulation of the cathodic impedance spectrum at different PEMFC temperatures using the cathodic impedance model (Eq. 6.14).

S_{O_2}	i_{cell} A/cm^2	T_a °C	p_a bar	T_{cell} °C	RH %	b_{O_2} V	δ_{GDL} μm	δ_{CCL} μm	D_{CCL} $\times 10^{-9}$m^2/s	D_{GDL} $\times 10^{-7}$m^2/s
2.5	0.11	37	1.7	40	85	0.0955	116.8	47	6.8335	1.9655
2.5	0.11	37	1.7	50	50	0.0934	116.8	47	6.8335	2.0604
2.5	0.11	37	1.7	70	20	0.0891	116.8	47	6.8335	2.2546

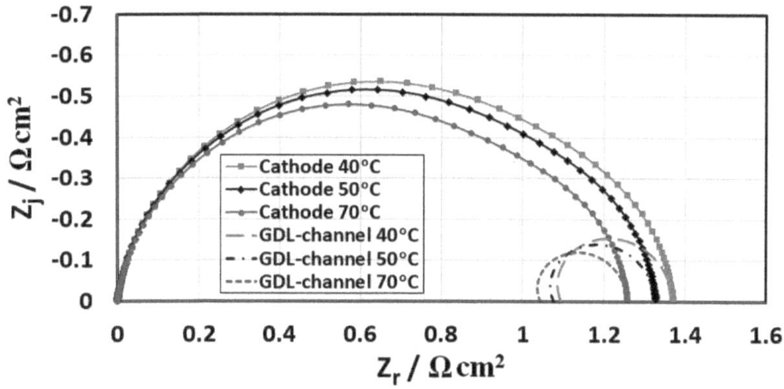

FIG. 6.19 Simulated effect of PEMFC temperature on the cathode impedance response and on the impedance response of the gas diffusion layer and channel (GDL-channel). The ohmic resistance is neglected $R_{ohm}=0$ and proton resistance in the cathode catalyst layer $R_P=1\times10^{-6}$ Ωcm^2 is considered very low in the simulated results generated by the cathodic impedance model, Eq. (6.14).

impedance responses of the cathode and GDL-channel with decreasing PEMFC temperature from 50°C (baseline) to 40°C is observed in Fig. 6.19. The increase of the impedance spectra is not only attributed to a decrease in PEMFC temperature from 50°C (baseline) to 40°C but also to an increase in the RH of air entering the cathode from 50% (baseline) to 85%. Previously in Chapter 5 Section 5.8, it was demonstrated that increasing the RH in the air entering the channel increases the magnitude of the impedance spectrum of the GDL-channel.

Fig. 6.20 shows the effect of PEMFC temperature on the CCL impedance spectrum considering fast (no loss) oxygen transport in the CCL. Fig. 6.20

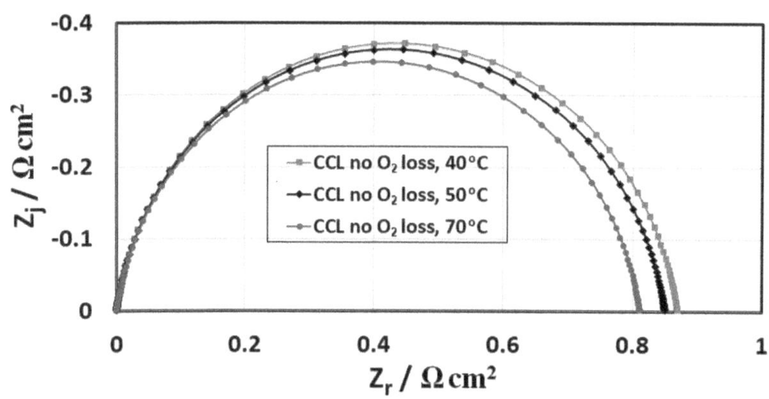

FIG. 6.20 Simulated effect of PEMFC temperature on the cathode catalyst layer (CCL) impedance response considering fast (no loss) oxygen transport. The ohmic resistance is neglected $R_{ohm}=0$ Ωcm^2 and proton resistance in the CCL $R_P=1\times10^{-6}$ Ωcm^2 is considered very low in the simulated results generated by the CCL impedance model (Eq. 6.7).

represents the impedance spectra of the CCL extracted from the cathodic impedance spectra shown in Fig. 6.19. The impedance spectra of the CCL shown in Fig. 6.20 were simulated using Eq. (6.7) and considering $Z_{O_2\_CCL}(y^*)=0$ where Eq. (6.7) comprises the charge transfer resistance $R_C(y^*)$. The diameter of the impedance spectra shown in Fig. 6.20 represents the charge transfer resistance which decreases with increasing temperature. The straight-line at high frequencies is not shown in the simulated impedance spectra of the CCL because the value of the proton resistance was considered to be very low $R_P=1\times10^{-6}\ \Omega\text{cm}^2$.

In Figs. 6.19 and 6.20, the effect of PEMFC temperature on the cathodic impedance spectrum had an effect on the GDL-channel impedance spectrum and impedance spectrum of the CCL. The effect of PEMFC temperature on the cathodic impedance spectrum at low current density operation could have a greater effect on the impedance of the CCL than on the GDL-channel impedance because the contribution of the GDL-channel impedance can be considered negligible at low current densities as demonstrated in Fig. 6.17. In Chapter 3 Section 3.9, the errors in interpreting EIS measurements were discussed. Boillat et al. [32] reported that the charge transfer resistance and activation overpotential become fully uncorrelated in EIS measurements carried out in the region where the Tafel approximation of the Butler-Volmer equation is valid. Therefore, the charge transfer resistance estimated from EIS measurements at a high current density cannot be considered to measure activation losses during the ORR. The effect of temperature on the charge transfer resistance to measure ORR activity in the CCL could be analysed in EIS measurements of PEMFCs carried out at low current density operation (near equilibrium conditions). In Chapter 3 Section 3.8.3, the effect of PEMFC temperature on EIS measurements was discussed. The EIS measurements shown in Fig. 3.27 from Chapter 3 were carried out in a PEMFC operated at low current density. A negligible effect of oxygen diffusion on the EIS measurements shown in Fig. 3.27 from Chapter 3 can be considered. Increasing PEMFC temperature decreased the magnitude of the impedance spectrum as shown in Fig. 3.27 from Chapter 3.

In the simulated cathodic impedance spectra shown in Fig. 6.19, the diffusion impedance of the GDL-channel is masked with the impedance of the CCL. Hence a single semicircle is observed in the cathodic impedance spectrum as shown in Fig. 6.19. This single semicircle feature in the cathodic impedance spectrum is attributed to a low value of the oxygen diffusion time constant in the GDL. The oxygen diffusion time constant in the GDL $\tau_{O_2\_GDL}=\delta_{GDL}{}^2/D_{GDL}$ defined in Eq. (5.2) from Chapter 5 related to the cathodic impedance spectra shown in Fig. 6.19 results in an order of magnitude of 10^{-2}s. For instance, at a PEMFC temperature of 70°C with $D_{GDL}=2.2546\times10^{-7}\text{m}^2/\text{s}$ and $\delta_{GDL}=116.8\times10^{-6}$ m, the oxygen diffusion time constant results in $\tau_{O_2\_GDL}=0.0605$ s. The separation between the charge transfer resistance during the ORR and oxygen diffusion resistance on PEMFC-EIS measurements represented through the diameter of two individual semicircles in the Nyquist plot has been reported when the order of magnitude of the oxygen diffusion time constant in the GDL was 10^{-1}s [16,33]. The oxygen diffusion time constant $\tau_{O_2\_GDL}$ increases with increasing oxygen diffusion distance δ_{GDL} in the GDL. The oxygen diffusion distance will depend on properties of the GDL such as permeability, porosity, tortuosity and hydrophobicity and operating conditions. Fig. 5.13 from Chapter 5 shows that oxygen diffusion below the channel rib yields an increase in the oxygen diffusion distance across the GDL.

The effect of increasing/decreasing PEMFC temperature can also be manifested in the ohmic resistance of the PEMFC. The ohmic resistance represents the opposition to flow of protons across the PEM and the opposition to flow of electrons in the GDL and FFP. The effect of the ohmic resistance on the cathodic impedance spectrum with increasing or decreasing PEMFC temperature was not simulated. The value of the ohmic resistance in the cathodic impedance model (Eq. 6.14) shifts the simulated cathodic impedance spectrum along the real axis Z_r of the Nyquist plot. Increasing the operating temperature of the PEMFC increases the proton conductivity in the PEM and reduces the ohmic resistance. Nevertheless, a severe increase in the heat generated during PEMFC operation could dehydrate the PEM and increase proton resistance [34]. An empirical equation has been reported by Mann et al. [35] to calculate the resistivity of a Nafion membrane. The empirical equation expressed in Eq. (6.29) could be implemented with the cathodic impedance model expressed in Eq. (6.14) to simulate the cathodic impedance spectrum from a PEMFC considering a Nafion membrane and operated at different temperatures.

$$r_M = \frac{181.6\left[1 + 0.03\left[\frac{I_{cell}}{A}\right] + 0.062\left[\frac{T_{cell}}{303}\right]^2\left[\frac{I_{cell}}{A}\right]^{2.5}\right]}{\left[\lambda - 0.634 - 3\left[\frac{I_{cell}}{A}\right]\right]\exp\left(4.18\left[\frac{T_{cell}-303}{T_{cell}}\right]\right)} \tag{6.29}$$

where A is the cell active area and λ is a fitted parameter influenced by the membrane preparation procedure, RH and stoichiometry of the hydrogen in the anode. The units of Eq. (6.29) are Ωcm. Note that Eq. (6.29) has been considered in Chapter 2 for the simulation of the polarisation curve shown in Fig. 2.8 considering the steady-state PEMFC model reported by Correa et al. [36].

6.6 Simulation of the cathodic impedance spectrum at different oxygen stoichiometry

Oxygen transport limitations can be reduced with increasing oxygen stoichiometry during PEMFC operation. It has been considered that optimal conditions during PEMFC operation can be achieved when the PEMFC is operated with an air stoichiometry of 2.5 and 3 [27]. A high oxygen stoichiometry during PEMFC operation homogenises the profile of the current distribution along the length of the channel of the FFP. The effect of oxygen stoichiometry on the current distribution along the channel length has been discussed in Section 5.5.3 from Chapter 5. The effect of oxygen stoichiometry on the cathodic impedance spectrum can be simulated using the cathodic impedance model represented in Eq. (6.14). The parameters from Table 6.4 are considered a baseline and were estimated from cathodic EIS measurements carried out in a PEMFC operated with an oxygen stoichiometry of $S_{O_2}=2.5$. The considered values of the parameters Y and P of the CPE in the cathodic impedance model (Eq. 6.14) were similar to the values shown in Table 6.3. The ohmic resistance is neglected $R_{ohm}=0\ \Omega cm^2$ and the proton resistance in the CCL $R_P=1\times10^{-6}$ Ωcm^2 is considered very low in the simulated results generated by the cathodic impedance model, Eq. (6.14). The simulation of the impedance response of the cathode at low oxygen stoichiometry $S_{O_2}<2$ can be achieved by considering Eq. (6.15) in the third term on the right-hand side of Eq. (6.14). Eq. (6.15) considers the integral of the local impedance along the channel length. In addition, a local Tafel slope b_{O_2} and local oxygen diffusion coefficient of the GDL along the length of the channel would be required in the cathodic model [37] to simulate the cathodic impedance response at low oxygen stoichiometry $S_{O_2}<2$. In this section, a qualitative analysis and simulation of the cathodic impedance response at low oxygen stoichiometry is carried out. Therefore, at $S_{O_2}<2$, Eq. (6.12) $Z_{CCL+GDL-ch}(y^*)$ instead of Eq. (6.15) $Z_{T\_CCL+GDL-ch}$ will be considered in the cathodic impedance model (Eq. 6.14) as a first approximation. This approximation will be considered to reduce the computational effort during the simulation. The local impedance of the cathode considers the position y^* of the channel length where the current density distribution is equal to the total current density of the PEMFC $i_F(y^*)=i_{cell}$. The current density distribution as a function of the oxygen stoichiometry is represented in Eq. (5.44) from Chapter 5. At different oxygen stoichiometries the position y^* of the channel length for $i_F(y^*)=i_{cell}$ will change. The position y^* with changing oxygen stoichiometry has been considered in the cathodic impedance model (Eq. 6.14). For instance, at a stoichiometry of $S_{O_2}=1.5$ the local current density is equal to the cell current density considering $y^*=0.45\ell_c$ as such $i_F(0.45\ell_c)=0.11\ \text{A/cm}^2$. The Tafel slope, oxygen diffusion coefficient and oxygen diffusion distance in the CCL and GDL are all considered to be constant along the length of the channel and during the simulations of the cathodic impedance response at low and high oxygen stoichiometry.

Fig. 6.21 shows the effect of increasing oxygen stoichiometry $S_{O_2}=4$ and decreasing oxygen stoichiometry $S_{O_2}=1.5$ on the cathodic impedance spectrum with respect to the baseline of $S_{O_2}=2.5$. The cathodic impedance spectrum decreases with increasing oxygen stoichiometry and vice versa. Fig. 6.21 also shows the effect of increasing and decreasing oxygen stoichiometry on the CCL impedance spectrum. The simulated impedance spectrum of the CCL shown in Fig. 6.21 considers oxygen diffusion impedance and also decreases with increasing oxygen stoichiometry and vice versa. The impedance spectrum of the CCL is simulated using Eq. (6.7). The oxygen diffusion coefficient in the

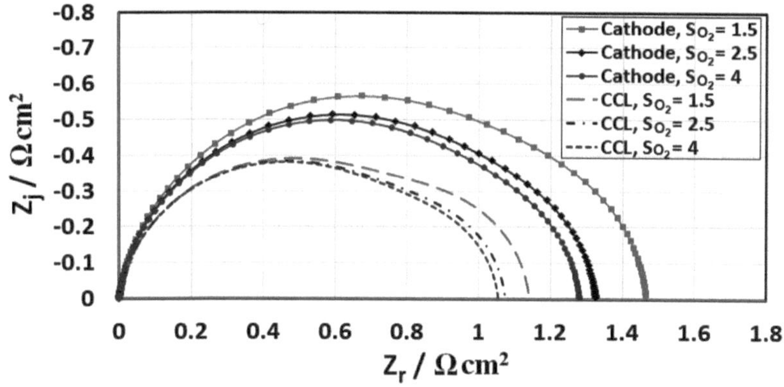

FIG. 6.21 Simulated effect of oxygen stoichiometry on the impedance spectrum of the cathode and on the impedance spectrum of the cathode catalyst layer (CCL). The impedance spectrum of the CCL considers charge transfer resistance and oxygen diffusion impedance. The ohmic resistance is neglected $R_{\text{ohm}} = 0\ \Omega\text{cm}^2$ and proton resistance in the CCL $R_P = 1 \times 10^{-6}\ \Omega\text{cm}^2$ is considered very low in the simulated results generated by the cathodic impedance model, Eq. (6.14). The parameters shown in Table 6.4 are considered a baseline.

GDL was considered constant $D_{\text{GDL}} = 2.0604 \times 10^{-7}\text{m}^2/\text{s}$ in the simulation results shown in Fig. 6.21. Wang and Nguyen [38] reported an increase in the liquid saturation level in the GDL with decreasing oxygen stoichiometry. The effect of oxygen stoichiometry influences the water saturation w_s on the diffusion coefficient of the GDL expressed in Eq. (6.28). Therefore, a decrease in the diffusion coefficient in the GDL with decreasing oxygen stoichiometry is expected. The decrease in oxygen diffusion coefficient in the GDL with decreasing oxygen stoichiometry has been demonstrated in Table 5.2 from Chapter 5. The effect of oxygen stoichiometry on oxygen diffusion coefficient was not considered in the simulation results shown in Fig. 6.21. A bigger effect of oxygen stoichiometry on the simulated impedance spectra shown in Fig. 6.21, (e.g. greater increase in magnitude of the impedance spectra with decreasing oxygen stoichiometry), would have resulted if the oxygen diffusion coefficient in the GDL, which was considered in the cathodic impedance model, was also influenced by the oxygen stoichiometry.

The difference between the diameter of the impedance spectra of the cathode and CCL shown in Fig. 6.21 represents the diameter of the impedance spectra of the GDL-channel as shown in Fig. 6.22. The simulated impedance spectra of the GDL-channel are shifted along the real component Z_r by the sum of charge transfer resistance and oxygen transport resistance in the CCL, $R_C(y^*) + R_{O_2\_\text{CCL}}(y^*)$. The simulated impedance spectrum of the GDL-channel is generated by Eq. (6.13) and considers steady-state oxygen concentration in the channel $Z_{\text{GDL}-\text{ch}_{ss}}(y^*)$ (Eq. 6.1) and oscillating oxygen concentration in the channel during AC conditions $Z_{\text{GDL}-\text{ch}_{ac}}(y^*)$ (Eq. 6.3). The diameter of the impedance spectrum of the GDL-channel decreases with increasing oxygen

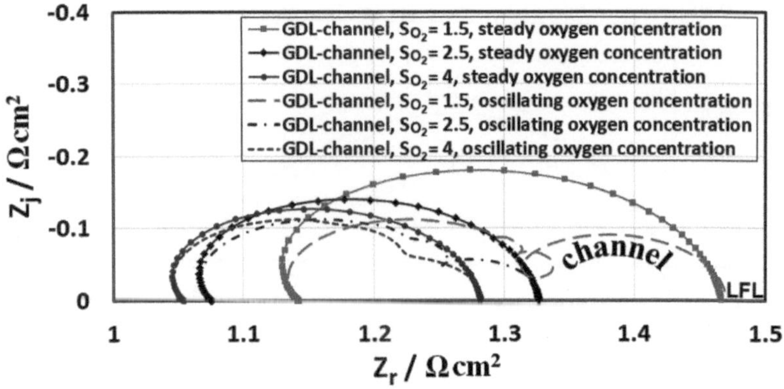

FIG. 6.22 Simulated effect of oxygen stoichiometry on the impedance spectrum of the gas diffusion layer and channel (GDL-channel). The simulated impedance response of the GDL-channel considers steady-state oxygen concentration in the channel and oscillating oxygen concentration in the channel during AC conditions. The simulated impedance spectra of the GDL-channel are shifted along the real component Z_r by the sum of charge transfer resistance and oxygen transport resistance in the CCL, $R_C(y^*)+R_{O_2\_CCL}(y^*)$. The parameters shown in Table 6.4 are considered a baseline in Eq. (6.13).

stoichiometry. The separation of the GDL and channel impedance responses can be simulated through the impedance model of the GDL-channel considering oscillating oxygen concentration in the channel during AC conditions $Z_{GDL-ch_{ac}}(y^*)$ (Eq. 6.3). The contribution of the impedance response of the channel manifested by a second semicircle at the lowest frequencies in the overall impedance spectrum of the GDL-channel is reduced with increasing oxygen stoichiometry. The increase of the impedance response of the channel at low stoichiometry can be related to an increase of the amplitude of the oscillating oxygen concentration along the channel during AC conditions [39]. In Chapter 5, Fig. 5.15, it was demonstrated that there is a difference in the value of the real component Z_r at the low frequency limit (LFL) $\omega \to 0$ between the impedance spectra of the GDL-channel considering steady-state oxygen concentration in the channel and oscillating oxygen concentration in the channel during AC conditions. The difference was attributed to the effect of the amplitude of the sinusoidal current density on the impedance spectrum of the GDL-channel considering oscillating oxygen concentration in the channel during AC conditions. The simulation results shown in Fig. 6.22 generated by the impedance model of the GDL-channel Eq. (6.13) with Eq. (6.3) $Z_{GDL-ch_{ac}}(y^*)$ considered the amplitude of the oscillating contribution of the sinusoidal current density with a value equal to 5% of the DC current density. Fig. 6.22 shows a negligible difference of the real component Z_r in the LFL between the impedance spectra of the GDL-channel considering steady-state oxygen concentration in the channel and oscillating oxygen concentration in the channel during AC conditions. In the next section, the effect of the amplitude of the oscillating contribution of the sinusoidal current density on the impedance spectrum of the GDL-channel at high current densities will be discussed.

6.7 Simulation of the cathodic impedance spectrum at high current density

The effect of the oxygen stoichiometry on the cathodic impedance spectrum can be more noticeable at high current density operation. Fig. 6.23 shows the simulated impedance spectrum of the cathode at a current density of $i_{cell}=0.5$ A/cm^2 considering the parameters shown in Table 6.4 as a baseline in Eq. (6.14). The ohmic resistance is neglected $R_{ohm}=0$ Ωcm^2 and the proton resistance in the CCL $R_P=1\times10^{-6}$ Ωcm^2 is considered very low in the simulated results generated by the cathodic impedance model, Eq. (6.14). The considered values of the parameters Y and P of the CPE in the cathodic impedance model (Eq. 6.14) were similar to the values shown in Table 6.3. The simulated impedance spectra shown in Fig. 6.23 at a current density of $i_{cell}=0.5$ A/cm^2 and oxygen stoichiometry of $S_{O_2}=2.5$ can be compared with the simulated impedance spectra shown in Fig. 6.11 at $i_{cell}=0.11$ A/cm^2 and $S_{O_2}=2.5$. The charge transfer resistance represented through the diameter of the impedance spectrum labelled as 'CCL no O_2 loss' decreases with increasing current density as shown in the simulated impedance spectra from Figs 6.11 and 6.23. The oxygen diffusion impedance in the CCL represented through the impedance spectrum labelled as 'O_2 loss in CCL' increases with increasing current density. The impedance spectrum of the GDL-channel also increases with increasing current density, the simulated results are generated by Eq. (6.13) and considers steady-state oxygen concentration in the channel $Z_{GDL-ch_{ss}}(y^*)$

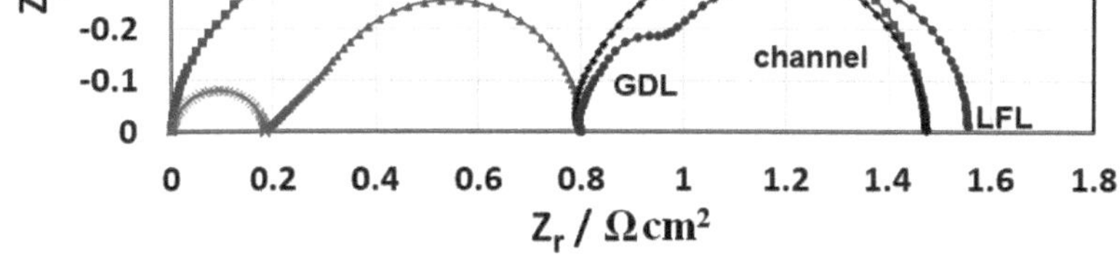

FIG. 6.23 Simulated cathodic impedance response considering a current density of $i_{cell}=0.5$ A/cm^2 and oxygen stoichiometry of $S_{O_2}=2.5$. The ohmic resistance is neglected $R_{ohm}=0$ Ωcm^2 and proton resistance in the cathode catalyst layer $R_P=1\times10^{-6}$ Ωcm^2 is considered very low in the simulated results generated by the cathodic impedance model, Eq. (6.14). The parameters shown in Table 6.4 are considered a baseline.

(Eq. 6.1) and oscillating oxygen concentration in the channel during AC conditions $Z_{GDL-ch_{ac}}(y^*)$ (Eq. 6.3). Fig. 6.23 shows that the impedance spectrum of the GDL-channel considering oscillating oxygen concentration in the channel during AC conditions, labelled as 'GDL-ch, oscillating O_2', results in a higher value of the real component Z_r at the LFL at $\omega \rightarrow 0$ than the LFL value from the impedance spectrum of the GDL-channel considering steady-state oxygen concentration in the channel labelled as 'GDL-ch, steady O_2'. This increase of the LFL value is attributed to the effect of the amplitude of the oscillating contribution of the sinusoidal current density which drifts the LFL value away from its steady-state value. The impedance spectrum labelled as 'GDL-ch, oscillating O_2' shown in Fig. 6.23 and generated by Eq. (6.13) and $Z_{GDL-ch_{ac}}(y^*)$ (Eq. 6.3) considered the amplitude of the oscillating contribution of the sinusoidal current density with a value equal to 5% of the DC current density. The effect of the amplitude of the oscillating contribution of the sinusoidal current density has a higher effect at high current density as demonstrated in the increase of the second semicircle of the impedance spectrum of the GDL-channel attributed to oscillating oxygen concentration in the channel at 0.5 A/cm^2 (Fig. 6.23) with respect to the simulation results at 0.11 A/cm^2 shown in Fig. 6.13 at $S_{O_2}=2.5$. In Chapter 5, Fig. 5.15, it was demonstrated that effect of the amplitude of the oscillating contribution of the sinusoidal current density on the impedance spectrum of the GDL-channel can be reduced by considering a lower amplitude in the oscillating contribution of the sinusoidal current density or by increasing the oxygen stoichiometry.

Fig. 6.24 shows the effect of increasing oxygen stoichiometry on the cathodic impedance spectrum at high current density. Comparing the simulated impedance spectra shown in Figs. 6.23 and 6.24, the diameter of the impedance

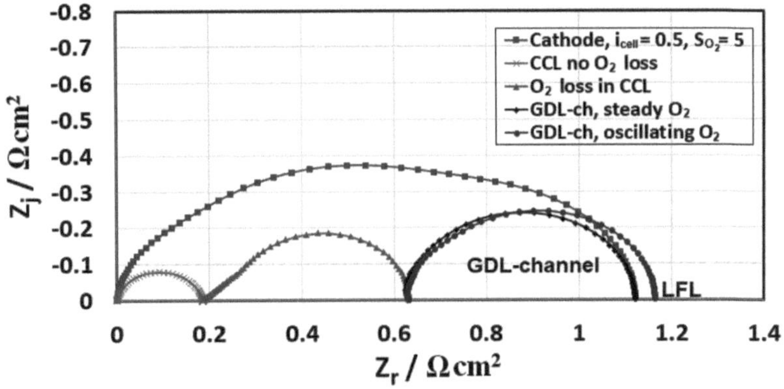

FIG. 6.24 Simulated cathodic impedance response considering a current density of $i_{cell}=0.5$ A/cm^2 and oxygen stoichiometry of $S_{O_2}=5$. The ohmic resistance is neglected $R_{ohm}=0$ Ωcm^2 and proton resistance in the cathode catalyst layer $R_P=1\times10^{-6}$ Ωcm^2 is considered very low in the simulated results generated by the cathodic impedance model, Eq. (6.14). The parameters shown in Table 6.4 are considered a baseline.

spectra is reduced with increasing oxygen stoichiometry $S_{O_2}=5$, but with an exception of the charge transfer resistance (diameter of the impedance spectrum labelled 'CCL no O_2 loss'). The charge transfer resistance remained constant because the considered charge transfer resistance in the mathematical treatment (Eq. 6.11) is not a function of oxygen stoichiometry. Weiß et al. [40] reported that a change in oxygen stoichiometry causes a change in partial pressure of oxygen and hence a change in exchange current density. The difference between the LFL value at $\omega \rightarrow 0$ from the impedance spectra labelled 'GDL-ch, steady O_2' and 'GDL-ch, oscillating O_2', is also reduced with increasing oxygen stoichiometry as shown in Fig. 6.24. In addition, the contribution of the impedance response of the channel on the GDL-channel impedance response represented through a second loop at low frequencies is reduced with increasing oxygen stoichiometry.

References

[1] M. Kim, et al., Effects of anode flooding on the performance degradation of polymer electrolyte membrane fuel cells, J. Power Sources 266 (2014) 332–340.

[2] J.N. Schwämmlein, P.J. Rheinländer, Y. Chen, K.T. Freyer, H.A. Gasteiger, Anode aging during PEMFC start-up and shut-down: H_2-air fronts vs voltage cycles, J. Electrochem. Soc. 165 (2018) F1312–F1322.

[3] R. Makharia, M.F. Mathias, D.R. Baker, Measurement of catalyst layer electrolyte resistance in PEFCs using electrochemical impedance spectroscopy, J. Electrochem. Soc. 152 (2005) A970–A977.

[4] X. Yuan, H. Wang, J.C. Sun, J. Zhang, AC impedance technique in PEM fuel cell diagnosis—a review, Int. J. Hydrog. Energy 32 (2007) 4365–4380.

[5] S.M.R. Niya, M. Hoorfar, Study of proton exchange membrane fuel cells using electrochemical impedance spectroscopy technique—a review, J. Power Sources 240 (2013) 281–293.

[6] S. Asghari, A. Mokmeli, M. Samavati, Study of PEM fuel cell performance by electrochemical impedance spectroscopy, Int. J. Hydrog. Energy 35 (2010) 9283–9290.

[7] S.A. Niknam, M. Mortazavi, A.D. Santamaria, Signature analysis of two-phase flow pressure drop in proton exchange membrane fuel cell flow channels, Result. Eng. 5 (2020), 100071.

[8] A. El-kharouf, R. Steinberger-Wilckens, The effect of clamping pressure on gas diffusion layer performance in polymer electrolyte fuel cells, Fuel Cells 15 (2015) 802–812.

[9] A. Kulikovsky, Why impedance of the gas diffusion layer in a PEM fuel cell differs from the Warburg finite-length impedance? Electrochem. Commun. 84 (2017) 28–31.

[10] MathWorks®. https://uk.mathworks.com/help/optim/ug/lsqcurvefit.html. last access June 2022.

[11] J. Song, M.Z. Bazant, Effects of nanoparticle geometry and size distribution on diffusion impedance of battery electrodes, J. Electrochem. Soc. 160 (2013) A15–A24.

[12] M. Orazem, B. Tribollet, Electrochemical Impedance Spectroscopy, second ed., Wiley, Hoboken, NJ, 2017.

[13] A. Kulikovsky, M. Eikerling, Analytical solutions for impedance of the cathode catalyst layer in PEM fuel cell: layer parameters from impedance spectrum without fitting, J. Electroanal. Chem. 691 (2013) 13–17.

[14] S. Cruz-Manzo, P. Greenwood, Study of oxygen diffusion in the cathode catalyst layer and gas diffusion layer for polymer electrolyte fuel cells with EIS, J. Electroanal. Chem. 892 (2021), 115270.

[15] S. Cruz-Manzo, R. Chen, A generic electrical circuit for performance analysis of the fuel cell cathode catalyst layer through electrochemical impedance spectroscopy, J. Electroanal. Chem. 694 (2013) 45–55.

[16] D. Malevich, E. Halliop, B.A. Peppley, J.G. Pharoah, K. Karan, Investigation of charge-transfer and mass-transport resistances in PEMFCs with microporous layer using electrochemical impedance spectroscopy, J. Electrochem. Soc. 156 (2009) B216–B224.

[17] Z. Tang, et al., Recent progress in the use of electrochemical impedance spectroscopy for the measurement, monitoring, diagnosis and optimization of proton exchange membrane fuel cell performance, J. Power Sources 468 (2020), 228361.

[18] N. Fouquet, C. Doulet, C. Nouillant, D. Dauphin-Tanguy, B. Ould-Bouamama, Model based PEM fuel cell state-of-health monitoring via AC impedance measurements, J. Power Sources 159 (2006) 905–913.

[19] I. Pivac, B. Šimić, F. Barbir, Experimental diagnostics and modeling of inductive phenomena at low frequencies in impedance spectra of proton exchange membrane fuel cells, J. Power Sources 365 (2017) 240–248.

[20] J. Harper, M. Rust, B. Sayers, A. Savage, High Frequency, High Current Impedance Spectroscopy: Experimental Protocols Enabling Measurement up to 1MHz at High Current Densities, Solartron Analytical, Farnborough, 2004. TB/ANALYTICAL/001.

[21] Y. Bultel, K. Wiezell, F. Jaouen, P. Ozil, G. Lindbergh, Investigation of mass transport in gas diffusion layer at the air cathode of a PEMFC, Electrochim. Acta 51 (2005) 474–488.

[22] F. Jaouen, G. Lindbergh, G. Sundholm, Investigation of mass-transport limitations in the solid polymer fuel cell cathode I. Mathematical model, J. Electrochem. Soc. 149 (2002) A437–A447.

[23] A. Kulikovsky, A fast low-current model for impedance of a PEM fuel cell cathode at low air stoichiometry, J. Electrochem. Soc. 164 (2017) F911–F915.

[24] M. Eikerling, Water management in cathode catalyst layers of PEM fuel cells: a structure-based model, J. Electrochem. Soc. 153 (2006) E58–E70.

[25] T. Romero-Castañón, L.G. Arriaga, U. Cano-Castillo, Impedance spectroscopy as a tool in the evaluation of MEA's, J. Power Sources 118 (2003) 179–182.

[26] J.X. Wang, F.A. Uribe, T.E. Springer, J. Zhang, R.R. Adzic, Intrinsic kinetic equation for oxygen reduction reaction in acidic media: the double Tafel slope and fuel cell applications, Faraday Discuss. 140 (2009) 347–362.

[27] X. Yan, M. Hou, L. Sun, D. Liang, Q. Shen, H. Xu, P. Ming, B. Yi, AC impedance characteristics of a 2 kW PEM fuel cell stack under different operating conditions and load changes, Int. J. Hydrog. Energy 32 (2007) 4358–4364.

[28] A. Iranzo, S.J. Navas, F. Rosa, M.R. Berber, Determination of time constants of diffusion and electrochemical processes in polymer electrolyte membrane fuel cells, Energy 221 (2021), 119833.

[29] A. El-kharouf, T.J. Mason, D.J.L. Brett, B.G. Pollet, Ex-situ characterisation of gas diffusion layers for proton exchange membrane fuel cells, J. Power Sources 218 (2012) 393–404.

[30] J.H. Nam, M. Kaviany, Effective diffusivity and water-saturation distribution in single- and two-layer PEMFC diffusion medium, Int. J. Heat Mass Transf. 46 (2003) 4595–4611.

[31] L. Xing, Y. Xu, Ž. Penga, Q. Xu, H. Su, F. Barbir, A segmented fuel cell unit with functionally graded distributions of platinum loading and operating temperature, Chem. Eng. J. 406 (2021) 126889.

[32] P. Boillat, F.N. Büchi, L. Gubler, T.J. Schmidt, Use and misuse of electrochemical impedance spectroscopy (EIS) in fuel cell research, in: ECS meeting abstracts, MA2019-02, 2019, p. 1435.

[33] M.A. Rubio, A. Urquia, R. Kuhn, S. Dormido, Electrochemical parameter estimation in operating proton exchange membrane fuel cells, J. Power Sources 183 (2008) 118–125.

[34] T. Ous, C. Arcoumanis, The formation of water droplets in an air-breathing PEMFC, Int. J. Hydrog. Energy 34 (2009) 3476–3487.

[35] R.F. Mann, J.C. Amphlett, M.A.I. Hooper, H.M. Jensen, B.A. Peppley, P.R. Roberge, Development and application of a generalised steady-state electrochemical model for a PEM fuel cell, J. Power Sources 86 (2000) 173–180.

[36] J.M. Correa, F.A. Farret, L.N. Canha, An analysis of the dynamic performance of proton exchange membrane fuel cells using an electrochemical model, in: IECON'01. 27th Annual Conference of the IEEE Industrial Electronics Society (Cat. No.37243), Denver, CO, USA, vol. 1, 2001, pp. 141–146.

[37] T. Reshetenko, A. Kulikovsky, A model for local impedance: validation of the model for local parameters recovery from a single spectrum of PEM fuel cell, J. Electrochem. Soc. 166 (2019) F431–F439.

[38] X. Wang, T.V. Nguyen, An experimental study of the liquid water saturation level in the cathode gas diffusion layer of a PEM fuel cell, J. Power Sources 197 (2012) 50–56.

[39] G. Maranzana, J. Mainka, O. Lottin, J. Dillet, A. Lamibrac, A. Thomas, S. Didierjean, A proton exchange membrane fuel cell impedance model taking into account convection along the air channel: on the bias between the low frequency limit of the impedance and the slope of the polarization curve, Electrochim. Acta 83 (2012) 13–27.

[40] A. Weiß, S. Schindler, S. Galbiati, M.A. Danzer, R. Zeis, Distribution of relaxation times analysis of high-temperature PEM fuel cell impedance spectra, Electrochim. Acta 230 (2017) 391–398.

Chapter 7

Impedance analysis on the individual cells of a PEMFC stack

7.1 EIS in PEMFC stacks

Electrochemical impedance spectroscopy (EIS) is a powerful technique that can be applied in situ for diagnosis of the performance of polymer electrolyte membrane fuel cell (PEMFC) stacks. EIS measurements carried out in PEMFC stacks allow the study of the factors that can yield a reduction in performance during PEMFC stack operation. The state of health in PEMFC stacks has been monitored through EIS measurements [1–5]. The impedance response of the whole PEMFC stack does not provide information about the factors that limit the performance in the individual cells of the stack. Not much work has been done in the study of the impedance response of individual cells within a large PEMFC stack, and the results have clearly shown that the impedance response of the cells changes at different locations in the stack. The change in impedance of individual cells within the stack could be attributed to the nonuniform distribution of current along the PEMFC stack. Therefore, in order to improve the performance of a PEMFC stack, it is important to know and to optimise the current distribution within each cell. The current distribution in the individual cells of a PEMFC stack has been analysed through magnetic field measurements [6].

Much of the work conducted in fuel cell research has limited the EIS technique to a single cell due to the requirement of setting up of a high-cost power source and limitations in the experimental set-up, for example some commercial load banks operating at high currents do not have good frequency response. Commercial PEMFC stacks are equipped with a control system for safe operation and optimal performance. It has been suggested [3] that the effect of all system components on the PEMFC stack must be taken into consideration during EIS measurements. The disadvantage of this method is that inaccurate quantitative data at high frequencies can be obtained by intrusive signals from the system components. Cimenti et al. [7] reported that if a shunt resistor is

Electrochemical Phenomena in the Cathode Impedance Spectrum of PEM Fuel Cells.
https://doi.org/10.1016/B978-0-323-90607-4.00006-9
Copyright © 2022 Elsevier Inc. All rights reserved.

connected with a PEMFC stack to measure the current, a capacitive arc at high frequencies (>1 kHz) in EIS measurements is observed. They concluded that the high frequency arc is likely an experimental artefact caused by neglecting the frequency dependence of the shunt resistor. Some works have limited the PEMFC-EIS study to 1 kHz to avoid complications from such high frequency external artefacts [8]. Yuan et al. [5] carried out EIS measurements in a 6-cell H_2/air PEMFC stack with an active area of 280 cm^2 as shown in Fig. 7.1. The range of frequencies was from 20 kHz down to 0.1 Hz. The EIS measurements were carried out at different currents and featured a bigger loop from 1 kHz to 0.1 Hz which decreased with increasing current from 10 to 70 A. The reduction of the loop was attributed to a decrease of the charge transfer resistance during the redox reaction with increasing current. Thereafter, the bigger loop increased from 90 to 190 A and was attributed to an increase in oxygen transport limitations. The small loop from 20 to 1 kHz was attributed to an artefact of the measurement system because the considered load does not have an accurate response to the sinusoidal perturbation at high frequencies.

The measurement of the individual cells of a PEMFC stack could provide an insight into the nonhomogeneous performance of the individual cells across the PEMFC stack [9,10]. Diard et al. [11] reported a method to estimate the impedance of the individual cells of a 4-cell PEMFC stack. The method is able to calculate the impedance of the individual cells assuming that the reference electrode of the measurement system can be located in the graphite plates of different cells of the stack. The use of a multichannel frequency response analyser (FRA) with control of the DC current allows the measurement of the simultaneous impedance response of the individual cells and whole stack

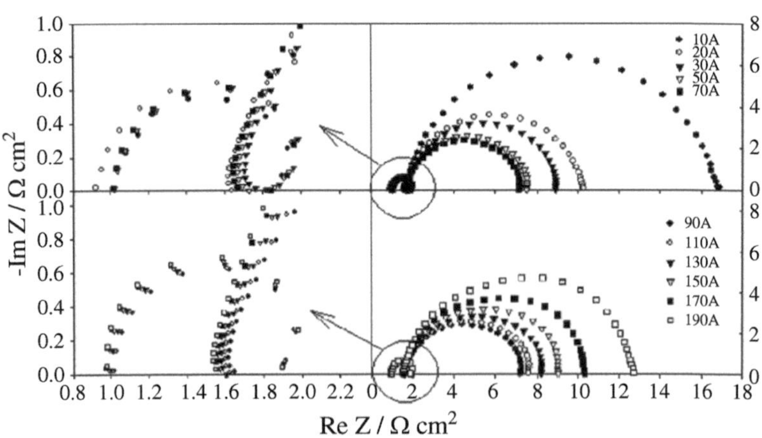

FIG. 7.1 EIS measurements carried out in a 6-cell PEMFC stack with active area of 280 cm^2. *(Reproduced from X. Yuan, J.C. Sun, M. Blanco, H. Wang, J. Zhang, D.P. Wilkinson, AC impedance diagnosis of a 500W PEM fuel cell stack, part I: stack impedance, J. Power Sources 161, (2006) 922, with permission of Elsevier. Copyright Elsevier.)*

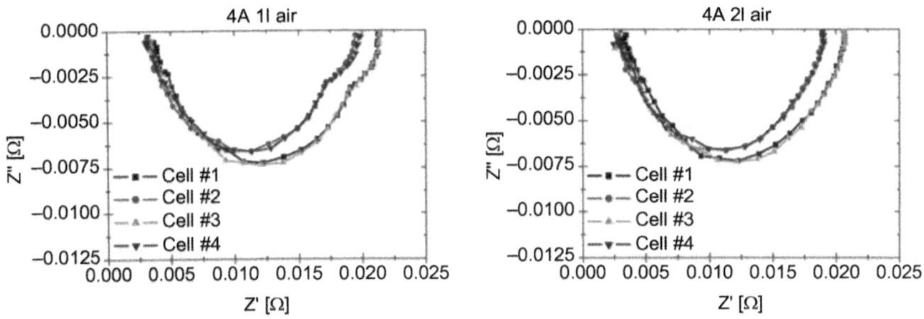

FIG. 7.2 EIS measurements carried out in the individual cells of a PEMFC stack. The PEMFC stack consisted of four cells with active area of 53 cm^2 and was operated at 4 A and at different air flow rate, left: 1 L/min, right: 2 L/min. *(Reproduced from A. Hakenjos, M. Zobel, J. Clausnitzer, C. Hebling, Simultaneous electrochemical impedance spectroscopy of single cells in a PEM fuel cell stack, J. Power Sources 154 (2006) 362, with permission of Elsevier. Copyright Elsevier.)*

[9,12]. Fig. 7.2 shows EIS measurements carried out in the individual cells of a 4-cell PEMFC stack at two different air flow rates [9]. At an air flow rate of 1 L/min, an overlapped small semicircle at low frequencies can be observed which can be attributed to oxygen transport limitations. The semicircle at low frequencies disappears with increasing air flow rate (2 L/min). Fig. 7.2 shows a difference in the impedance spectra of the cells which can be attributed to a different cell performance in the PEMFC stack. Zhang et al. [13] studied the inconsistent performance of the individual cells of a 30-cell PEMFC stack through EIS measurements and concluded that the operating strategy of the PEMFC stack plays an important role to protect vulnerable cells, avoid severe fault states, and reduce inconsistency in cell performance, and improve the durability of the PEMFC stack.

7.2 Analysis of EIS measurements featuring an inductive loop at low frequencies

EIS measurements with positive imaginary components at low frequencies are coined inductive loops. Inductive loops at low frequencies have been attributed to side reactions with intermediate species [14], carbon monoxide poisoning [15], and water transport characteristics [16]. A literature review on low-frequency inductive loops of EIS measurements in PEMFCs has been carried out by Pivac and Barbir [17]. EIS measurements carried out in the individual cells comprising a PEMFC stack can feature inductive loops at low frequencies. Fig. 7.3 shows EIS measurements carried out in the individual cells of a PEMFC stack from a Ballard Nexa PEMFC system [3]. The impedance spectra of the individual cells were measured at 10 A DC and comprise inductive loops at low frequencies. The impedance spectrum of the cell near the air inlet (cell 1) does not present an inductive loop. Dale et al. [3] attributed the formation of the

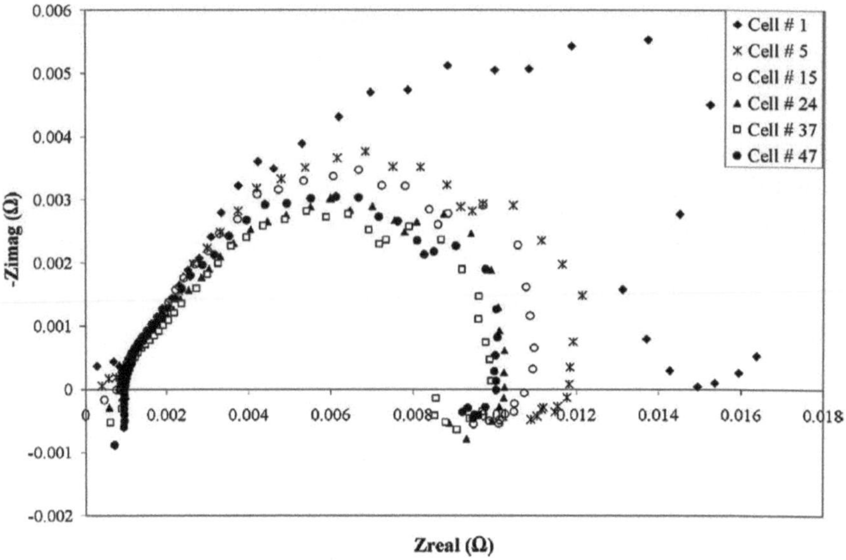

FIG. 7.3 EIS measurements carried out in the individual cells of a PEMFC stack from a Ballard Nexa PEMFC system at 10 A DC. *(Reproduced from N.V. Dale, M.D. Mann, H. Salehfar, A.M. Dhirde, T. Han, AC impedance study of a proton exchange membrane fuel cell stack under various loading conditions, J. Fuel Cell Sci. Technol. 7 (2010) 031010-8, with permission of ASME. All rights reserved.)*

low-frequency inductive loops in the impedance spectra to side reactions with intermediate species during the PEMFC reactions.

If the EIS measurements of the individual cells of a PEMFC stack do not feature inductive loops at low frequencies and the contribution of the anode in the PEMFC impedance response is neglected; then it would be possible to apply the cathodic modelling framework presented previously in Chapter 6 with the EIS measurements to estimate the contribution of the cathode catalyst layer (CCL), gas diffusion layer (GDL) and air channels in the overall impedance spectrum of the PEMFC. The application of the cathodic modelling architecture presented previously in Chapter 6 with individual EIS measurements of the cells comprising a PEMFC stack (impedance spectra with no inductive loops at low frequencies and neglecting anode contribution) would allow the assessment of the nonhomogeneous cathodic performance of a PEMFC stack.

To date, there is no a generic physics-based impedance model developed from fundamental electrode theory considering the electrochemical phenomena in the different layers comprising the PEMFC and considering all the possible phenomena that yield the formation of the inductive loops in the impedance spectrum at low frequencies. Although different physics-based impedance models have been proposed [18–21] and applied with EIS measurements in PEMFCs to relate the low-frequency inductive loops with different processes of the PEMFC, the use of equivalent electrical circuits [22–27] is still an acceptable method. These estimate the charge transfer resistance during the redox reaction and the mass transport resistance from EIS measurements of PEMFCs featuring inductive loops at low frequencies. In this chapter, an analysis of the charge transfer resistance and mass transport resistance in the EIS measurements from the individual cells of an open-cathode PEMFC stack featuring low-frequency inductive loops will be carried out through the application of three electrical circuits reported in the literature [25–27] with the EIS measurements.

7.2.1 EIS measurements in an open-cathode PEMFC stack

Open-cathode PEMFCs are fed with air from ambient conditions through forced convection and the use of fans or blowers. Open-cathode PEMFCs require a simplified balance of plant in which cooling, humidification and air circulation subsystems are not required. The performance of open-cathode PEMFCs can be affected at different ambient temperatures, for example, at winter low subzero air temperatures and summer high relative humidities (RHs). Open-cathode PEMFCs can also be exposed under a range of atmospheric compositions which can include sulphur dioxide, nitrous oxides, and ionic contamination, as well as fuel impurities that can cause irreversible damage to the PEMFC [28]. Other critical contaminants at which open-cathode PEMFCs may be exposed are from battlefield gases in military application [29]. Uniform current density in PEMFC stacks can be achieved at a high stoichiometry of gases, but for open-cathode PEMFCs with dry inlet air, this can be a challenge. Kurnia et al. [30] discussed the progress carried out in open-cathode PEMFCs and point-out the advantages and disavtangaes from open-cathode PEMFCs and closed-cathode PEMFCs. In addition, the performance, design, challenges and future directions of open-cathode PEMFCs are reported in the study of Kurnia et al. [30].

In this chapter, EIS measurements carried out in a commercial 4-cell open-cathode PEMFC stack with an active area of $16\,cm^2$ are considered. The membrane electrode assemblies (MEAs) were made of Nafion 211 with platinum loadings of $0.4\,mg/cm^2$ and carbon black for the electrodes. The GDLs were made of carbon felt with $200\,\mu m$ width. Whilst the flow field channels have a thickness of 5 mm and are made of FU4369 HT material. Two 5 V DC fans for air supply and cooling are attached to the open-cathode PEMFC stack. The open-cathode PEMFC stack was operated at ambient temperature 22°C and considered a through-flow mode on the hydrogen side. Flow rate of hydrogen in the PEMFC stack was kept constant during all the experiments with a stoichiometry of 2. The hydrogen supplied was dry. Simultaneous EIS measurements can be carried out in the 4-cell open-cathode stack through the use of a multichannel FRA. Fig. 7.4 shows a five-channel multichannel FRA (Z#106 WonATech Co) to carry out simultaneous EIS measurements in the individual cells and whole PEMFC stack. The multichannel FRA (Z#106 WonATech Co)

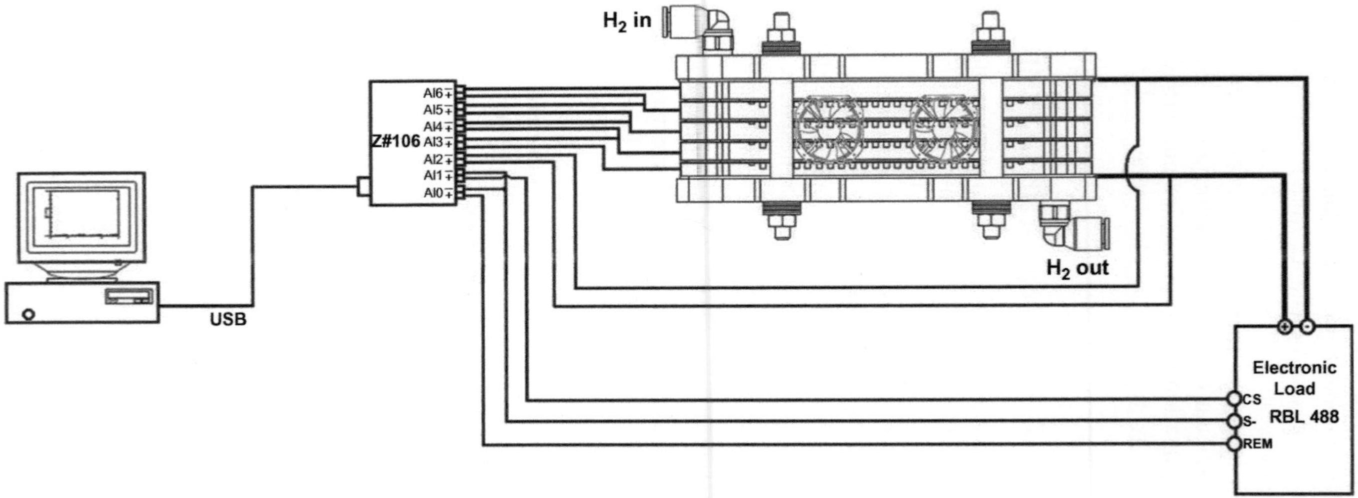

FIG. 7.4 Multichannel frequency response analyser for EIS measurements in the open-cathode PEMFC stack.

is operated with a programmable load unit (Dynaload RBL488) in galvanostatic mode. Note that this chapter does not aim to demonstrate the best testing practise for obtaining homogeneous performance in the individual cells of an open-cathode PEMFC stack.

The use of the programmable load unit (Dynaload RBL488) shown in Fig. 7.4 allows the measurement of the steady-state polarisation curve of the 4-cell open-cathode PEMFC stack. The load unit (Dynaload RBL488) can be controlled through MATLAB software to measure the polarisation curve. Fig. 7.5 shows the polarisation curve of the 4-cell open-cathode PEMFC stack. EIS measurements in the open-cathode PEMFC stack have been carried out at two current densities, 0.1875 and 0.4375 A/cm^2. The considered frequency range is from 10 kHz down to 0.1 Hz. The considered amplitude of the sinusoidal perturbation during the EIS measurements was 5% of the DC current.

Fig. 7.6 shows the EIS measurements carried out in the whole open-cathode PEMFC stack at two different current densities. The impedance spectra with

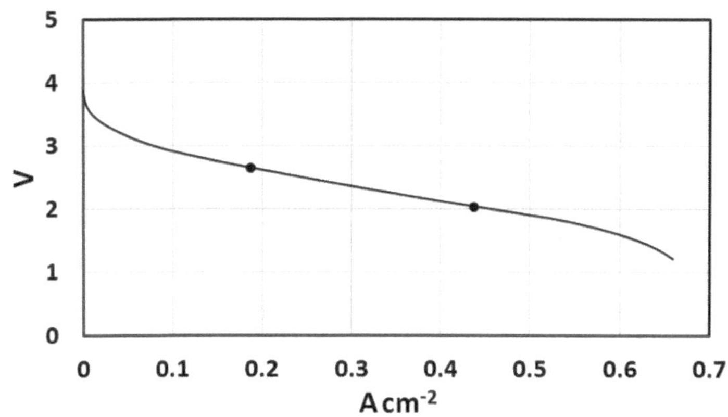

FIG. 7.5 Polarisation curve carried out in the 4-cell open-cathode PEMFC stack.

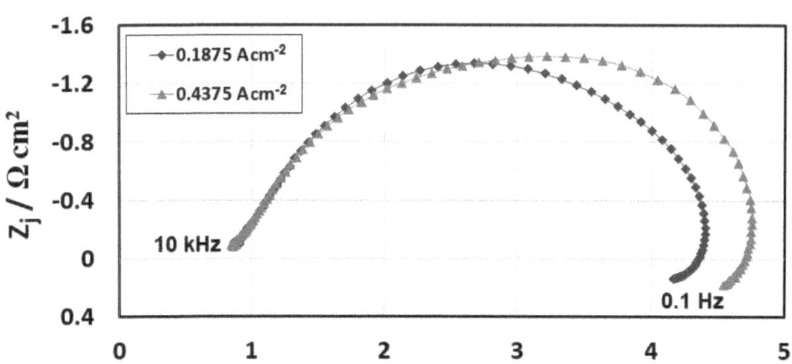

FIG. 7.6 Impedance spectra measured from the whole open-cathode PEMFC stack.

negative imaginary components $-Z_j$ are comprised of two overlapped semicircles. The first semicircle at high-low frequencies is attributed to kinetics of the redox reaction and the second semicircle at low frequencies is attributed to oxygen transport limitations. It would be expected that the first semicircle at high-low frequencies decreases with increasing current density and the second semicircle at low frequencies increases with increasing current density. The increase of the impedance spectrum measured at 0.4375 A/cm^2 with respect to the impedance spectrum measured at 0.1875 A/cm^2 is mainly attributed to an increase in oxygen transport limitations. At the highest frequency of 10 kHz, the impedance spectra shown in Fig. 7.6 present roughly the same value of the real component Z_r or ohmic resistance which suggests that the polymer electrolyte membrane (PEM) was completely hydrated when the impedance spectrum was measured at the current density of 0.4375 A/cm^2. Therefore, the water produced by the oxygen reduction reaction (ORR) with increasing current density hinders the transport of oxygen through the GDL to reach the active sites in the CCL. The inductive loops at low frequencies can be related to side reactions with intermediate species (e.g. hydrogen peroxide, platinum oxide) during the ORR [14] or water transport mechanisms in the electrode [19].

The impedance spectra of the individual cells of the PEMFC stack at the two different current densities are shown in Fig. 7.7. Fig. 7.7 shows that the impedance spectra of the individual cells at the same current density are different. The difference between the impedance spectra in the cells could be a consequence of different water, temperature and air distribution across the PEMFC stack [31]. Based on the EIS measurements shown in Fig. 7.7, it is considered that cell 2 presents a higher charge transfer resistance than the rest of the cells. The impedance spectrum of cell 2 at 0.1875 A/cm^2 as shown in Fig. 7.7b presents a value of $Z_j \approx -0.3\ \Omega\text{cm}^2$ at $Z_r = 0.45\ \Omega\text{cm}^2$. The impedance spectra of the rest of the cells at 0.1875 A/cm^2 present a value of $Z_j \approx -0.25\ \Omega\text{cm}^2$ at $Z_r = 0.45\ \Omega\text{cm}^2$ as shown in Fig. 7.7a, c, and d. The more negative imaginary component of impedance spectrum of cell 2 $Z_j \approx -0.3\ \Omega\text{cm}^2$ with respect to the impedance spectra of the rest of the cells $Z_j \approx -0.25\ \Omega\text{cm}^2$ at $Z_r = 0.45\ \Omega\text{cm}^2$ can be attributed to low electrocatalytic activity of cell 2. The EIS measurements were carried out during short term operation of the PEMFC stack therefore, the reduction of performance of cell 2 could be associated with a defective MEA [13] in cell 2 from the manufacturer and not a reduction of electrochemical surface area during long-term PEMFC operation [32]. The EIS measurements shown in Fig. 3.39 from Chapter 3 correspond to the same open-cathode PEMFC stack presented in this section and were carried out at 0.3125 A/cm^2. The EIS measurements of the open-cathode PEMFC stack complied with linearity and stability properties for Kramers–Kronig (K–K) transformation (real component to imaginary component and vice versa). This was demonstrated in Fig. 3.39 from Chapter 3 in which a comparison between K–K transformed data and the EIS measurements of the open-cathode PEMFC stack at 0.3125 A/cm^2 with a sinusoidal signal amplitude of 5% of the DC current was carried out.

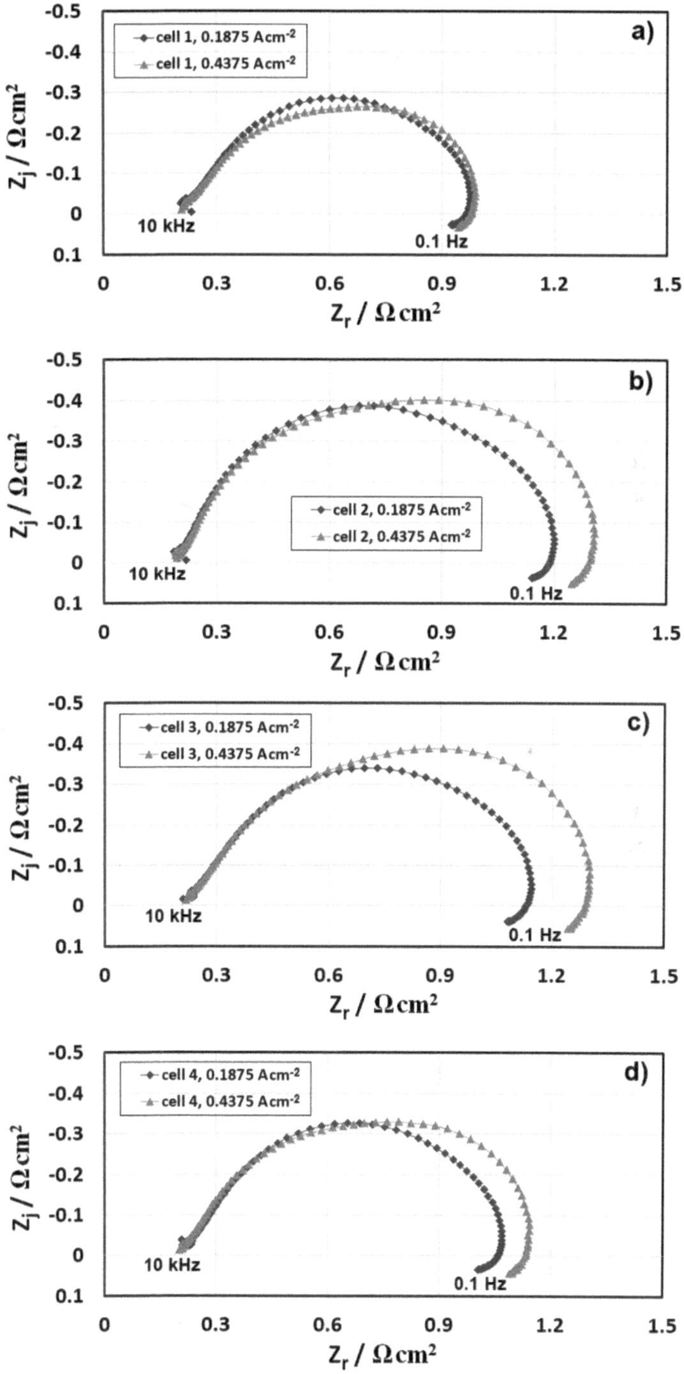

FIG. 7.7 Impedance spectra measured in the individual cells of the open-cathode PEMFC stack, (a) cell 1, (b) cell 2, (c) cell 3, and (d) cell 4.

7.2.2 Electrical circuits

Three electrical circuits [25–27] are considered to estimate the charge transfer resistance and mass transport resistance from the EIS measurements of the individual cells of the open-cathode PEMFC stack. The considered electrical circuits can reproduce the inductive loop at low frequencies although there are other reported electrical circuits that can analyse EIS measurements in PEMFCs featuring inductive loops at low frequencies as discussed in the literature review carried out by Pivac and Barbir [17].

Pivac et al. [25] reported the electrical circuit shown in Fig. 7.8 to analyse EIS measurements carried out in a single PEMFC featuring inductive loops at low frequencies. The inductor element in the electrical circuit configuration shown in Fig. 7.8 represents water absorption and water desorption at the polymer surface within the catalyst layer. The electrical circuit shown in Fig. 7.8 reproduces low-frequency inductive loops attributed to water transport mechanisms in the catalyst layer.

The impedance of the electrical circuit shown in Fig. 7.8 can be represented through the following equation:

$$Z_{\mathrm{EEC\_P}} = R_0 + \frac{R_1 L_1 j\omega}{L_1 j\omega[1 + R_1 C_1 j\omega] + R_1} + \frac{R_2}{1 + R_2 C_2 j\omega} + \frac{R_3}{1 + R_3 C_3 j\omega} + \frac{R_4 L_4 j\omega}{L_4 j\omega[1 + R_4 C_4 j\omega] + R_4} \quad (7.1)$$

where ω is the angular frequency, j is the imaginary component of a complex number, R_0 is the ohmic resistance in the PEM, the parameters R_1, R_2, C_1, C_2, L_1 are related to electro-diffusion phenomena in the anode and the parameters R_3, R_4, C_3, C_4, L_4 are related to electro-diffusion phenomena in the cathode.

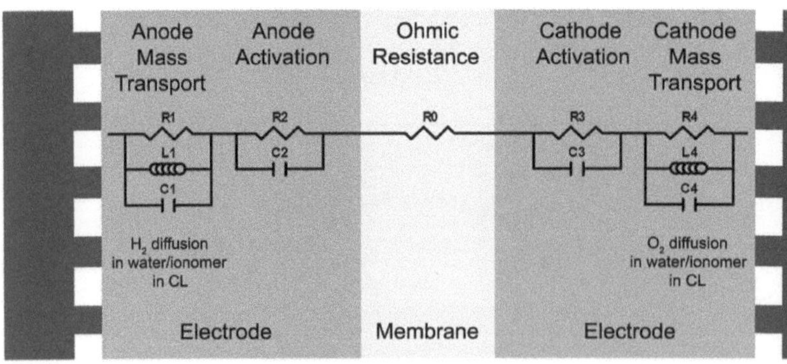

FIG. 7.8 Equivalent electrical circuit reported by Pivac et al. [25] to analyse EIS measurements in single PEMFCs with inductive loops at low frequencies. *(Reproduced from I. Pivac, B. Šimić, F. Barbir, Experimental diagnostics and modeling of inductive phenomena at low frequencies in impedance spectra of proton exchange membrane fuel cells, J. Power Sources 365 (2017) 242, with permission of Elsevier. Copyright Elsevier.)*

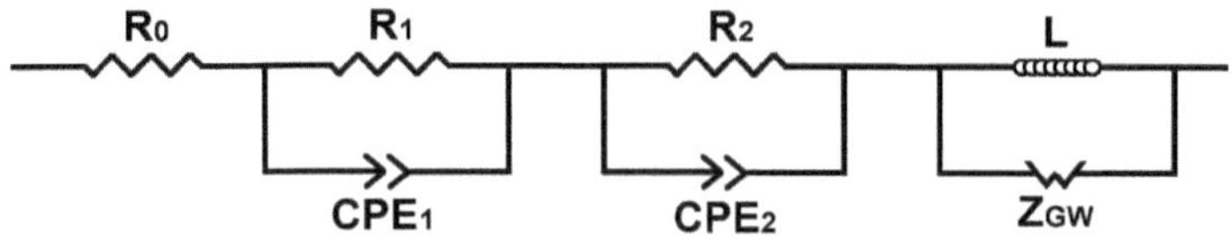

FIG. 7.9 Equivalent electrical circuit reported by Dhirde et al. [26] to analyse EIS measurements in a PEMFC stack with inductive loops at low frequencies.

Further information about the electrical circuit shown in Fig. 7.8 can be found in the study of Pivac et al. [25].

Dhirde et al. [26] proposed the electrical circuit shown in Fig. 7.9 to analyse EIS measurements carried out in a PEMFC stack featuring inductive loops at low frequencies. An inductor element connected in parallel with the generalised finite-length Warburg impedance was considered to reproduce the low-frequency inductive loops in the EIS measurements. Dhirde et al. [26] concluded that the formation of the low-frequency inductive loops was attributed to side reactions with intermediate species during the redox reaction or nonstationary behaviour of the PEMFC during EIS tests.

The impedance of the electrical circuit shown in Fig. 7.9 can be represented through the following equation:

$$Z_{\mathrm{EEC\_D}} = R_0 + \frac{R_1}{1 + R_1 Y_1 [j\omega]^{P_1}} + \frac{R_2}{1 + R_2 Y_2 [j\omega]^{P_2}} + \frac{Z_{\mathrm{GW}} L j\omega}{Z_{\mathrm{GW}} + L j\omega} \tag{7.2}$$

where R_0 represents the ohmic resistance in the PEMFC stack, R_1 is the charge transfer resistance in the anode, R_2 is the charge transfer resistance in the cathode, $Y_{1,2}$ and $P_{1,2}$ are parameters of a constant phase element (CPE) to account for the nonhomogeneous charge distribution in the electrode/membrane interfaces (the impedance of a CPE is represented in Eq. 1.63 from Chapter 1), L is an inductor element related to side reactions with intermediate species during the redox reaction. The electrical circuit also shown in Fig. 7.9 considers the generalised finite-length Warburg impedance [33] represented as:

$$Z_{\mathrm{GW}} = \frac{R_{\mathrm{GW}}}{[\tau_{\mathrm{GW}} j\omega]^{\varphi_{\mathrm{GW}}}} \tanh\left([\tau_{\mathrm{GW}} j\omega]^{\varphi_{\mathrm{GW}}}\right) \tag{7.3}$$

where R_{GW} is the generalised Warburg resistance, τ_{GW} is time constant of the generalised Warburg impedance, φ_{GW} is the phase constant of the generalised Warburg impedance. In Chapter 1 Section 1.4.3, the impedance response of the finite-length Warburg impedance was discussed. The generalised finite-length Warburg impedance represented in Eq. (7.3) differs from the finite-length Warburg impedance represented in Eq. (1.55) from Chapter 1 where the term φ_{GW} represented in Eq. (7.3) is different from a value of 0.5 in Eq. (1.55). The finite-length Warburg impedance represented in Eq. (1.55) from Chapter 1 has been commonly applied to EIS measurements of PEMFCs to estimate mass transport resistance [4].

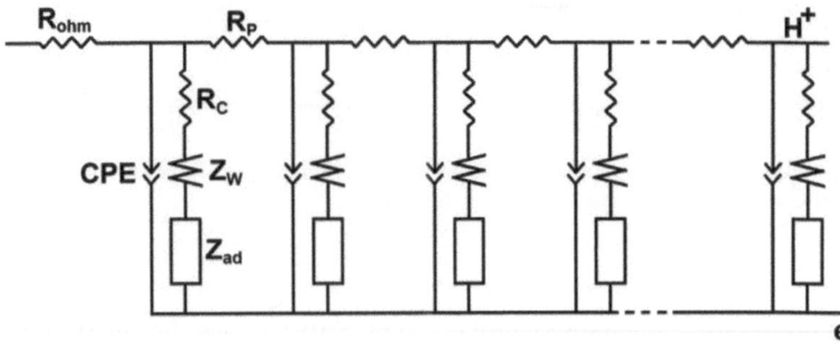

FIG. 7.10 Transmission-line model reported by Cruz-Manzo et al. [27] representing the impedance of the PEMFC with side reactions with intermediate species during the oxygen reduction reaction.

Cruz-Manzo et al. [27] reported the transmission-line model (TLM) shown in Fig. 7.10 to simulate low-frequency inductive loops of PEMFC-EIS measurements. The TLM comprised distributed resistors, CPEs, and finite-length Warburg impedances. The TLM shown in Fig. 7.10 represents the impedance of the PEMFC considering ohmic resistance R_{ohm} in the PEMFC, proton resistance R_P in the CCL, charge transfer resistance R_C during the ORR, oxygen diffusion impedance Z_W represented as a finite-length Warburg impedance, CPE for non-homogeneous charge distribution in the CCL and the impedance Z_{ad} attributed to side reactions with intermediate species during the ORR. The impedance of the TLM shown in Fig. 7.10 can be represented through Eq. (7.4) but neglects the contribution of the impedance response of the anode on the PEMFC impedance spectrum.

$$Z_{TLM\_C} = R_{ohm} + \frac{R_P \coth\left(\sqrt{R_P\left[\frac{1}{R_C + Z_W + Z_{ad}} + Y[j\omega]^P\right]}\right)}{\sqrt{R_P\left[\frac{1}{R_C + Z_W + Z_{ad}} + Y[j\omega]^P\right]}} \tag{7.4}$$

Where the finite-length Warburg impedance Z_W can be expressed through Eq. (1.55) from Chapter 1, Y and P are parameters related to the CPE. The impedance Z_{ad} consisted of a resistor R_{ad} connected in parallel with a CPE_{ad}, the impedance Z_{ad} which was attributed to side reactions with intermediate species during the ORR was connected in series with the charge transfer resistance R_C and finite-length Warburg impedance Z_W. The impedance Z_{ad} can be represented as:

$$Z_{ad} = \frac{R_{ad}}{1 + R_{ad} Y_{ad} [j\omega]^{P_{ad}}} \tag{7.5}$$

The parameters R_{ad} and Y_{ad} are negative values to simulate the inductive loop of the impedance spectrum of the PEMFC at low frequencies. Klotz

[34] demonstrated that inductive loops at low frequencies can be reproduced by electrical circuits comprising a negative capacitance and a negative resistance. If the impedance attributed to side reactions with intermediate species during the ORR is neglected in Eq. (7.4) $Z_{ad}=0$, the second term on the right-hand side of Eq. (7.4) is analogous to the impedance of the CCL represented in Eq. (4.103) from Chapter 4.

7.2.3 Fitting process

The electrical circuits reported in the literature [25–27] and shown in Figs 7.8–7.10 will be applied to the EIS measurements of the open-cathode PEMFC stack. The electrical circuits can be fitted to the EIS measurements using ZView software (Scribner Associates Inc.) and MATLAB software. The fitting process is carried out through a nonlinear regression method such as the Levenberg–Marquardt algorithm.

Eq. (7.4) represents the impedance of the TLM shown in Fig. 7.10. The impedance response of the TLM can be fitted to the EIS measurements of the open-cathode PEMFC stack using MATLAB software. The TLM shown in Fig. 7.10 cannot be constructed in ZView software because the TLM defined in ZView software neglects the components Z_W and Z_{ad} shown in Fig. 7.10. The fitting process of the TLM represented in Eq. (7.4) with the EIS measurements is carried out in a two-step fitting process:

1. Eq. (7.4) neglecting the impedance $Z_{ad}=0$ is first fitted to the EIS measurements. During the first-step fitting process, the EIS measurements with positive imaginary components at low frequencies (inductive loop) are neglected as shown in Fig. 7.11.
2. Eq. (7.4) considering the impedance Z_{ad} is fitted to the impedance spectrum featuring low-frequency inductive loop as shown in Fig. 7.11. During the second step, a fixed value of the charge transfer resistance R_C and ohmic

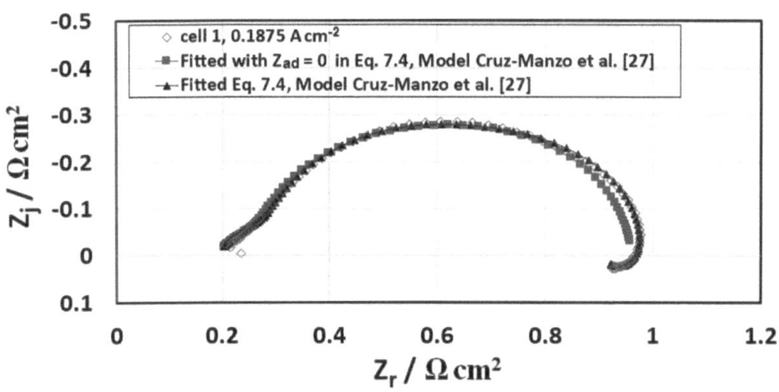

FIG. 7.11 Two-step fitting process to estimate the parameters of the transmission-line model represented in Eq. (7.4) and reproduce the inductive loop at low frequencies.

resistance R_{ohm} is considered. The same value of R_C and R_{ohm} can be considered from the first step. The value of the rest of the parameters calculated from the first-step fitting process are considered as initial values during the second-step fitting process.

The aforementioned two-step fitting process will allow the estimation of the values of the parameters from the TLM shown in Fig. 7.10 and the simulation of the low-frequency inductive loop as shown in Fig. 7.11. The fitting process in MATLAB software is carried out through the function *lsqcurvefit* defined in MATLAB software [35]. This function solves nonlinear curve-fitting (data-fitting) problems in a least-squares sense using the Levenberg–Marquardt algorithm. The fitting process of an impedance model with PEMFC-EIS measurements in MATLAB software has been discussed in Section 6.2 from Chapter 6. A good quality fit is obtained when the real and imaginary relative residuals between the simulated and measured impedance data has a minimum value. The average of the real and imaginary relative residuals is calculated through Eqs (6.16) and (6.17) from Chapter 6. The average of the real and imaginary relative residuals can also be considered to evaluate the quality of the fitting results of the TLM represented in Eq. (7.4) with the EIS measurements of the open-cathode PEMFC stack. The sum of Eqs (6.16) and (6.17) $\Delta Z = \Delta Z_r + \Delta Z_j$ from Chapter 6 to evaluate the goodness of the fit of Eq. (7.4) with the EIS measurements as shown in Fig. 7.11 resulted in $\Delta Z = 0.0328$. A good quality fit is obtained when the real and imaginary relative residuals have a minimum value. The script written in MATLAB software to carry out the simulations shown in Fig. 7.11 can be found in Appendix A.

The impedance response of the electrical circuits shown in Figs 7.8 and 7.9 will be fitted to the EIS measurements of the open-cathode PEMFC stack using ZView software. Fig. 7.12 shows the fitting process of the impedance response of the electrical circuit shown in Fig. 7.8 with EIS measurements using ZView software. Fig. 7.12 also shows that the estimation of the values of the parameters of the electrical circuit shown in Fig. 7.8 are carried out through a single-step fitting process with no fixed-values in the parameters. A single-step fitting process with no fixed-values in the parameters is also considered for the estimation of the values of the parameters of the electrical circuit shown in Fig. 7.9 using ZView software. The goodness of fit in ZView software is evaluated through chi-squares values. In Section 3.3.4 from the book of Orazem and Tribollet [36], an example of the analysis of the chi-squared test for goodness of fit is considered, the example considers a Voigt circuit fitted to EIS measurements.

The electrical circuits shown in Figs 7.8 and 7.9 were fitted to the EIS measurements of the open-cathode PEMFC stack using ZView software. The script written in MATLAB software to fit the TLM shown in Fig. 7.10 with the EIS measurements and using a two-step fitting process is shown in Appendix A. Fig. 7.13 shows that the three electrical circuits reported in the literature [25–27] and shown in Figs 7.8–7.10 can reproduce the EIS measurements of

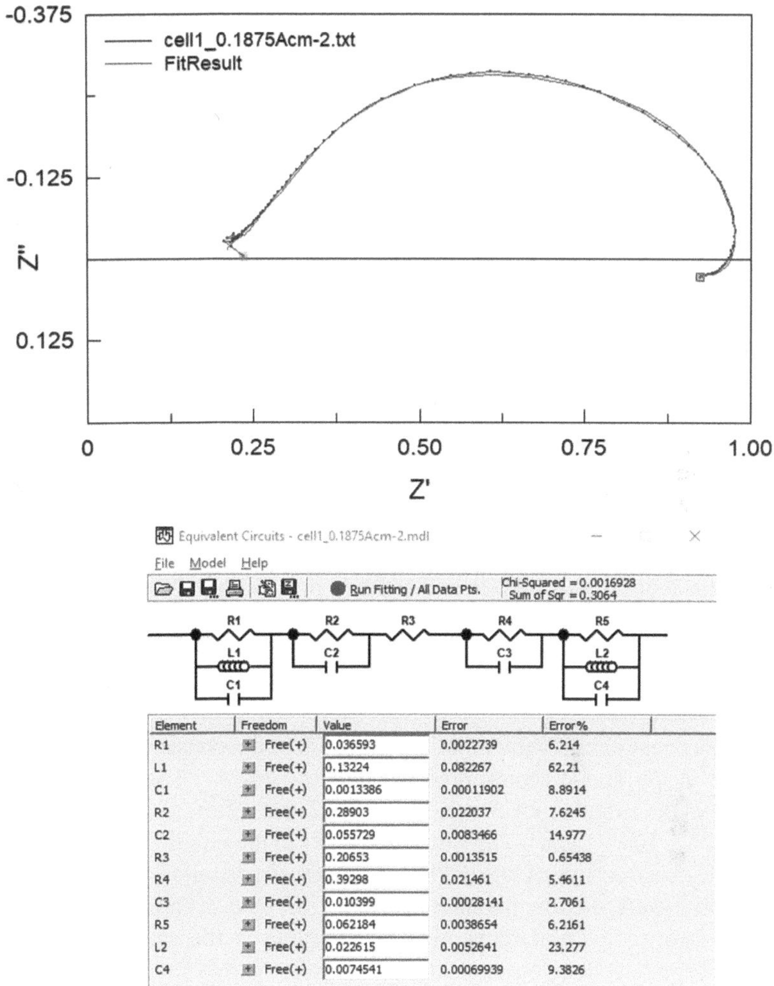

FIG. 7.12 Fitting process applied to the electrical circuit shown in Fig. 7.8 reported by Pivac et al. [25] with EIS measurements of cell 1 of the open-cathode PEMFC stack at 0.1875 A/cm^2 and using ZView software (Scribner Associates Inc.).

cell 1 at 0.4375 A/cm^2. The electrical circuits were also able to reproduce the EIS measurements of the rest of the cells of the open-cathode PEMFC stack at 0.1875 and 0.4375 A/cm^2, although the fitting results are not shown.

7.2.4 Impedance analysis

Tables 7.1–7.3 show the estimated and resulting parameters from fitting the impedance responses of the electrical circuits shown in Figs 7.8–7.10 with

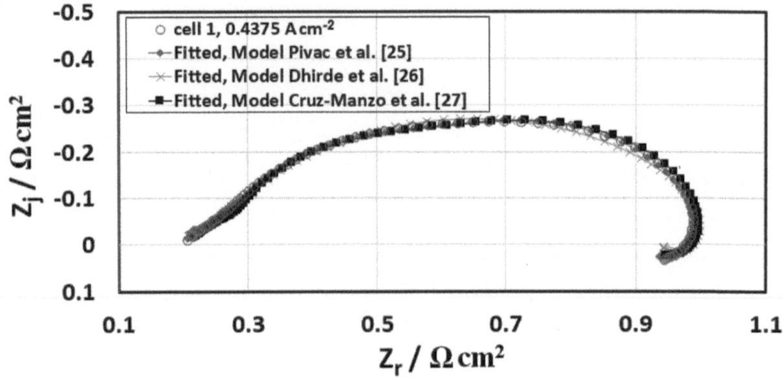

FIG. 7.13 Comparison between impedance response generated by the circuits shown in Figs 7.8–7.10 reported in the literature [25–27] and EIS measurements of cell 1 of the open-cathode PEMFC stack at 0.4375 A/cm^2.

the EIS measurements of the individual cells of the open-cathode PEMFC stack shown in Fig. 7.7. The electrical circuit shown in Fig. 7.8 estimated the charge transfer resistances in the anode and cathode represented through the resistors R_2 and R_3 with the same order of magnitude. The electrical circuit shown in Fig. 7.8 also estimated the mass transport resistance in the anode and cathode represented through the resistors R_1 and R_4 with the same order of magnitude as shown in Table 7.1.

The electrical circuit shown in Fig. 7.9 estimated a constant charge transfer resistance in the anode represented through resistor R_1. The charge transfer resistance in the anode is not shown in Table 7.2, but resulted in a constant value of $R_1 = 1.21 \times 10^{-10}$ Ωcm^2. The low value of R_1 in the circuit shown in Fig. 7.9 makes the contribution of the CPE$_1$ connected in parallel with the resistor R_1 negligible therefore, the parameters shown in Table 7.2 attribute the impedance spectrum to the electrochemical phenomena in the cathode. The TLM shown in Fig. 7.10 does not consider parameters related to the impedance of the anode.

Fig. 7.14 shows the charge transfer resistance R_C during the redox reaction represented in Tables 7.1–7.3 and estimated by the electrical circuits reported by Pivac et al. [25], Dhirde et al. [26], and Cruz-Manzo et al. [27] at a current density of 0.1875 A/cm^2. The charge transfer resistance R_C estimated by the electrical circuit reported by Pivac et al. [25] shown in Fig. 7.8 is calculated from Table 7.1 by considering $R_C = R_2 + R_3$. The charge transfer resistance R_C estimated by the electrical circuit reported by Dhirde et al. [26] shown in Fig. 7.9 is calculated from Table 7.2 as $R_C = R_1 + R_2$ and considering that $R_1 = 1.21 \times 10^{-10}$ Ωcm^2. Fig. 7.14 shows that both electrical circuits shown in Figs 7.8 and 7.9 reported by Pivac et al. [25] and Dhirde et al. [26] estimated close values of the charge transfer resistance R_C in the EIS measurements of the cells comprising the open-cathode PEMFC stack at 0.1875 A/cm^2. The charge

TABLE 7.1 Parameters estimated by fitting the impedance response of the electrical circuit reported by Pivac et al. [25] shown in Fig. 7.8 with the individual EIS measurements of the open-cathode PEMFC stack (Fig. 7.7).

0.1875 A/cm^2	R_0 Ωcm^2	R_1 Ωcm^2	L_1 Hcm^2	C_1 F/cm^2	R_2 Ωcm^2	C_2 F/cm^2	R_3 Ωcm^2	C_3 F/cm^2	R_4 Ωcm^2	L_4 Hcm^2	C_4 F/cm^2
Cell 1	0.206	0.036	0.132	0.0013	0.289	0.055	0.392	0.0103	0.062	0.0226	0.0074
Cell 2	0.190	0.028	0.027	0.0012	0.382	0.045	0.552	0.0077	0.055	0.0245	0.0087
Cell 3	0.222	0.034	0.078	0.0013	0.378	0.041	0.446	0.0085	0.074	0.0295	0.0060
Cell 4	0.183	0.051	0.135	0.0002	0.371	0.033	0.408	0.0079	0.063	0.0214	0.0042
0.4375 A/cm^2	R_0 Ωcm^2	R_1 Ωcm^2	L_1 Hcm^2	C_1 F/cm^2	R_2 Ωcm^2	C_2 F/cm^2	R_3 Ωcm^2	C_3 F/cm^2	R_4 Ωcm^2	L_4 Hcm^2	C_4 F/cm^2
Cell 1	0.204	0.034	0.007	0.0015	0.417	0.051	0.295	0.0131	0.058	0.0639	0.0084
Cell 2	0.187	0.035	0.013	0.0015	0.618	0.035	0.421	0.0105	0.062	0.0478	0.0121
Cell 3	0.217	0.032	0.039	0.0013	0.648	0.030	0.346	0.0107	0.072	0.0429	0.0063
Cell 4	0.199	0.031	0.031	0.0012	0.521	0.037	0.348	0.0108	0.058	0.0271	0.0076

TABLE 7.2 Parameters estimated by fitting the impedance response of the electrical circuit reported by Dhirde et al. [26] shown in Fig. 7.9 with the individual EIS measurements of the open-cathode PEMFC stack (Fig. 7.7).

0.1875 A/cm^2	R_0 Ωcm^2	R_2 Ωcm^2	Y_2 s^P/Ωcm^2	P_2	R_{GW} Ωcm^2	τ_{GW} s	φ_{GW}	L Hcm^2
Cell 1	0.182	0.738	0.0211	0.835	0.087	5.316E-4	0.303	0.0321
Cell 2	0.106	1.019	0.0158	0.833	0.111	1.464E-5	0.263	0.0526
Cell 3	0.191	0.882	0.0188	0.825	0.115	1.634E-3	0.248	0.0421
Cell 4	0.151	0.851	0.0170	0.828	0.109	1.07E-4	0.261	0.0356
0.4375 A/cm^2	R_0 Ωcm^2	R_2 Ωcm^2	Y_2 s^P/Ωcm^2	P_2	R_{GW} Ωcm^2	τ_{GW} s	φ_{GW}	L Hcm^2
Cell 1	0.119	0.827	0.0456	0.733	0.149	5.95E-4	0.129	0.0414
Cell 2	0.033	1.205	0.0271	0.771	0.177	2.917E-6	0.209	0.0669
Cell 3	0.196	1.041	0.0322	0.784	0.179	0.0124	0.281	0.0568
Cell 4	0.068	1.019	0.0335	0.750	0.166	6.635E-6	0.163	0.0538

TABLE 7.3 Parameters estimated by fitting the impedance response of the TLM reported by Cruz-Manzo et al. [27] shown in Fig. 7.10 with the individual EIS measurements of the open-cathode PEMFC stack (Fig. 7.7).

0.1875 A/cm^2	R_{ohm} Ωcm^2	R_P Ωcm^2	R_C Ωcm^2	Y s^P/Ωcm^2	P	R_W Ωcm^2	τ_W s	R_{ad} Ωcm^2	Y_{ad} s^{Pad}/Ωcm^2	P_{ad}
Cell1	0.1840	0.3027	0.5220	0.0098	0.961	0.1907	0.0334	−0.0887	−5.8658	0.819
Cell2	0.1781	0.1829	0.8056	0.0094	0.920	0.2102	0.0453	−0.1456	−3.8788	0.705
Cell3	0.1982	0.322	0.5819	0.0069	0.983	0.2798	0.0334	−0.1159	−5.124	0.803
Cell4	0.1842	0.3027	0.561	0.0065	0.993	0.2335	0.0289	−0.0999	−5.3875	0.839
0.4375 A/cm^2	R_{ohm} Ωcm^2	R_P Ωcm^2	R_C Ωcm^2	Y s^P/Ωcm^2	P	R_W Ωcm^2	τ_W s	R_{ad} Ωcm^2	Y_{ad} s^{Pad}/Ωcm^2	P_{ad}
Cell1	0.2001	0.2699	0.384	0.0097	0.990	0.3606	0.046	−0.1161	−5.3202	0.707
Cell2	0.1752	0.2146	0.5652	0.007	0.988	0.5412	0.0447	−0.1655	−4.4975	0.716
Cell3	0.2011	0.2907	0.4927	0.0072	0.987	0.5541	0.0404	−0.184	−5.1836	0.687
Cell4	0.1982	0.2423	0.461	0.0075	0.994	0.4509	0.0421	−0.1408	−5.0542	0.725

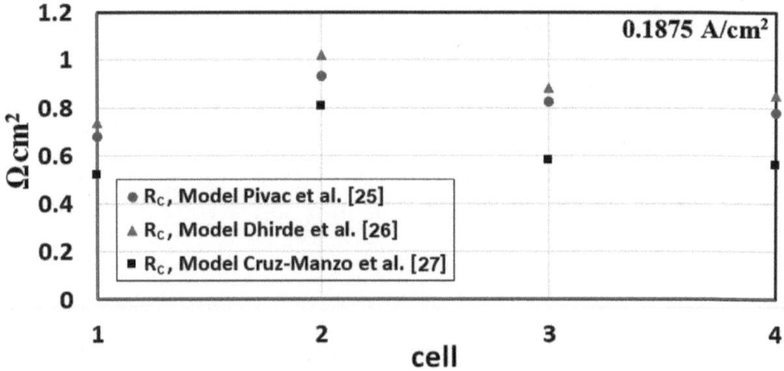

FIG. 7.14 Estimation of the charge transfer resistance R_C at 0.1875 A/cm^2 through the electrical circuits reported in the literature (Figs 7.8–7.10) and the EIS measurements of the four cells comprising the open-cathode PEMFC stack.

transfer resistance R_C during the ORR estimated by the TLM shown in Fig. 7.10 reported by Cruz-Manzo et al. [27] is lower than the charge transfer resistance R_C estimated by the electrical circuits shown in Figs 7.8 and 7.9 reported by Pivac et al. [25] and Dhirde et al. [26]. The resulting value of the charge transfer resistance R_C estimated by the TLM and shown in Fig. 7.14 can also be related to the fact that the TLM considers the proton resistance R_P in the CCL which is represented through a 45-degree straight line on the high frequency region of the impedance spectrum. The effect of the proton resistance on the impedance spectrum of the CCL has been discussed in Chapter 4. The value of the charge transfer resistance R_C for cell 2 at a current density of 0.1875 A/cm^2 resulted in a higher resistance value than the rest of the cells as shown in Fig. 7.14. As previously discussed in Section 7.2.1, the more negative imaginary component of the impedance spectrum of cell 2 $Z_j \approx -0.3$ Ωcm^2 (impedance spectrum at 0.1875 A/cm^2) with respect to the impedance spectra of the rest of the cells $Z_j \approx -0.25$ Ωcm^2 at $Z_r = 0.45$ Ωcm^2 can be attributed to low electrocatalytic activity of cell 2. The increase of charge transfer resistance in cell 2 can be due to a defective MEA in the cell.

Fig. 7.15 shows the results for the charge transfer resistance R_C at 0.4375 A/cm^2 estimated by the electrical circuits reported by Pivac et al. [25], Dhirde et al. [26] and Cruz-Manzo et al. [27]. The charge transfer resistance R_C estimated by the electrical circuits reported by Pivac et al. [25] and Dhirde et al. [26] increased with increasing current density from 0.1875 to 0.4375 A/cm^2 as shown in Figs 7.14 and 7.15. Yuan et al. [5] reported an increase of the charge transfer resistance with increasing current density when fitting the impedance response of an electrical circuit with the impedance spectra of a PEMFC stack. The EIS measurements of the PEMFC stack reported by Yuan et al. [5] are shown in Fig. 7.1. The increase of charge transfer resistance with increasing current density in the study of Yuan et al. [5] was related to an

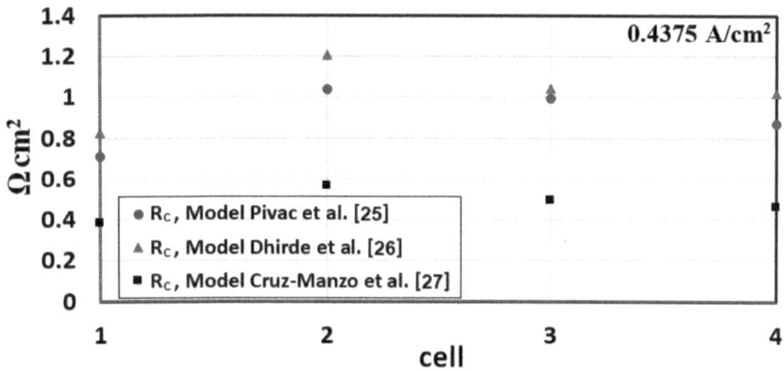

FIG. 7.15 Estimation of the charge transfer resistance R_C at 0.4375 A/cm^2 through the electrical circuits reported in the literature (Figs 7.8–7.10) and the EIS measurements of the four cells comprising the open-cathode PEMFC stack.

effective charge transfer resistance containing mass transport phenomena. It seems that it is difficult to separate and estimate the electrochemical phenomena in impedance spectra of a PEMFC that comprises overlapped electro-diffusion time constants by only relying on a parametric fitting analysis.

The value of the charge transfer resistance R_C at 0.4375 A/cm^2 estimated by the TLM reported by Cruz-Manzo et al. [27] resulted in a lower resistance value than the charge transfer resistances R_C estimated by the electrical circuits reported by Pivac et al. [25] and Dhirde et al. [26] as shown in Fig. 7.15. The TLM was fitted considering a two-step fitting process. In the first-step the low-frequency inductive loop in the EIS measurements is neglected and the charge transfer resistance R_C is estimated. Thereafter, in the second-step, the low-frequency inductive loop in the EIS measurements is considered in the fitting process and the value of the charge transfer resistance R_C is considered with a fixed value. The two-step fitting process to estimate the values of the parameters of the TLM from EIS measurements has been discussed in Section 7.2.3.

The total electro-diffusion resistance of the PEMFC can be calculated through the sum of the charge transfer resistance R_C and mass transport resistance R_m. Therefore, from Fig. 7.15, it is easy to predict that the mass transport resistance estimated by the TLM shown in Fig. 7.10 is higher than the mass transport resistances estimated by the electrical circuits shown in Figs 7.8 and 7.9. A lower value of the charge transfer resistance R_C estimated by the electrical circuits reported by Pivac et al. [25] and Dhirde et al. [26] with increasing current density could result if a two-step fitting process with a fixed value of R_C is considered.

Fig. 7.16 shows the mass transport resistance R_m represented in Tables 7.1–7.3 estimated by the electrical circuits reported by Pivac et al. [25], Dhirde et al. [26], Cruz-Manzo et al. [27] at a current density of 0.1875 A/cm^2. The mass

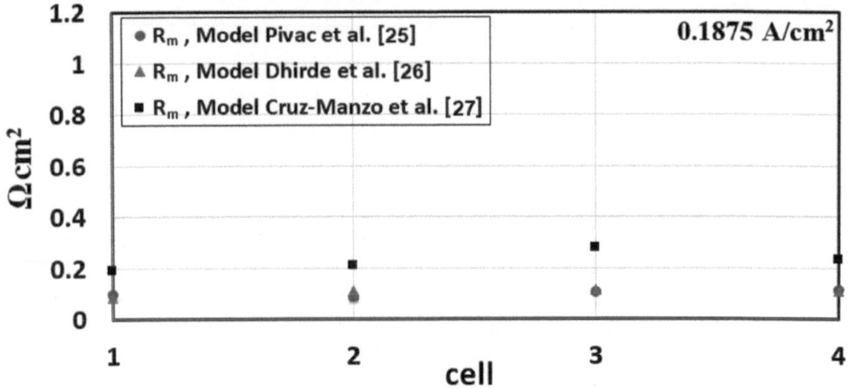

FIG. 7.16 Estimation of the mass transport resistance R_m at 0.1875 A/cm^2 through the electrical circuits reported in the literature (Figs 7.8–7.10) and the EIS measurements of the four cells comprising the open-cathode PEMFC stack.

transport resistance R_m estimated by the electrical circuit reported by Pivac et al. [25] is calculated from Table 7.1 considering $R_m=R_1+R_4$. The mass transport resistance R_m estimated by the electrical circuit reported by Dhirde et al. [26] represents the resistance of the generalised finite-length Warburg impedance R_m-R_{GW} shown in Table 7.2. The mass transport resistance R_m estimated by the TLM reported by Cruz-Manzo et al. [27] represents the resistance of the finite-length Warburg impedance $R_m=R_W$ shown in Table 7.3. Close values of the mass transport resistance R_m were predicted by the electrical circuits reported by Pivac et al. [25] and Dhirde et al. [26] at 0.1875 A/cm^2 as shown in Fig. 7.16. The mass transport resistance R_m predicted by the TLM reported by Cruz-Manzo et al. [27] resulted in a higher value than the mass transport resistance R_m predicted by the electrical circuits reported by Pivac et al. [25] and Dhirde et al. [26]. This can be attributed to the estimated low value of the charge transfer resistance R_C of the TLM at 0.1875 A/cm^2 as shown in Fig. 7.14.

Fig. 7.17 shows the mass transport resistances R_m represented in Tables 7.1–7.3 estimated by the three electrical circuits at 0.4375 A/cm^2. Close results for the mass transport resistance R_m represented in Tables 7.1 and 7.2 estimated by the electrical circuits reported by Pivac et al. [25] and Dhirde et al. [26] are shown in Fig. 7.17. Fig. 7.17 also shows that the mass transport resistance R_m estimated by the TLM (Fig. 7.10) reported by Cruz-Manzo et al. [27] resulted in higher values with respect to the mass transport resistance R_m estimated by the other two circuits (Figs 7.8 and 7.9). The higher mass transport resistance R_m is a consequence of a lower estimated charge transfer resistance R_C at 0.4375 A/cm^2 as shown in Fig. 7.15 because the total electro-diffusion resistance of the PEMFC can be calculated through the sum of the charge transfer resistance R_C and mass transport resistance R_m.

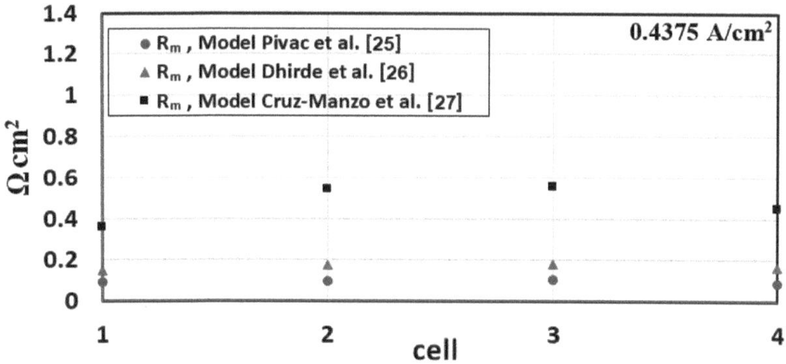

FIG. 7.17 Estimation of the mass transport resistance R_m at 0.4375 A/cm^2 through the electrical circuits reported in the literature (Figs 7.8–7.10) and the EIS measurements of the four cells comprising the open-cathode PEMFC stack.

7.2.5 Limitations with electrical circuits

The electrical circuits can reproduce the impedance response of the PEMFC and can be considered to qualitatively compare the performance of the PEMFC represented through its impedance response at different operating conditions. The parameters previously shown in Tables 7.1–7.3 with their respective electrical circuits shown in Figs 7.8–7.10 can reproduce the impedance spectra of the individual cells comprising the open-cathode PEMFC stack. Therefore, the electrical circuits can be represented through differential equations in the time-domain to simulate the electrochemical dynamic response of the PEMFC stack. In Chapter 2 Section 2.4, it was demonstrated that the dynamic output voltage of the PEMFC during a step change in current can be simulated through the differential equation representing an electrical circuit composed of resistors and a capacitor. The dynamic output voltage of the individual cells of the open-cathode PEMFC stack in the time-domain could be simulated through the differential equations representing the electrical circuit shown in Fig. 7.8 reported by Pivac et al. [25] and considering the parameters from Table 7.1. The electrical circuits can only be applied to EIS measurements of PEMFC stacks or individual cells at different operating conditions to empirically estimate the impedance response.

The increase of the number of parameters in an electrical circuit during the fitting process between the impedance response of the electrical circuit and PEMFC-EIS measurements increases the ambiguity whether or not the parameters truly represent the electrochemical phenomena that are occurring during PEMFC operation. The results shown in Figs 7.14–7.17 demonstrates that depending on how the electrical circuit is applied/fitted to the EIS measurements, it can yield a different estimation of the parameters. The results from Figs 7.14–7.17 also demonstrates that the electrical circuits calculate different time constants, e.g. $\tau_C = R_C C_{dl}$ with C_{dl} as the double-layer capacitance, which are distributed over the overall impedance spectrum at different frequencies. The double-layer capacitance can be calculated from the parameters of the CPE as discussed in Chapter 4 Section 4.5.3.

The electrical circuits represent resistive, capacitive, and inductive components and relate the electrical components with electro-diffusion phenomena of the PEMFC. However, the electrical circuits cannot predict the change of the impedance response of the PEMFC at different operating conditions. Therefore the parameters from Tables 7.1–7.3 and the equations representing the impedance of the electrical circuits expressed in Eqs (7.1)–(7.5) are not able to predict and simulate the change of the magnitude of the impedance spectrum with changing operating conditions (e.g. PEMFC temperature). The electrical circuits shown in Figs 7.8 and 7.9 comprise inductors to simulate the inductive loops of the EIS measurements at low frequencies. The TLM shown in Fig. 7.10 considers a negative CPE to simulate the inductive loops at low frequencies. Other electrical circuits that analyse EIS measurements in PEMFCs featuring low-frequency inductive loops are reported in the literature review carried out by Pivac and Barbir [17]. The following section will discuss the electrochemical phenomena of the PEMFC that yield the formation of inductive loops in PEMFC-EIS measurements at low frequencies.

7.3 Analysis of inductive loops at low frequencies in PEMFC-EIS measurements

Inductive loop formation at low frequencies in EIS measurements of PEMFCs has been attributed to the formation of hydrogen peroxide [14] and platinum oxide during the ORR [18]. The inductive loop at low frequencies has also been related to water transport mechanisms in the PEM [16] and water vapour transport in the CCL [37]. In this section, a brief discussion about the formation of the low-frequency inductive loops in PEMFC-EIS measurements attributed to hydrogen peroxide, platinum oxide, and water vapour transport during the ORR in the CCL will be discussed.

7.3.1 Hydrogen peroxide formation

The formation of hydrogen peroxide H_2O_2 considering the ORR as a two-step electron reaction has been considered as a cause of the formation of the inductive loops at low frequencies in EIS measurements of PEMFCs [14]. Crossover of hydrogen from the anode to the cathode facilitates the reaction of oxygen and hydrogen at the cathode, generating hydroxyl, and hydroperoxyl radicals which react further to produce hydrogen peroxide during the ORR [14]. Inaba et al. [38] studied the effect of agglomeration of Pt/C on H_2O_2 formation during the ORR. In Chapter 2, Section 2.3.4.1 the faradaic current density considering the formation of H_2O_2 during the ORR was presented. In Section 10.4.1 of the book of Orazem and Tribollet [36], the formation of the low-frequency inductive loop in the impedance response of an electrode considering a hypothetical reaction sequence was demonstrated. The hypothetical reaction was considered as:

$$M \rightarrow X^{+} + e^{-} \tag{7.6}$$

$$X^{+} \rightarrow P^{2+} + e^{-} \tag{7.7}$$

where the adsorption reaction of the intermediate X^+ is represented by a surface coverage γ_{X^+}. Orazem and Tribollet [36] defined the charge transfer resistance associated with the reactions (7.6) and (7.7) as:

$$R_{t,M} = \frac{b_M}{K_M[1 - \overline{\gamma}_X] \exp\left(\overline{V}/b_M\right)} \tag{7.8}$$

$$R_{t,X} = \frac{b_X}{K_X \overline{\gamma}_X \exp\left(\overline{V}/b_X\right)} \tag{7.9}$$

where $b_{M,X}$ is the Tafel slope, $K_{M,X}$ is the rate constant and $\overline{\gamma}_X$ is the steady-state fractional surface coverage. The impedance response of the electrode considering the hypothetical reactions (7.6) and (7.7) demonstrated that an inductive loop is formed at low frequencies when the fractional surface coverage γ_X is small and a capacitive loop at low frequencies is formed when fractional surface coverage γ_X is large (Figs 10.10 and 10.11 from the book of Orazem and Tribollet [36]). Roy et al. [14] developed an impedance model of the PEMFC considering the adsorption reaction of hydrogen peroxide in a two-step ORR, the impedance model was able to simulate the inductive loop at low frequencies in the impedance spectrum of the PEMFC. The charge transfer resistance associated with the oxygen reaction neglecting oxygen transport limitation reported by Roy et al. [14] was analogous to the charge transfer resistance represented in Eq. (7.8). The charge transfer resistance of the hydrogen peroxide reaction was analogous to Eq. (7.9) as such:

$$R_{t,O_2} = \frac{b_{O_2}}{K_{O_2}\left[1 - \overline{\gamma}_{H_2O_2}\right] \exp\left(\overline{V}/b_{O_2}\right)} \tag{7.10}$$

$$R_{t,H_2O_2} = \frac{b_{H_2O_2}}{K_{H_2O_2} \overline{\gamma}_{H_2O_2} \exp\left(\overline{V}/b_{H_2O_2}\right)} \tag{7.11}$$

An impedance model of the PEMFC to simulate the formation of the low-frequency inductive loop attributed to the reaction of hydrogen peroxide during the two-step ORR has also been reported by Cruz-Manzo et al. [39]. The impedance model reported by Cruz-Manzo et al. [39] considered that the formation of hydrogen peroxide is governed by the Langmuir isotherm [40]. The impedance model was also able to decouple the contribution of the charge transfer resistance in the oxygen reaction, charge transfer resistance in the hydrogen peroxide reaction and oxygen transport limitations. Similar to Eq. (7.10), the impedance model reported by Cruz-Manzo et al. [39] considered the charge transfer resistance in the oxygen reaction as a function of steady-state fractional surface coverage of hydrogen peroxide $\left(1-\overline{\gamma}_{H_2O_2}\right)$. The impedance model of the PEMFC reported by Cruz-Manzo et al. [39] to simulate inductive loops at low frequencies attributed to hydrogen peroxide formation in the PEMFC impedance spectrum is represented as:

$$Z_{PEMFC} = R_{ohm} + Z_{anode} + Z_{cathode_{O_2,H_2O_2}} \tag{7.12}$$

where R_{ohm} is the total ohmic resistance of the PEMFC, Z_{anode} is the impedance associated with the hydrogen oxidation reaction (HOR) in the anode represented through the impedance of a parallel resistor-capacitor, and $Z_{cathode_{O_2,H_2O_2}}$ is the impedance during a two-step ORR in the cathode represented as:

$$Z_{cathode_{O_2,H_2O_2}} = \left[\frac{1}{\frac{1}{R_{C,O_2}+Z_{O_2}} + \frac{1}{R_{C,H_2O_2}} + A\varphi_{H_2O_2} + Y[j\omega]^P}\right] \tag{7.13}$$

where

$$A\varphi_{H_2O_2} = \frac{1}{[\tau_{H_2O_2} j\omega + 1] R\varphi_{H_2O_2}} \tag{7.14}$$

with

$$\tau_{H_2O_2} = \frac{zF\beta}{i_{O_2,ss} + i_{H_2O_2,ss}} \tag{7.15}$$

and

$$R\varphi_{H_2O_2} = \frac{R_{C,H_2O_2}[R_{C,O_2}+Z_{O_2}][i_{O_2,ss}+i_{H_2O_2,ss}]}{[R_{C,H_2O_2}-R_{C,O_2}-Z_{O_2}][i_{H_2O_2,ss}-i_{O_2,ss}]} \tag{7.16}$$

where β is the maximum coverage of hydrogen peroxide, F is the Faraday constant, $i_{O_2,ss}$ is the steady-state current density for the oxygen reaction, $i_{H_2O_2,ss}$ is the steady-state current density for the hydrogen peroxide reaction, R_{C,O_2} is the charge transfer resistance associated with the oxygen reaction, R_{C,H_2O_2} is the charge transfer resistance associated with the hydrogen peroxide reaction, $\tau_{H_2O_2}$ is the time constant for hydrogen peroxide formation, Z_{O_2} is the diffusion impedance for oxygen transport limitations in the CCL and is represented though the finite-length Warburg impedance. A detailed derivation of the cathodic impedance model represented in Eq. (7.13) can be found in the study of Cruz-Manzo et al. [39].

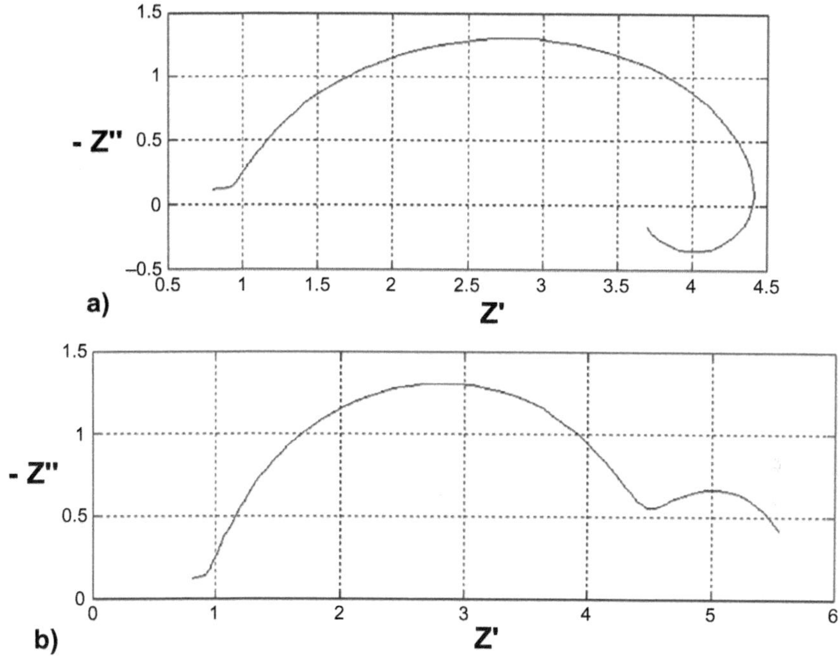

FIG. 7.18 Effect of charge transfer resistance in the hydrogen peroxide reaction on the impedance spectrum of the PEMFC at low frequencies, (a) $R_{C,H_2O_2} > R_{C,O_2} + Z_{O_2}$, (b) $R_{C,H_2O_2} < R_{C,O_2} + Z_{O_2}$. *(Reproduced from S. Cruz-Manzo, R. Chen, P. Greenwood, An impedance model for analysis of EIS of polymer electrolyte fuel cells under hydrogen peroxide formation in the cathode, J. Electroanal. Chem. 745 (2015) 32, with permission of Elsevier. Copyright Elsevier.)*

Fig. 7.18a shows that the impedance model reported by Cruz-Manzo et al. [39] simulates an inductive loop in the impedance spectrum when the charge transfer resistance in the hydrogen peroxide reaction is greater than the sum of the charge transfer resistance in the oxygen reaction and the oxygen diffusion impedance $R_{C,H_2O_2} > R_{C,O_2} + Z_{O_2}$. Fig. 7.18b shows that a capacitive loop is formed at low frequencies when the relation $R_{C,H_2O_2} < R_{C,O_2} + Z_{O_2}$ is satisfied. For a low fractional surface coverage of hydrogen peroxide $\gamma_{H_2O_2}$ there is an increase in charge transfer resistance for hydrogen peroxide formation R_{C,H_2O_2} and the formation of the inductive loop at low frequencies is apparent in the impedance spectrum.

The electrical circuits shown in Figs 7.8–7.10 simulate impedance spectra featuring an inductive loop at low frequencies through inductive elements or a negative capacitive element, but the electrical components are independent from each other. The impedance model represented through Eq. (7.12) demonstrates that there is an interaction between the parameters of the model to simulate the low-frequency inductive loop. The analytical parameter $A\varphi_{H_2O_2}$ represented in Eq. (7.14) simulates the low-frequency inductive loop and is a

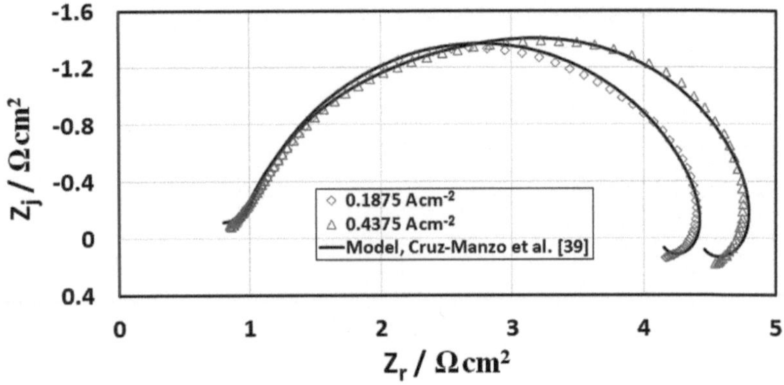

FIG. 7.19 Comparison between EIS measurements in a 4-cell open-cathode PEMFC stack and impedance response generated by the impedance model [39] expressed in Eq. (7.12) considering hydrogen peroxide formation H_2O_2 in a two-step oxygen reduction reaction.

function of the parameters that simulate the impedance spectrum with negative imaginary components such as charge transfer resistance in the oxygen reaction and oxygen diffusion impedance. Therefore, a change in the parameters that simulate the impedance spectrum with negative imaginary components also yield a change in the inductive loop at low frequencies.

Cruz Manzo et al. [39] analysed the EIS measurements of the whole open-cathode PEMFC stack described previously in Section 7.2.1 through the impedance model considering hydrogen peroxide formation during a two-step ORR expressed in Eq. (7.12). Fig. 7.19 shows a comparison between the EIS measurements carried out over the whole open-cathode PEMFC stack and the simulated impedance response generated by Eq. (7.12) at two different current densities.

Fig. 7.19 shows that even though the impedance model of the PEMFC considering hydrogen peroxide formation can simulate inductive loops at low frequencies, it is not possible to accurately reproduce the low-frequency inductive loops from the EIS measurements. The discrepancy between the measured and the simulated low-frequency inductive loops shown in Fig. 7.19 can be attributed to the presence of other electrochemical phenomena with time constants represented at the lowest frequencies.

7.3.2 Platinum oxide formation

Cruz-Manzo et al. [41] proposed an impedance model of the PEMFC that not only considers hydrogen peroxide formation, but also platinum oxide formation during a two-step ORR. Similar to Eq. (7.12), the impedance model considered the sum of the ohmic resistance R_{ohm} of the PEMFC with the impedance Z_{anode} associated with the HOR in the anode and with the impedance of the cathode. The impedance of the cathode in the impedance model of the PEMFC considering hydrogen peroxide formation and platinum oxide formation during a two-step ORR is represented as:

$$Z_{\text{cathode}_{O_2,H_2O_2,PtO}} = \left[\frac{1}{\frac{1}{R_{C,O_2} + Z_{O_2}} + \frac{1}{R_{C,H_2O_2}} + A\varphi_{H_2O_2} + \frac{1}{[\tau_{PtO} j\omega + 1] R_{PtO}} + Y[j\omega]^P} \right] \quad (7.17)$$

where R_{C,O_2} is the charge transfer resistance for the oxygen reaction, Z_{O_2} is the finite-length Warburg impedance for the impedance of oxygen diffusion, R_{C,H_2O_2} is the charge transfer resistance for the hydrogen peroxide reaction, R_{PtO} is the resistance of the platinum oxide formation, $A\varphi_{H_2O_2}$ is an analytical parameter expressed in Eq. (7.14) and with

$$\tau_{PtO} = \frac{L}{R_{PtO}} = \frac{zF\lambda}{i_{Pt,f,ss} + i_{Pt,b,ss}} \quad (7.18)$$

and

$$R_{PtO} = \frac{[R_{Pt,f} R_{Pt,b}][i_{Pt,f,ss} + i_{Pt,b,ss}]}{[R_{Pt,f} + R_{Pt,b}][i_{O_2,ss} + i_{H_2O_2,ss}]} \quad (7.19)$$

where τ_{PtO} is the time constant for platinum oxide formation and λ is the maximum coverage of platinum oxide. A detailed derivation of the cathodic impedance model represented in Eq. (7.17) can be found in the study of Cruz-Manzo et al. [41].

In the study by Cruz-Manzo et al. [41], the impedance response generated by the impedance model of the PEMFC considering the impedance of the cathode with hydrogen peroxide formation and platinum oxide formation was fitted to the EIS measurements of the whole open-cathode PEMFC stack described previously in Section 7.2.1. Fig. 7.20 shows a comparison between the EIS

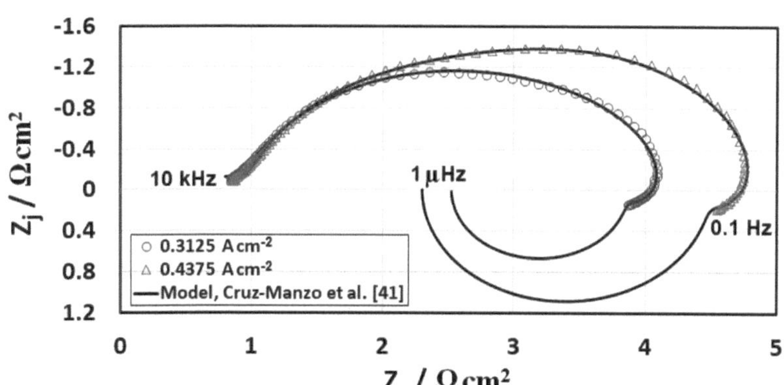

FIG. 7.20 Comparison between EIS measurements in a 4-cell open-cathode PEMFC stack and impedance response generated by the impedance model [41] considering hydrogen peroxide H_2O_2 formation and platinum oxide PtO formation in a two-step oxygen reduction reaction.

measurements and simulated impedance spectra generated by the PEMFC impedance model considering hydrogen peroxide formation and platinum oxide formation during the ORR. Low-frequency inductive loops of PEMFC-EIS measurements should converge with the value of the DC polarisation resistance estimated from the slope of the polarisation curve as the frequency is reduced to a value near zero. The resistive value Z_r of the impedance spectra as the frequency is decreasing $\omega \to 0$ should be equal to the value of the polarisation resistance estimated from the slope of the polarisation curve. For instance, the value of the polarisation resistance estimated from the slope of the polarisation curve shown in Fig. 7.5 at 0.3125 A/cm^2 is $R_{pol} = 2.52\ \Omega cm^2$, this estimation of the polarisation resistance has been demonstrated in Fig. 3.20 from Chapter 3. Further information about the modelling validation procedure from Fig. 7.20 can also be found in the study of Cruz-Manzo et al. [41].

A dead-ended test configuration of the gas reactants allows a gradual water accumulation condition in a PEMFC. EIS measurements carried out in a PEMFC with a dead-ended configuration of the gas reactants during water accumulation can be found in the study of Cano-Castillo et al. [42]. Fig. 7.21 shows a comparison between EIS measurements carried out in a single 5 cm^2 H_2/O_2 PEMFC during gradual water accumulation and the impedance response predicted by the impedance model of the PEMFC considering the impedance of the cathode with hydrogen peroxide formation and platinum oxide formation during the ORR (Eq. 7.17). The EIS measurements featured inductive loops at low frequencies as shown in Fig. 7.21. The low-frequency inductive loops from the EIS measurements shown in Fig. 7.21 were attributed to electrochemical mechanisms during the ORR and not to instability during the gradual water accumulation as the low-frequency inductive loops complied with K–K properties. More details about the K–K analysis for the EIS tests

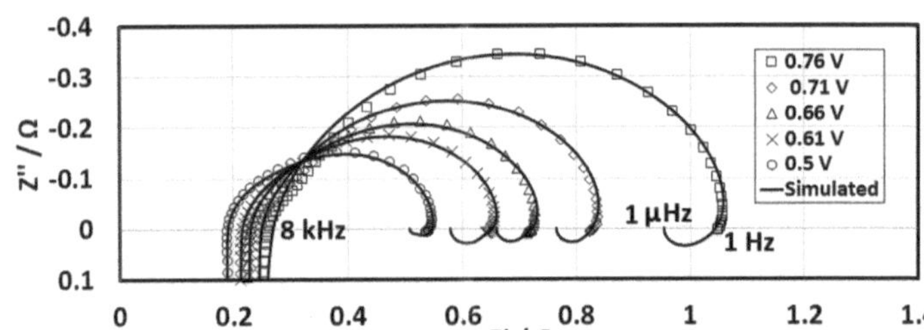

FIG. 7.21 Comparison between measured and simulated EIS spectra at different cell voltage. EIS measurements carried out in a single 5 cm^2 H_2/O_2 PEMFC operated with a dead-ended configuration of the gas reactants. The simulated impedance spectrum is generated with an impedance model of the PEMFC considering the impedance of the cathode with hydrogen peroxide formation and platinum oxide formation during the oxygen reduction reaction [41]. *(Reproduced from S. Cruz-Manzo, U. Cano-Castillo, P. Greenwood, Impedance study on estimating electrochemical mechanisms in a polymer electrolyte fuel cell during gradual water accumulation, Fuel Cells 19 (2019) 78, with permission of John Wiley & Sons. Copyright John Wiley & Sons.)*

carried out during the water accumulation in the PEMFC shown in Fig. 7.21 can be found in the study of Cruz-Manzo et al. [43]. The experimental low-frequency inductive loops were validated by reducing the frequency during model simulation until the simulated inductive loop converged with the DC polarisation resistance calculated from the slope of the polarisation curve. The modelling validation procedure from Fig. 7.21 can also be found in the study of Cruz-Manzo et al. [43]. The low-frequency inductive loops were mainly attributed to platinum oxide formation which increased with increasing water accumulation during PEMFC operation.

Figs 7.20 and 7.21 demonstrated that the impedance model considering platinum oxide formation during the ORR was able to predict low-frequency inductive loops in EIS measurements of PEMFCs operated at different conditions. The hypothesis of platinum oxide formation for the prediction of the low-frequency inductive loop in the impedance spectrum of the PEMFC was found to be in accordance with the frequency response of electrochemical mechanisms during EIS measurements in PEMFCs.

Other physics based-impedance models have been proposed to predict low-frequency inductive loops of EIS measurements of PEMFCs attributed to platinum oxide formation during the ORR. The impedance response of the model that considers the formation of platinum oxide as reported by Setzler and Fuller [18] agreed with EIS measurements carried out in a H_2/air PEMFC operated at 80°C and 75% RH in the gas reactants as shown in Fig. 7.22. In addition, Fig. 7.22 shows that the simulated impedance spectrum generated by the impedance model reported by Setzler and Fuller [18] was extrapolated to converge the low-frequency limit of the impedance spectrum with the DC polarisation resistance estimated from the polarisation curve. The authors concluded that the growth of the platinum oxide layer in the CCL which yields the formation of the inductive loop at low frequencies should not be ignored in EIS modelling.

7.3.3 Water transport in the electrode

The hypothesis that different processes during PEMFC operation yield the formation of low-frequency inductive loops during EIS measurements should not be ruled out. Inductive loops at low frequencies in PEMFC-EIS measurements have also been attributed to water transport mechanisms in the PEMFC. Schiefer et al. [44] reported EIS measurements in a H_2/O_2 PEMFC operated with different RH and keeping the same humidity level in both inlet gas reactants. The size of the inductive loops at low frequencies was reduced with increasing RH. Kosakian et al. [19] related low-frequency inductive loops to the water absorption/desorption kinetics in the electrolyte of the catalyst layers using a single-phase 2D PEMFC impedance model. The authors demonstrated that a faster interfacial transport of water yields a lower water content in the electrolyte and stronger capacitive behaviour in the catalyst layer. Niya et al. [24] developed an impedance model of the PEMFC to understand electrochemical

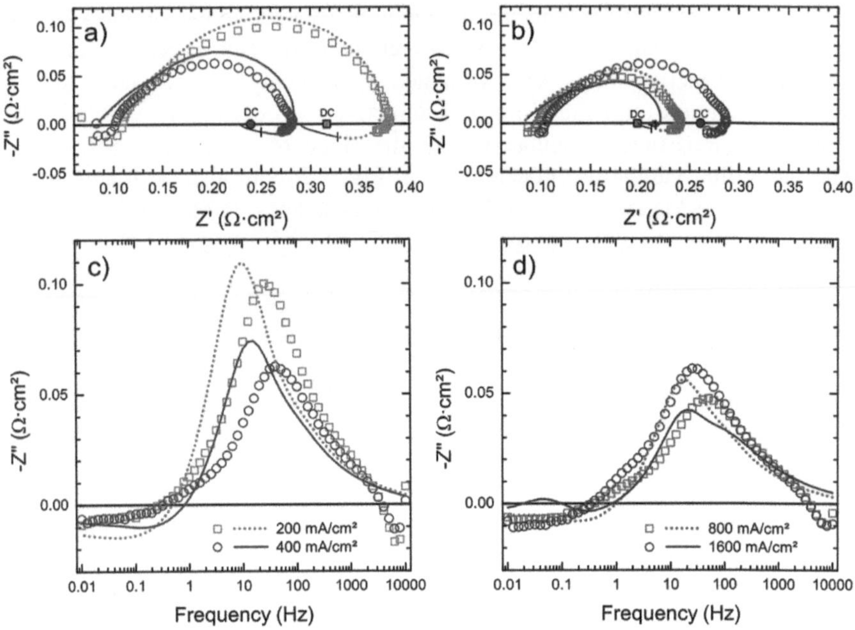

FIG. 7.22 Comparison between EIS measurements carried out in a H_2/air PEMFC operated at 80°C and 75% relative humidity in the gas reactants and simulated impedance spectra reported by Setzler and Fuller [18]. Outline symbols: experimental; points labelled DC: DC polarisation resistance; lines: model; thin black lines: extended model results to 1 μHz. (a) Low current density Nyquist plot, (b) high current density Nyquist plot, (c) low current density imaginary component Bode plot, (d) high current density imaginary component Bode plot. *(From B.P. Setzler, T.F. Fuller, A physics-based impedance model of proton exchange membrane fuel cells exhibiting low-frequency inductive loops, J. Electrochem. Soc. 162 (2015) F526. Figure published under license of Creative Commons Attribution 4.0 License (CC BY).)*

phenomena that yield the formation of the inductive loops at low frequencies. Niya et al. [24] concluded that the low-frequency inductive loop is inversely proportional to the water transport coefficient in the GDL. When the water concentration increases in the GDL, there is a reduction of the inductive loop as shown in Fig. 7.23.

7.3.4 Water vapour diffusion in the cathode catalyst layer

Water management in the CCL is a complicated process as different water transport mechanisms are involved such as water vapour condensation and liquid water evaporation, membrane/ionomer absorption and desorption, water transport through the PEM via electroosmotic drag, back diffusion, and hydraulic permeation. The formation of inductive loops at low frequencies in EIS measurements of PEMFCs can be attributed to water vapour transport and ionomer swelling/shrinking (hydration/dehydration) in the CCL [37].

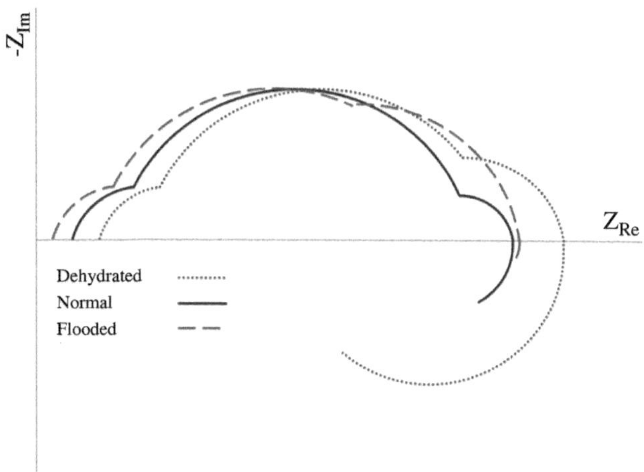

FIG. 7.23 Effect of water transport coefficient in the GDL on the impedance spectrum of the PEMFC. When the GDL is flooded a decrease in the inductive loop is observed. *(Reproduced from S.M.R. Niya, R.K. Phillips, M. Hoorfar, Process modeling of the impedance characteristics of proton exchange membrane fuel cells, Electrochim. Acta 191 (2016) 603, with permission of Elsevier. Copyright Elsevier.)*

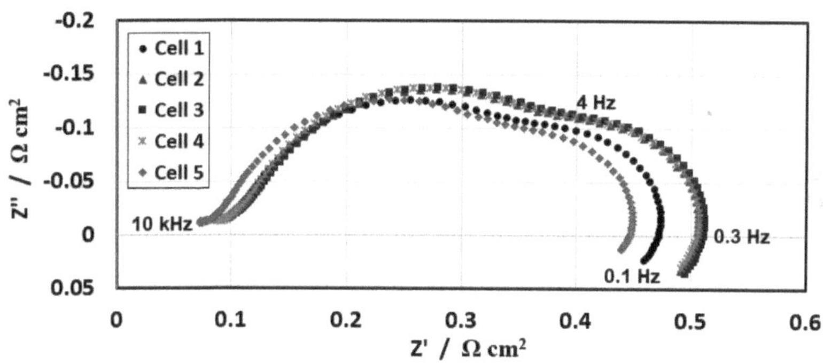

FIG. 7.24 EIS measurements carried out on the individual cells of a 5-cell open-cathode PEMFC stack operated at 0.47 A/cm^2. *(Reproduced from S. Cruz-Manzo, P. Greenwood, Low frequency inductive loop in EIS measurements of an open-cathode polymer electrolyte fuel cell stack. Impedance of water vapour diffusion in the cathode catalyst layer, J. Electroanal. Chem. 900 (2021) 115733/4, with permission of Elsevier. Copyright Elsevier.)*

In the study of Cruz-Manzo and Greenwood [37], EIS measurements were carried out on the individual cells of a 5-cell open-cathode PEMFC stack with an active area of 65 cm^2. The EIS measurements featured inductive loops at low frequencies and the size of the inductive loops were indirectly proportional to the ohmic resistance of the PEMFC. Fig. 7.24 shows the EIS measurements carried out in the individual cells of the open-cathode PEMFC stack at 0.47 A/cm^2 reported in the study of Cruz-Manzo and Greenwood [37]. The open-cathode PEMFC stack was operated at 50°C and considered a through-flow mode on the hydrogen side. The EIS measurements shown in Fig. 7.24 were carried out with a multichannel FRA (Z#106 WonATech Co) and programmable load unit (Dynaload RBL488) in galvanostatic mode. The amplitude of the signal perturbation during EIS measurements was kept to 5% of the DC current to comply with the linearity property of EIS analysis and comply with K–K properties. Further information about the experimental set-up for the EIS measurements shown in Fig. 7.24 can be found in the study of Cruz-Manzo and Greenwood [37].

Fig. 7.24 shows an overlapped second semicircle at low frequencies (e.g. impedance spectrum of cell 3 from 4 Hz to 0.3 Hz) attributed to an increase in oxygen transport limitation in the CCL-GDL interface which hinders the transport of oxygen towards the active sites in the CCL. The size of the impedance spectra of cell 1 and 5 are similar. This similitude can also be related to the EIS measurement at the minimum imaginary component $-Z''$ (the highest point of the impedance spectrum of cell 1 and 5) shown in Fig. 7.24. The similitude in the minimum imaginary component $-Z''$ or similar highest point of the impedance spectrum of cells 1 and 5 is related to a similar value in double-layer capacitance.

Fig. 7.24 shows that the impedance spectra of cells 2, 3, and 4 present the same features but with increased size than the impedance spectra for cells 1 and 5. It is possible to propose different hypotheses for the difference in impedance spectra of the cells comprising the open-cathode PEMFC stack. The increase in the impedance spectra for cells 2, 3, and 4 can be attributed to a reduction in water content with increasing air flow rate [45,46]. The temperature of the PEMFC stack was measured at a single point (middle cell), but a gradient in temperature across the PEMFC stack is expected. The temperature improves the kinetics of the ORR, improves proton conduction in the PEM and yields a reduction in the size of the impedance spectrum [47]. Nevertheless, a severe increase in the heat generated during PEMFC operation could dehydrate the PEM and increase proton resistance [48]. In addition, the diffusion of water vapour increases with temperature [49], and water vapour swells the CCL electrolyte and increases proton conductivity in the CCL and PEM [50]. The impedance spectrum of cell 5 was similar to the impedance spectrum of cell 1, but the impedance spectrum of cell 5 is shifted towards the left on the real axis Z' of the Nyquist plot (low ohmic resistance) and presents a lower size of inductive loop. In the open-cathode PEMFC stack reported in the study of Cruz-Manzo and Greenwood [37], from which the impedance spectra shown in Fig. 7.24 were measured, cell 1 is located at the top and cell 5 at the bottom of the stack configuration. Hydrogen flows from cell 1 to cell 5 and carries water vapour and liquid water produced during the ORR and may have increased the water concentration in cell 5. Gravity's effect on water can also change the performance of PEMFC stacks [51,52]. Silva et al. [53] carried out an in situ evaluation of degradation in an eight-cell open-cathode PEMFC stack and reported greater water accumulation in the last cells (bottom of the stack).

The impedance spectrum of cell 5 is shifted towards the left of the real axis Z' in the Nyquist plot and presents the smallest inductive loop as shown in Fig. 7.24. This can be related to a higher water concentration in the MEA of cell 5. It can be assumed that the inductive loops can be related to water concentration in the CCL and the size of the inductive loops is inversely proportional to the water concentration present in the CCL. The value of the ohmic resistance is inversely proportional to water concentration in the PEM.

Based on relation between the ohmic resistance and inductive loop in the EIS measurements shown in Fig. 7.24, it is proposed that water transport plays an important role in the formation of the inductive loops at low frequencies. Cruz-Manzo and Greenwood [37] developed an impedance model of the PEMFC neglecting the anode contribution and considering that water vapour diffusion in the CCL yielded the formation of the inductive loop at low frequencies as such:

$$Z_{\text{PEMFC-cathode}} = R_{\text{ohm}} + \frac{R_{\text{P}}\coth(\beta)}{\beta} \tag{7.20}$$

with

$$\beta = \sqrt{R_{\text{P}}\left[\frac{1}{R_{\text{C}} + Z_{\text{O}_2} + Z_{\text{Wv}}} + Y[j\omega]^{P}\right]} \tag{7.21}$$

$$Z_{\text{Wv}} = -G_{\text{Wv}}\frac{j\omega\tau_{\text{Wv}}\coth\left(\sqrt{j\omega\tau_{\text{Wv}}}\right) - \sqrt{j\omega\tau_{\text{Wv}}}}{\left[\frac{1-\cosh\left(\sqrt{j\omega\tau_{\text{Wv}}}\right)}{\sinh\left(\sqrt{j\omega\tau_{\text{Wv}}}\right)}\right][1 + z\in]} \tag{7.22}$$

$$G_{\text{Wv}} = \frac{Z_{\text{Fw}}}{1 + Z_{\text{Fw}}j\omega C_{\text{Wv}}} \tag{7.23}$$

and

$$\tau_{\text{Wv}} = \frac{{\delta_{\text{Wv}}}^{2}}{D_{\text{H}_2\text{O}}} \tag{7.24}$$

where R_{P} is the proton resistance in the electrolyte of the CCL, R_{C} is the charge transfer resistance during the ORR, R_{ohm} is the ohmic resistance in the PEMFC and accounts for the opposition to the flow of protons and electrons in the PEM, GDL, and flow field plate (FFP), $\in$ is the number of water molecules that are dragged from anode to cathode per hydrogen ion, Z_{O_2} is the oxygen diffusion impedance in the CCL and is represented through the finite-length Warburg impedance. Y and P are parameters related to a CPE and account for the non-homogeneous accumulation of charge in the Nafion/Pt/carbon interface in the CCL, without taking into account the water vapour adsorption in the multiphase interface in the CCL. C_{Wv} accounts for the capacitance for accumulation of water vapour in the multiphase network interface (Pt/Nafion, carbon/Nafion, Pt/water, carbon/water, and Nafion/pore interface) of the CCL, Z_{Fw} is the faradaic impedance associated with the binary oxygen-water vapour transport in the CCL during the ORR, δ_{Wv} is the diffusion distance for water vapour in the CCL, and D_{H_2O} is the effective diffusion coefficient of water vapour in the CCL. A detailed derivation of the cathodic impedance model represented in Eq. (7.20) can be found in the study of Cruz-Manzo and Greenwood [37].

The impedance model of the PEMFC represented in Eq. (7.20) was fitted to the EIS measurements of the open-cathode PEMFC stack using MATLAB software. The fitting was carried out in a two-step process. First the impedance model (Eq. 7.20) neglecting the parameter $Z_{Wv}=0$ was fitted to the EIS measurements. Thereafter the full expression (Eq. 7.20) was fitted to the EIS measurements considering the inductive loops at low frequencies as shown in Fig. 7.25. Further information about the fitting process for the results shown Fig. 7.25 can be found in the study of Cruz-Manzo and Greenwood [37]. Fig. 7.25 shows that the impedance model of the PEMFC considering water vapour diffusion in the CCL is able to reproduce the EIS measurements of the open-cathode PEMFC stack. A lower size of inductive loop in the impedance spectrum is related to a higher water vapour concentration in the CCL and hence, an increase in capacitance in the multiphase network interface (Pt/Nafion, carbon/Nafion, Pt/water, carbon/water, and Nafion/pore interface). The size of the inductive loop and value of ohmic resistance in the EIS measurements can indicate the hydration state in the MEA. Cruz-Manzo and Greenwood [37] concluded that extensive work needs to be carried out to simulate and study the difference in the impedance spectra from the individual cells of the open-cathode PEMFC stack using a physics-based impedance model that considers different transport mechanisms for water in a liquid and vapour state as well as oxygen, hydrogen and temperature distributions across the CCL, GDL, PEM, and FFP.

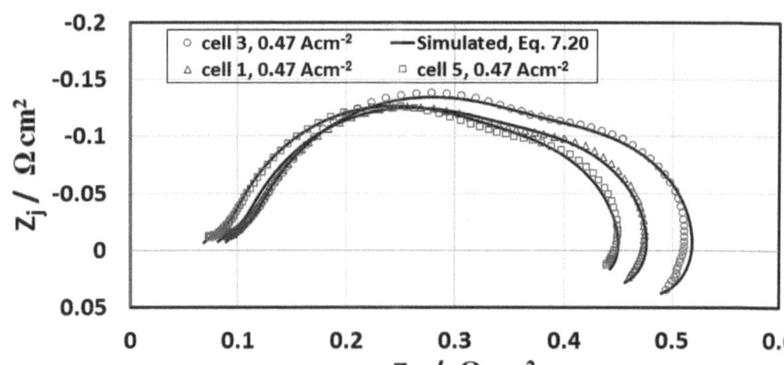

FIG. 7.25 Comparison between simulated (Eq. 7.20) and measured impedance spectra for cells 1, 3, and 5 of the open-cathode PEMFC stack reported by Cruz-Manzo and Greenwood [37] at 0.47 A/cm^2.

Inductive loops of PEMFC-EIS measurements at low frequencies should converge with the value of the DC polarisation resistance estimated from the slope of the polarisation curve as the frequency is reduced to a value near zero. Nevertheless, it may not be feasible to carry out EIS measurements at very low frequencies (the EIS test may take too long) to assess the performance of a PEMFC stack operated in a real-world application. As discussed in previous sections, the formation of low-frequency inductive loops in PEMFC-EIS measurements can be related to different phenomena. Previously in Section 7.3.2, it was proposed that the formation of inductive loops at the lowest frequencies in EIS measurements carried out in an open-cathode PEMFC stack was attributed to platinum oxide formation during the ORR. Fig. 7.20 shows that the formation of the low-frequency inductive loop was also attributed to hydrogen peroxide formation which was overlapped with the platinum oxide formation during the ORR. The platinum oxide formation was related to EIS measurements lower than 0.1 Hz as shown in Fig. 7.20. In Fig. 7.25, the formation of the low-frequency inductive loop is associated with water vapour diffusion in the CCL and shows that the inductive loop simulated by the impedance model represented in Eq. (7.20) is not extrapolated to lower frequencies, e.g. 1 μHz, to validate the inductive loop with the DC polarisation resistance estimated from the polarisation curve. The inductive loops from the EIS measurements shown in Fig. 7.25 at frequencies lower than 0.1 Hz may be associated with a different process (e.g. platinum oxide formation) as discussed in Section 7.3.2 which is not considered in the impedance model represented through Eq. (7.20). Even though the inductive loop from PEMFC-EIS measurements may not be measured at very low frequency e.g. <0.1 Hz, it could still provide relevant information during the assessment of the performance of the individual cells in a PEMFC stack using EIS. Cruz-Manzo and Greenwood [37] demonstrated that the developed impedance model expressed in Eq. (7.20) is able to provide an insight into the relation between the trend of the inductive loops and the phase angle of the EIS measurements which can be associated with the hydration state in the MEA of the individual cells of the open-cathode PEMFC stack.

7.3.4.1 Relation between inductive loops and phase angle of EIS measurements

As previously discussed, the analysis of inductive loops at low frequencies from EIS measurements using an impedance model should be carried out at a very low frequency, e.g. 0.01 mHz, to validate the low-frequency resistive value of the EIS measurements with the DC polarisation resistance estimated from the polarisation curve. Nevertheless, it may not be feasible to carry out EIS measurements at very low frequencies to assess the performance of a PEMFC stack operated in a real-world application. It is possible to carry out an assessment of the individual cells of a PEMFC stack by analysing the trend of the EIS measurements at the lowest frequency considered.

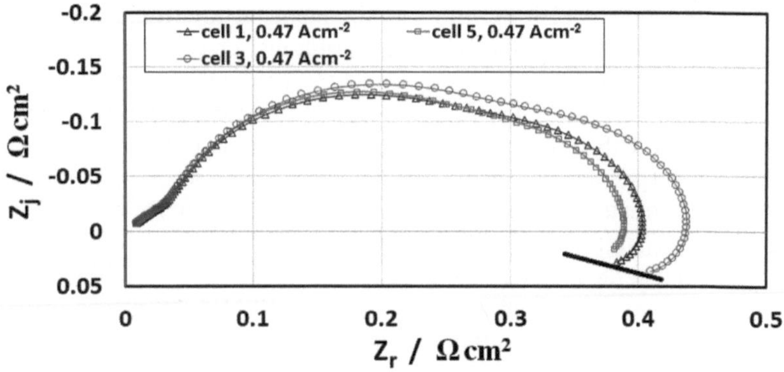

FIG. 7.26 Simulated impedance spectra of cells 1, 3, and 5 of the open-cathode PEMFC stack reported by Cruz-Manzo and Greenwood [37] at 0.47 A/cm^2 generated by Eq. (7.20) and neglecting the ohmic resistance $R_{ohm}=0$.

Fig. 7.26 shows the simulated impedance spectra generated by the impedance model represented in Eq. (7.20) and neglecting the ohmic resistance $R_{ohm}=0$. The simulated impedance spectra shown in Fig. 7.26 correspond to the simulated impedance spectra shown in Fig. 7.25, but shifted towards the left of the Nyquist plot. Fig. 7.26 shows that the inductive loop for cell 5 does not follow the same trend for the EIS measurement at the lowest frequency of 0.1 Hz and with respect to cells 1 and 3. This difference in inductive loop for cell 5 with respect to cells 1 and 3 can be attributed to an out of phase angle in the EIS measurements for cell 5. Cruz-Manzo and Greenwood [37] demonstrated that the time constant for water-vapour diffusion τ_{Wv} and water-vapour capacitance C_{Wv} in the impedance model represented in Eq. (7.20) predict the low-frequency inductive loop and hence a change in the overall spectrum which can be an indication that not only water vapour transport but also ionomer swelling/shrinking (hydration/dehydration) affect the size of the inductive loop at low frequencies. Further discussion about the change of parameters τ_{Wv} and C_{Wv} to simulate the impedance spectra of the individual cells of the open-cathode PEMFC stack can be found in the study of Cruz-Manzo and Greenwood [37].

Fig. 7.27 shows the phase angle of EIS measurements at 0.47 A/cm^2 for cells 1, 3, and 5 of the open-cathode PEMFC stack reported by Cruz-Manzo and Greenwood [37]. Fig. 7.27 clearly shows the difference in phase angle of cell 5 with respect to cells 1 and 3. The phase angle of EIS measurements becomes more negative with increasing water concentration in the electrode [54]. The results from Figs 7.25–7.27 demonstrate that a decrease in ohmic resistance, a decrease in size of the inductive loops and a change in phase angle in the EIS measurements with respect to the other cells can indicate the hydration state in the MEA of the open-cathode PEMFC stack.

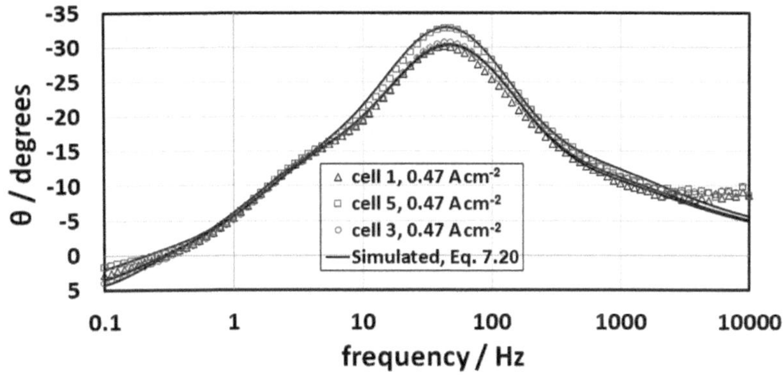

FIG. 7.27 Comparison between simulated (Eq. 7.20) and measured phase angle for cells 1, 3, and 5 of the open-cathode PEMFC stack reported by Cruz-Manzo and Greenwood [37] at 0.47 A/cm^2.

References

[1] W. Mérida, D.A. Harrington, J.M. Le Canut, G. McLean, Characterisation of proton exchange membrane fuel cell (PEMFC) failures via electrochemical impedance spectroscopy, J. Power Sources 161 (2006) 264–274.

[2] T. Kurz, A. Hakenjos, J. Kramer, M. Zedda, C. Argert, An impedance-based predictive control strategy for the state-of-health of PEM fuel cell stacks, J. Power Sources 180 (2008) 742–747.

[3] N.V. Dale, M.D. Mann, H. Salehfar, A.M. Dhirde, T. Han, AC impedance study of a proton exchange membrane fuel cell stack under various loading conditions, J. Fuel Cell Sci. Technol. 7 (2010), 031010.

[4] N. Fouquet, C. Doulet, C. Nouillant, G. Dauphin-Tanguy, B. Ould-Bouamama, Model based PEM fuel cell state-of-health monitoring via AC impedance measurements, J. Power Sources 159 (2006) 905–913.

[5] X. Yuan, J.C. Sun, M. Blanco, H. Wang, J. Zhang, D.P. Wilkinson, AC impedance diagnosis of a 500W PEM fuel cell stack, part I: stack impedance, J. Power Sources 161 (2006) 920–928.

[6] T. Nasu, Y. Matsushita, J. Okano, K. Okajima, Study of current distribution in PEMFC stack using magnetic sensor probe, J. Int. Counc. Electr. Eng. 2 (2012) 391–396.

[7] M. Cimenti, M. Tam, J. Stumper, High frequency artifacts in electrochemical impedance spectroscopy measurements on PEM fuel cells, Electrochem. Solid-State Lett. 12 (2009) B131–B134.

[8] M. Ciureanu, R. Roberge, Electrochemical impedance study of PEM fuel cells. Experimental diagnostics and modeling of air cathodes, J. Phys. Chem. B 105 (2001) 3531–3539.

[9] A. Hakenjos, M. Zobel, J. Clausnitzer, C. Hebling, Simultaneous electrochemical impedance spectroscopy of single cells in a PEM fuel cell stack, J. Power Sources 154 (2006) 360–363.

[10] X. Yuan, J.C. Sun, H. Wang, J. Zhang, AC impedance diagnosis of a 500W PEM fuel cell stack part II: individual cell impedance, J. Power Sources 161 (2006) 929–937.

[11] J.-P. Diard, N. Glandut, B.L. Gorrec, C. Montella, Impedance measurement of each cell of a 10 W PEMFC stack under load, J. Electrochem. Soc. 151 (2004) A2193–A2197.

[12] S. Cruz-Manzo, R. Chen, P. Greenwood, Analysis of the performance of an open-cathode polymer electrolyte fuel cell stack using simultaneous electrochemical impedance spectroscopy measurements, ECS Trans. 48 (2014) 47–58.

[13] X. Zhang, Y. Jiang, L. Huang, W. Chen, D. Brett, Inconsistent responses of cells on operating conditions in a 5 kW proton exchange membrane fuel cell stack, Electrochim. Acta 391 (2021), 138925.

[14] S.K. Roy, M.E. Orazem, B. Tribollet, Interpretation of low-frequency inductive loops in PEM fuel cells, J. Electrochem. Soc. 154 (2007) B1378–B1388.

[15] N. Wagner, M. Schulze, Change of electrochemical impedance spectra during CO poisoning of the Pt and Pt–Ru anodes in a membrane fuel cell (PEFC), Electrochim. Acta 48 (2003) 3899–3907.

[16] I.A. Schneider, M.H. Bayer, A. Wokaun, G.G. Scherer, Impedance response of the proton exchange membrane in polymer electrolyte fuel cells, J. Electrochem. Soc. 155 (2008) B783–B792.

[17] I. Pivac, F. Barbir, Inductive phenomena at low frequencies in impedance spectra of proton exchange membrane fuel cells—a review, J. Power Sources 326 (2016) 112–119.

[18] B.P. Setzler, T.F. Fuller, A physics-based impedance model of proton exchange membrane fuel cells exhibiting low-frequency inductive loops, J. Electrochem. Soc. 162 (2015) F519–F530.

[19] A. Kosakian, L.P. Urbina, A. Heaman, M. Secanell, Understanding single-phase water-management signatures in fuel cell impedance spectra: a numerical study, Electrochim. Acta 350 (2020), 136204.

[20] G.A. Futter, P. Gazdzicki, K.A. Friedrich, A. Latz, T. Jahnke, Physical modeling of polymer-electrolyte membrane fuel cells: understanding water management and impedance spectra, J. Power Sources 391 (2018) 148–161.

[21] K. Wiezell, N. Holmström, G. Lindbergh, Studying low-humidity effects in PEFCs using EIS II. Modeling, J. Electrochem. Soc. 159 (2012) F379–F392.

[22] I. Pivac, D. Bezmalinović, F. Barbir, Catalyst degradation diagnostics of proton exchange membrane fuel cells using electrochemical impedance spectroscopy, Int. J. Hydrog. Energy 43 (2018) 13512–13520.

[23] S.M.R. Niya, M. Hoorfar, Process modeling of the ohmic loss in proton exchange membrane fuel cells, Electrochim. Acta 120 (2014) 193–203.

[24] S.M.R. Niya, R.K. Phillips, M. Hoorfar, Process modeling of the impedance characteristics of proton exchange membrane fuel cells, Electrochim. Acta 191 (2016) 594–605.

[25] I. Pivac, B. Šimić, F. Barbir, Experimental diagnostics and modeling of inductive phenomena at low frequencies in impedance spectra of proton exchange membrane fuel cells, J. Power Sources 365 (2017) 240–248.

[26] A.M. Dhirde, N.V. Dale, H. Salehfar, M.D. Mann, T.-H. Han, Equivalent electric circuit modeling and performance analysis of a PEM fuel cell stack using impedance spectroscopy, IEEE Trans. Energy Convers. 25 (2010) 778–786.

[27] S. Cruz-Manzo, R. Chen, An electrical circuit for performance analysis of polymer electrolyte fuel cell stacks using electrochemical impedance spectroscopy, J. Electrochem. Soc. 160 (2013) F1109–F1115.

[28] R. Mohtadi, W.-K. Lee, J.W.V. Zee, Assessing durability of cathodes exposed to common air impurities, J. Power Sources 138 (2004) 216–225.

[29] J.M. Moore, P.L. Adcock, J.B. Lakeman, G.O. Mepsted, The effects of battlefield contaminants on PEMFC performance, J. Power Sources 85 (2000) 254–260.

[30] J.C. Kurnia, B.A. Chaedir, A.P. Sasmito, T. Shamim, Progress on open cathode proton exchange membrane fuel cell: performance, designs, challenges and future directions, Appl. Energy 283 (2021), 116359.

[31] S. Chevalier, B. Auvity, J.C. Olivier, C. Josset, D. Trichet, M. Machmoum, Detection of cells state-of-health in PEM fuel cell stack using EIS measurements coupled with multiphysics modeling, Fuel Cells 14 (2014) 416–429.

[32] C. Bonnet, L. Franck-Lacaze, B.T. Huang, Y. Chatillon, G. Valentin, F. Lapicque, Aging of polymer electrolyte membrane fuel cells (PEMFC): general features and investigation of two typical examples, J. Appl. Electrochem. 42 (2012) 699–709.

[33] B.A. Boukamp, Derivation of a distribution function of relaxation times for the (fractal) finite length warburg, Electrochim. Acta 252 (2017) 154–163.

[34] D. Klotz, Negative capacitance or inductive loop?—a general assessment of a common low frequency impedance feature, Electrochem. Commun. 98 (2019) 58–62.

[35] MathWorks®. https://uk.mathworks.com/help/optim/ug/lsqcurvefit.html. last access, June 2022.

[36] M. Orazem, B. Tribollet, Electrochemical Impedance Spectroscopy, second ed., Wiley, Hoboken, NJ, 2017.

[37] S. Cruz-Manzo, P. Greenwood, Low frequency inductive loop in EIS measurements of an open-cathode polymer electrolyte fuel cell stack. Impedance of water vapour diffusion in the cathode catalyst layer, J. Electroanal. Chem. 900 (2021), 115733.

[38] M. Inaba, H. Yamada, J. Tokunaga, A. Tasaka, Effect of agglomeration of Pt/C catalyst on hydrogen peroxide formation, Electrochem. Solid-State Lett. 7 (2004) A474–A476.

[39] S. Cruz-Manzo, R. Chen, P. Greenwood, An impedance model for analysis of EIS of polymer electrolyte fuel cells under hydrogen peroxide formation in the cathode, J. Electroanal. Chem. 745 (2015) 28–36.

[40] J.-P. Diard, B. Gorrec, M.C. Le, Calculation, simulation and interpretation of electrochemical impedances: part 3. Conditions for observation of low frequency inductive diagrams for a two-step electron transfer reaction with an adsorbed intermediate species, J. Electroanal. Chem. 326 (1992) 13–36.

[41] S. Cruz-Manzo, C. Perezmitre-Cruz, P. Greenwood, R. Chen, An impedance model for analysis of EIS of polymer electrolyte fuel cells under platinum oxidation and hydrogen peroxide formation in the cathode, J. Electroanal. Chem. 771 (2016) 94–105.

[42] U. Cano-Castillo, A. Ortiz, S. Manzo, L.G. Arriaga, G. Orozco, Parameter changes during gradual flooding of a PEM fuel cell through EIS studies, ECS Trans. 3 (2006) 931–939.

[43] S. Cruz-Manzo, U. Cano-Castillo, P. Greenwood, Impedance study on estimating electrochemical mechanisms in a polymer electrolyte fuel cell during gradual water accumulation, Fuel Cells 19 (2019) 71–83.

[44] A. Schiefer, M. Heinzmann, A. Weber, Inductive low-frequency processes in PEMFC-impedance spectra, Fuel Cells 0 (2020) 1–8.

[45] T.E. Springer, T.A. Zawodzinski, S. Gottesfeld, Polymer electrolyte fuel cell model, J. Electrochem. Soc. 138 (1991) 2334–2342.

[46] Y.-S. Chen, H. Peng, D.S. Hussey, D.L. Jacobson, D.T. Tran, T. Abdel-Baset, M. Biernacki, Water distribution measurement for a PEMFC through neutron radiography, J. Power Sources 170 (2007) 376–386.

[47] X. Yan, M. Hou, L. Sun, D. Liang, Q. Shen, H. Xu, P. Ming, B. Yi, AC impedance characteristics of a 2 kW PEM fuel cell stack under different operating conditions and load changes, Int. J. Hydrog. Energy 32 (2007) 4358–4364.

[48] T. Ous, C. Arcoumanis, The formation of water droplets in an air-breathing PEMFC, Int. J. Hydrog. Energy 34 (2009) 3476–3487.

[49] P.W. Majsztrik, M.B. Satterfield, A.B. Bocarsly, J.B. Benziger, Water sorption, desorption and transport in Nafion membranes, J. Membr. Sci. 301 (2007) 93–106.

[50] L. Xing, X. Liu, T. Alaje, R. Kumar, M. Mamlouk, K. Scott, A two-phase flow and non-isothermal agglomerate model for a proton exchange membrane (PEM) fuel cell, Energy 73 (2014) 618–634.

[51] Y. Yi, T. Zheng-kai, Z. Zhi-gang, P. Mu, Gravity effect on the performance of PEM fuel cell stack with different gas manifold positions, Int. J. Energy Res. 36 (2012) 845–855.
[52] S. Chen, Y. Wu, Gravity effect on water discharged in PEM fuel cell cathode, Int. J. Hydrog. Energy 35 (2010) 2888–2893.
[53] R.A. Silva, T. Hashimoto, G.E. Thompson, C.M. Rangel, Characterization of MEA degradation for an open air cathode PEM fuel cell, Int. J. Hydrog. Energy 37 (2012) 7299–7308.
[54] N. Wagner, Electrochemical Impedance Spectroscopy From: PEM Fuel Cell Diagnostic Tools, CRC Press, 25 Aug 2011. Accessed on: June 2022 https://www.routledgehandbooks.com/doi/10.1201/b11100-5.

Appendix

A Impedance models in MATLAB software

Scripts to be executed in MATLAB software. The scripts shown below can only be executed if the toolbox 'Optimization Toolbox' is installed in MATLAB software. Further information about the 'Optimization Toolbox' in MATLAB software can be found in the following link: https://uk.mathworks.com/help/optim/index.html?s_tid=CRUX_lftnav, last access June 2022.

Note. The command lines containing indentations means that the command line was separated from the one above. In the MATLAB script, the command line with indentation should be placed with the command above.

A.1 Chapter 1

A.1.1 Script to reproduce Fig. 1.24

```
% Frequency range 100 kHz - 0.01 Hz
Frequency=logspace(-2,5,100);
fr=(sort(Frequency,'descend'))';
% Parameters of Randles circuit with inductor element, Eq. 1.66
R1=1;
R2=3;
Rw=1.5;
Tw=0.7;
Y=0.009;
P=0.75;
R3=10;
L=50;
% Angular frequency rad/s
w=2.*pi.*fr;
% Constant phase element
Yr=Y.*(w.*1i).^P;
% Warburg impedance, Eq. 1.55
Zw=Rw.*tanh(sqrt(1i.*w.*Tw))./(sqrt(1i.*w.*Tw));
% Inductor impedance, Eq. 1.54
Lz=1i.*w.*L;
% Impedance of Randles circuit with inductor element, Eq. 1.66
Z=R1+((R2+Zw).*(R3+Lz)./((R3+Lz).*(1+(R2+Zw).*Yr)+R2+Zw));
```

```
% Argument of the impedance of the Randles circuit with inductor
  element
Zarg=abs(Z);
% Phase angle of the impedance of the Randles circuit with
  inductor element
Zang=angle(Z);
% Real component of the impedance of the Randles circuit with
  inductor element
Z_r=Zarg.*cos(Zang);
% Imaginary component of the impedance of the Randles circuit
  with inductor element
Z_i=Zarg.*sin(Zang);
% Initial conditions for estimation of elements in Voigt circuit
x0=[0.1 0.1 0.01 0.1 0.01 0.1 0.01 0.1 0.01 0.1 0.01 0.1 0.01 0.1
  0.01];
xdata=fr;
% Initial conditions for estimation of real and imaginary
  residuals
Res=2;
Res_r=0;
Res_i=0;
Iteration=0;
% Maximum number of iterations during fitting process
limit=100;
% Conditional statement for real and imaginary residuals
while Res > 0.01
if Iteration==limit
break
end
% Functions defining elements in the Voigt circuit
f_Voi_0=@(x,xdata)(x(1)+(x(2)./(1+(x(2).*x(3).*2.*pi.*xdata.
  *1i))));
f_Voi_1=@(x,xdata)((x(4)./(1+(x(4).*x(5).*2.*pi.*xdata.*1i))));
f_Voi_2=@(x,xdata)((x(6)./(1+(x(6).*x(7).*2.*pi.*xdata.*1i))));
f_Voi_3=@(x,xdata)((x(8)./(1+(x(8).*x(9).*2.*pi.*xdata.*1i))));
f_Voi_4=@(x,xdata)((x(10)./(1+(x(10).*x(11).*2.*pi.*xdata.
  *1i))));
f_Voi_5=@(x,xdata)((x(12)./(1+(x(12).*x(13).*2.*pi.*xdata.
  *1i))));
% Function defining the impedance of the Voigt circuit, Eq. 1.67
f_Z=@(x,xdata)(f_Voi_0(x,xdata)+f_Voi_1(x,xdata)+f_Voi_2(x,
  xdata)+f_Voi_3(x,xdata)+f_Voi_4(x,xdata)+f_Voi_5(x,xdata));
% Function defining the argument of the impedance of the Voigt
  circuit
```

```
arg_m=@(x,xdata)abs(f_Z(x,xdata));
% Function defining the phase angle of the impedance of the Voigt
  circuit
ang_m=@(x,xdata)angle(f_Z(x,xdata));
% Function defining real component of the impedance of the Voigt
  circuit
real_m=@(x,xdata)arg_m(x,xdata).*cos(ang_m(x,xdata));
% Function defining imaginary component of impedance of the Voigt
  circuit
img_m=@(x,xdata)arg_m(x,xdata).*sin(ang_m(x,xdata));
% Non-linear regression
options=optimoptions('lsqcurvefit','Algorithm',
  'levenberg-marquardt');
lb = [];
ub = [];
xr=lsqcurvefit(real_m,x0,xdata,Z_r,lb,ub,options);
% Estimation of elements of the Voigt circuit
R0=abs(xr(1))
R1=abs(xr(2))
C1=abs(xr(3))
R2=abs(xr(4))
C2=abs(xr(5))
R3=abs(xr(6))
C3=abs(xr(7))
R4=abs(xr(8))
C4=abs(xr(9))
R5=abs(xr(10))
C5=abs(xr(11))
R6=-abs(xr(12))
C6=-abs(xr(13))
% Vector of elements of the Voigt circuit as initial condition for
  next iteration
x0 = [R0 R1 C1 R2 C2 R3 C3 R4 C4 R5 C5 R6 C6];
% Impedance of the Voigt circuit, Eq. 1.67
Z=R0+(R1./(1+(R1.*C1.*2.*pi.*fr.*1i)))+(R2./(1+(R2.*C2.*2.*pi.
  *fr.*1i)))+(R3./(1+(R3.*C3.*2.*pi.*fr.*1i)))+(R4./(1+(R4.*C4.
  *2.*pi.*fr.*1i)))+(R5./(1+(R5.*C5.*2.*pi.*fr.*1i)))+(R6./(1
  +(R6.*C6.*2.*pi.*fr.*1i)));
% Argument of the impedance of the Voigt circuit
arg=abs(Z);
% Phase angle of the impedance of the Voigt circuit
ang=angle(Z);
% Real component of the impedance of the Voigt circuit
mod=arg.*cos(ang);
```

```
% Imaginary component of the impedance of the Voigt circuit
Imge=arg.*sin(ang);
% Estimation of real and imaginary residuals
for n=1:numel(fr)
% Residual of real component
Jpr=abs((Z_r(n)-(mod(n)))./Zarg(n));
KpR=(Res_r+Jpr);
Res_r=Jpr;
kpr(n,:)=KpR;
% Residual of imaginary component
Jpi=abs((Z_i(n)-(Imge(n)))./Zarg(n));
KpI=(Res_i+Jpi);
Res_i=Jpi;
kpi(n,:)=KpI;
end
% Average sum of real and imaginary residuals
Res=(mean(kpi+kpr));
% Number of iterations
Iteration=Iteration+1;
end
%%%%%% Plotting results %%%%%%
% Complex-impedance (Nyquist) plot
figure (1)
plot(Z_r,-Z_i,'ok')
hold on
plot(mod,-Imge,'r')
xlabel('Z_r / \Omega','FontSize',14,'FontWeight','bold',
  'Color','k')
ylabel('- Z_j / \Omega','FontSize',14,'FontWeight','bold',
  'Color','k')
grid;
hleg1 = legend(' Randles with inductor','Voigt circuit');
hleg1=gca;
hleg1.FontSize = 16;
hleg1.FontWeight = 'bold';
axis([0 6 -0.8 1.2])
% Bode plot, angle
figure (2)
semilogx(fr,-(Zang*360)/(2*pi),'ok')
hold on
semilogx(fr,-(ang*360)/(2*pi),'r')
xlabel('Frequency / Hz','FontSize',14,'FontWeight','bold',
  'Color','k')
```

```
ylabel('Angle / degrees','FontSize',14,'FontWeight','bold',
  'Color','k')
grid;
hlegl = legend(' Randles with inductor','Voigt circuit');
hlegl=gca;
hlegl.FontSize = 16;
hlegl.FontWeight = 'bold';
% Bode plot, argument
figure (3)
semilogx(fr,Zarg,'ok')
hold on
semilogx(fr,arg,'r')
xlabel('Frequency / Hz','FontSize',14,'FontWeight','bold',
  'Color','k')
ylabel('Modulus / \Omega','FontSize',14,'FontWeight','bold',
  'Color','k')
grid;
hlegl = legend(' Randles with inductor','Voigt circuit');
hlegl=gca;
hlegl.FontSize = 16;
hlegl.FontWeight = 'bold';
clear all
```

See Fig. A.1.

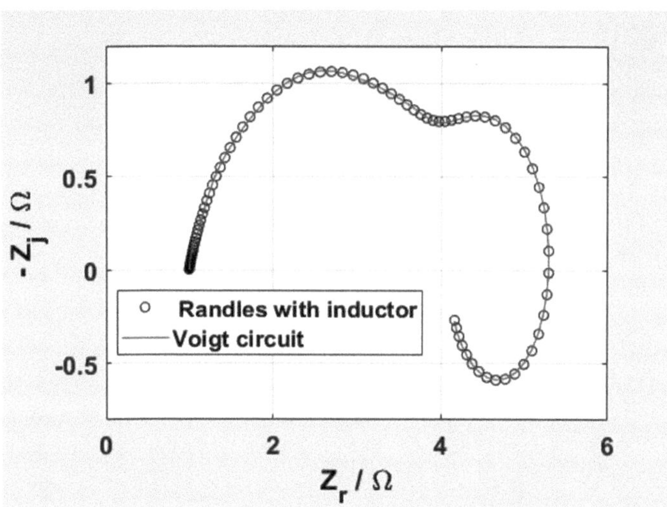

FIG. A.1 Plot in MATLAB resulting from the execution of script A.1.1.

A.1.2 Script to reproduce Fig. 1.25

```
% Frequency range 100 kHz - 0.01 Hz
Freq=logspace(-2,5,100);
freq=(sort(Freq,'descend'))';
% Parameters of the Randles circuit, Eq. 1.65
R1=1;
R2=3;
Rw=1.5;
Tw=0.7;
Y=0.009;
P=0.75;
% Angular frequency rad/s
w=2.*pi.*freq;
% Constant phase element
Yr=Y.*(w.*1i).^P;
% Warburg impedance, Eq. 1.55
Zw=Rw.*tanh(sqrt(1i.*w.*Tw))./(sqrt(1i.*w.*Tw));
% Impedance of the Randles circuit, Eq. 1.65
Z=R1+((R2+Zw)./(1+(R2+Zw).*Yr));
% Argument of the impedance of the Randles circuit
Zarg=abs(Z);
% Phase angle of the impedance of the Randles circuit
Zang=angle(Z);
% Real component of the impedance of the Randles circuit
re=Zarg.*cos(Zang);
% Imaginary component of the impedance of the Randles circuit
img=Zarg.*sin(Zang);
% Calculate the new frequency range, Eq. 1.41
minim=find(img==min(min(img)));
delta=2.*atan(freq./freq(minim))';
a=linspace(0,pi,1024);
%%%%% Imaginary to real transformation %%%%%
% Step 1 Interpolated imaginary component at negative and posi-
  tive frequencies
b= interp1(delta,img,a)';
c = -b;
d = c(end:-1:1);
e=cat(1,b,d);
e(isnan(e))=0;
% Step 2 FFT is applied to the imaginary interpolated data
f=fft(e);
% Step 3 The resulting data at positive frequencies are set to be
  zero
f(2:1025)=0;
```

```
% Step 4 The data at negative frequencies resulting from step 2
  are conjugated, doubled and concatenated with data from step 3
g = conj(f(1026:2048));
k=cat(1,2.*f,g);
% Step 5 The inverse FFT is applied and the transformed real com-
  ponents at positive frequencies are calculated
m=ifft(1i.*k,2048);
Re_trans=m(1:1024);
%%%%%% Real to imaginary transformation %%%%%%
% Step 1 Interpolated real component at negative and positive
  frequencies
b_real= interp1(delta,re,a)';
c_real = -b_real;
d_real = b_real(end:-1:1);
e_real=cat(1,b_real,d_real);
e_real(1024)=e_real(1);
e_real(1025)=e_real(1);
e_real(isnan(e_real))=[];
% Step 2 FFT is applied to the real interpolated data
f_img=fft(e_real);
% Step 3 Data at negative frequencies are set to be zero
f_img((size(e_real)/2)+1:size(e_real))=0;
% Step 4 The data at positive frequencies resulting from step 2
  are conjugated, doubled and concatenated with data from step 3
g_img = conj(f_img(1:size(e_real)/1)) ;
h_img = g_img(end:-1:1);
k_img=cat(1,2.*f_img,h_img);
% Step 5 The inverse FFT is applied and the transformed imaginary
  components at positive frequencies are calculated
m_img=ifft(1i.*k_img,2048);
rE_img=(-m_img);
Img_trans=rE_img(1:1024);
%%%%%% Plotting results %%%%%%
% The real component Zr of the EIS data is shifted to zero where
  the imaginary component reaches its minimum value
plot((re-re(minim)),-img,'ok','LineWidth',2,'MarkerSize',8)
hold on
ax = gca;
ax.FontSize = 16;
ax.FontWeight = 'bold';
% Transformed data considering only positive frequencies
plot(Re_trans(1:1024),Img_trans(1:1024),'r','LineWidth',3)
xlabel('Z_r / \Omega','FontSize',18,'FontWeight','bold',
  'Color','k')
```

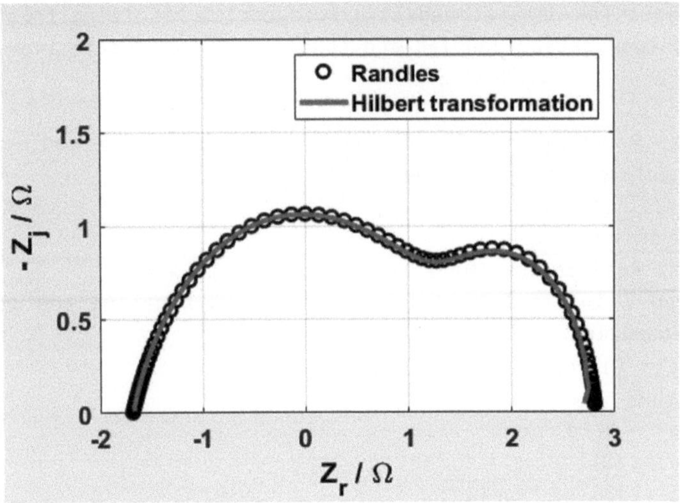

FIG. A.2 Plot in MATLAB resulting from the execution of script A.1.2.

```
ylabel('- Z_j / \Omega','FontSize',18,'FontWeight','bold',
  'Color','k')
axis([-2 3 0 2])
grid;
hleg1 = legend('Randles','Hilbert transformation');
hleg1.FontSize = 14;
clear all
```

See Fig. A.2.

A.2 Chapter 2

A.2.1 Script to reproduce Fig. 2.8

```
% Parameters of the PEMFC model
I=0.09:0.1:75.7;       % Range of electrical current, A
T=343.15;              % Temperature, K
A=50.6;                % Active area, cm^2
L=17.8e-03;            % Membrane thickness, cm
co2=1e-04;             % Oxygen concentration, mol/cm^3
ch2=1e-04;             % Hydrogen concentration, mol /cm^3
po2=1;                 % Oxygen pressure, atm
ph2=1;                 % Hydrogen pressure, atm
g1=-0.948;             % Activation overpotential coefficients
g2=0.00286+0.0002*log(A)+4.3e-05*log(ch2);
g3=7.6e-05;
g4=-1.93e-04;
```

```
lambda=23;              % Parameter related to ohmic overpotential
B=0.016;                % Parameter related to concentration overpo-
                        tential, V
Jmax=1.5;               % Limit current density, A/cm^2
J=I./A;                 % Current density, A/cm^2
% Reversible potential
Nerst=1.229-(0.85e-03)*(T-298.15)+(4.31e-05)*T*(log(ph2)
  +0.5*log(po2));
% Activation overpotential
nact=-(g1+g2*T+g3*T*log(co2)+g4.*T.*log(I));
% Ohmic Overpotential
Pm=(181.6*(1+0.03.*(I./A)+0.062*(T/303)^2.*(I./A).^2.5))./
  ((lambda-0.634-3*(I./A)).*exp(4.18*(T-303)/T));
Rm=(Pm.*L)./A;
nohmic=Rm.*I;
% Mass transport overpotential
ncon=-B.*log(1-(J./Jmax));
% Voltage output in PEMFC
V=Nerst-nact-nohmic-ncon;
% Plotting polarisation curve
plot(J,V,'k','LineWidth',2');
title('Polarisation curve');
xlabel('Current density / A cm^-^2','FontSize',12,'FontWeight',
  'bold','Color','k');
ylabel('PEMFC voltage / V','FontSize',12,'FontWeight','bold',
  'Color','k');
axis([0 1.5 0.4 1.4])
hold on
%%% Fitting empirical equation, Eq. 2.37
xdata=J';                      % Current density
x0 =[0.5 0.5 0.5 0.5 0.5];     % Initial conditions
% Defining a function for empirical equation, Eq. 2.37
f_V=@(x,xdata)(x(1)-x(2).*log10(xdata)-x(3).*xdata-(x(4).*exp(x
  (5).*xdata)));
% Curve fitting
options=optimoptions('lsqcurvefit','Algorithm',
  'levenberg-marquardt');
lb = [];
ub = [];
% V is the voltage of the polarisation curve to be predicted
x = lsqcurvefit(f_V,x0,xdata,V',lb,ub,options);
% Estimated parameters
E0=abs(x(1));
b=abs(x(2));
```

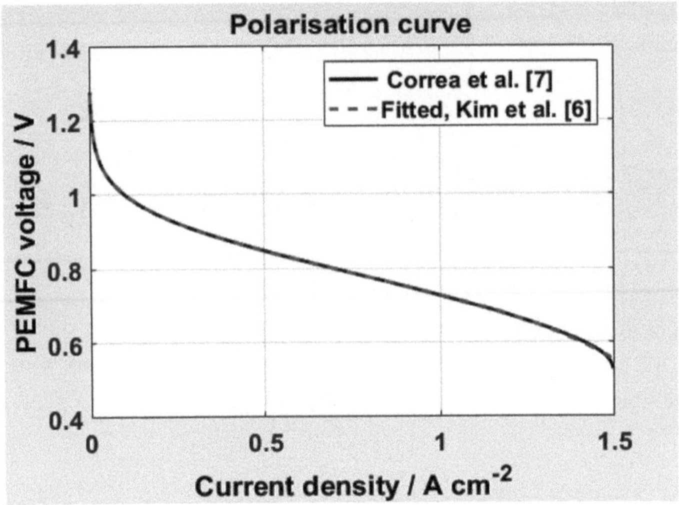

FIG. A.3 Plot in MATLAB resulting from the execution of script A.2.1.

```
r=abs(x(3));
m=abs(x(4));
d=abs(x(5));
% Empirical equation, Eq. 2.37
Vs=E0-b.*log10(J)-r.*J-(m.*exp(d.*J));
% Plotting simulation of empirical equation
plot(J,Vs,'--r','LineWidth',2')
grid;
hleg1=legend(' Correa et al. [7]','Fitted, Kim et al. [6]');
hleg1=gca;
hleg1.FontSize = 16;
hleg1.FontWeight = 'bold';
clear all
```

See Fig. A.3.

A.3 Chapter 4

A.3.1 Script to reproduce Fig. 4.15

```
% Frequency range 10 MHz - 0.1 Hz
Frequency=logspace(-1,7,100);
f=(sort(Frequency,'descend'))';
%%% Parameters of the transmission line model (TLM), Eq. 4.73
% Charge transfer resistance
Rc=0.6;
```

```
% Double-layer capacitance
C=0.0243;
% Thickness of cathode catalyst layer, cm
tcl=100*4.5e-06;
% Proton conductivity
Sp=1e-03;
% Electron conductivity
Se=3788e-03;
% Proton resistivity
Zp=1/Sp;
% Electron resistivity
Ze=1/Se;
% Parallel proton and electron resistivities
RHF=(Zp*Ze)/(Zp+Ze);
% Impedance between upper and lower rails of TLM
ZF=Rc./(1+(Rc.*C*1i.*2.*pi.*f));
% Impedance of the TLM, Eq. 4.73
k=sqrt(ZF.*tcl./(Zp+Ze));
Z_TLM=RHF.*(tcl+(2.*k./(sinh(tcl./k))))+k.*coth(tcl./k).*
  ((Zp^2)+(Ze^2))./(Zp+Ze);
% Modulus of the impedance of the TLM
arg=abs(Z_TLM);
% Phase angle of the impedance of the TLM
ang=angle(Z_TLM);
% Real component of the impedance of the TLM
Zr_TLM=arg.*cos(ang);
% Imaginary component of the impedance of the TLM
Zim_TLM=arg.*sin(ang);
% Complex-impedance plot
plot(Zr_TLM,-Zim_TLM,'-^r','LineWidth',1);
grid;
xlabel('Z_r / \Omega cm^2','FontSize',16,'FontWeight','bold',
  'Color','k')
ylabel('- Z_j / \Omega cm^2','FontSize',16,'FontWeight','bold',
  'Color','k')
axis([0 0.8 0 0.4])
hleg1 = legend(' Case 2');
hleg1=gca;
hleg1.FontSize = 16;
hleg1.FontWeight = 'bold';
clear all
```

See Fig. A.4.

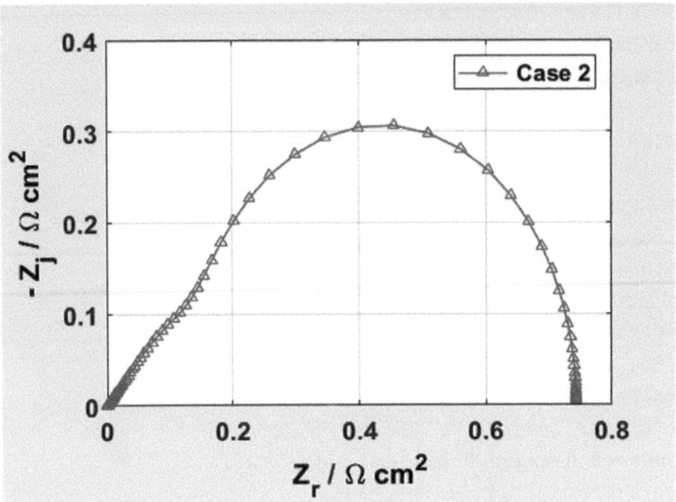

FIG. A.4 Plot in MATLAB resulting from the execution of script A.3.1.

A.4 Chapter 5

A.4.1 Script to estimate diffusion parameters defined in Table 5.2

```
% Frequency range 1000 Hz - 0.001 Hz
Frequency=logspace(-3,3,100);
f=(sort(Frequency,'descend'))';
% Channel inlet temperature
T=273.15+20;
% Channel inlet pressure
p=101325*1;
% Air relative humidity
RH=0;
% Oxygen Stoichiometry
so2=2.3;
% Channel depth, m
hc=0.001;
% Current density, A/cm^2
I=0.125;
% Cell active area, cm^2
Ac=100;
% Warburg resistance, 10^-3 ohms
Rsg=7.969;
% Warburg time constant, s
Tsg=0.698;
```

```
% Tafel slope, V
b=0.165;
% Gas constant
R=8.3145;
% Faraday constant
F=96485;
% Coefficients for calculation of vapour saturation pressure
aa=-5800.2206;
bb=1.3914993;
cc=-0.048640239;
dd=0.41764768E-04;
ee=-0.14452093E-07;
ff=6.5459673;
% Saturation pressure of vapour, Eq. 5.23
Pvs=exp((aa/T)+bb+(cc*T)+(dd*T^2)+(ee*T^3)+ff*log(T));
% Humidity molar ratio, Eq. 5.22
H=RH*Pvs/(p-RH*Pvs);
% Channel length, it can be any value
Lc=1;
% Calculation of inlet current density in channel
ydata=0:0.001:1;
y0=I;
IY_S=@(y,ydata)(y(1).*exp(-ydata.*y(1)./(I.*so2)));
I_in=@(y,ydata)(mean(IY_S(y,ydata)));
% Inlet current density in channel, A/cm^2
jo=lsqcurvefit(I_in,y0,ydata,I);
% Fraction of channel length for SO2=2.3 icell=0.125 A/cm^2
y=0.476*Lc;
% Oxygen concentration at inlet of channel, mol/m^3
c_ref=p./(5.*(H+1).*R.*T);
% Velocity of oxygen in channel, Eq. 5.17, m/s
Vc=so2.*(I.*100.^2).*Lc./(4.*F.*hc.*c_ref);
% Inlet stoichiometry ratio
Lpe=4.*F.*c_ref.*Vc.*hc./(Lc.*jo*100.^2);
% Current density distribution, Eq. 5.26, A/cm^2
Iy=jo.*exp(-y./(Lc.*Lpe));
% Molar oxygen concentration, Eq. 5.28
Ms_A=exp(-y./(Lpe.*Lc));
% Oxygen concentration in channel, Eq. 5.27
MCCT_s=Ms_A./((Ms_A./5)+0.8+H);
CCT_S=(p.*MCCT_s./(5.*R.*T));
% Initial conditions for estimation of diffusion parameters
x0=[5e-04 2e-06];
```

```
xdata=f;
% Initial conditions for estimation of real and imaginary
  residuals
Res=2;
Res_r=0;
Res_i=0;
Iteration=0;
% Maximum number of iterations during fitting process
limit=100;
% Conditional statement for real and imaginary residuals
while Res > 0.0001
if Iteration==limit
break
end
% Function representing oxygen concentration at CCL/GDL inter-
  face, Eq. 5.43
GC=@(x,xdata)((Iy.*(100.^2).*x(1)./(4.*F.*x(2))));
C_gdl_S=@(x,xdata)(CCT_S-GC(x,xdata));
% Function representing oxygen transport resistance, Eq. 5.42
Rgdl_dc=@(x,xdata)((100.^2).*(b.*x(1))./(4.*96485.*x(2).*
  (C_gdl_S(x,xdata))));
% Function representing oxygen diffusion time constant, Eq. 5.2
T_GDL=@(x,xdata)(x(1).*x(1)./x(2));
% Function representing local GDL-channel impedance, Eq. 5.41,
  10^-3 ohms
Z_dc=@(x,xdata)(((1000./Ac).*Rgdl_dc(x,xdata)).*(tanh((sqrt
  (-1).*2.*pi.*xdata.*T_GDL(x,xdata)).^0.5))./((sqrt(-1).*2.*pi.
  *xdata.*T_GDL(x,xdata)).^0.5));
% Function representing argument from local GDL-channel impedance
arg_m=@(x,xdata)abs(Z_dc(x,xdata));
% Function representing phase angle from local GDL-channel
  impedance
ang_m=@(x,xdata)angle(Z_dc(x,xdata));
% Function representing real component from local GDL-channel
  impedance
real_m=@(x,xdata)arg_m(x,xdata).*cos(ang_m(x,xdata));
% Function representing imaginary component from local
  GDL-channel impedance
img_m=@(x,xdata)arg_m(x,xdata).*sin(ang_m(x,xdata));
% Finite-length Warburg impedance, Eq. 5.1, 10^-3 ohms
Zw=((Rsg.*tanh((sqrt(-1).*2.*pi.*f.*Tsg).^0.5))./((sqrt(-1).*2.
  *pi.*f.*Tsg).^0.5));
```

```
% Argument of finite-length Warburg impedance
argw=abs(Zw);
% Angle phase of finite-length Warburg impedance
angw=angle(Zw);
% Real component of finite-length Warburg impedance
modw=argw.*cos(angw);
% Imaginary component of finite-length Warburg impedance
Imgew=argw.*sin(angw);
% Non-linear regression
options=optimoptions('lsqcurvefit','Algorithm',
  'levenberg-marquardt');
lb = [];
ub = [];
xr=lsqcurvefit(real_m,x0,xdata,modw,lb,ub,options);
xim=lsqcurvefit(img_m,x0,xdata,Imgew,lb,ub,options);
% Estimated GDL diffusion distance, m
lgdl=(abs(xr(1))+abs(xim(1)))/2;
% Estimated GDL diffusion coefficient, m^2/s
Dg=(abs(xr(2))+abs(xim(2)))/2 ;
% Vector of diffusion parameters as initial condition for next
  iteration
x0=[lgdl Dg];
%%% Simulation of GDL-channel impedance model considering
  steady-state oxygen concentration in channel
% Oxygen diffusion time constant in GDL, Eq. 5.2
T_gdl=(lgdl^2)/Dg;
% Oxygen concentration at GDL/CCL interface, Eq. 5.43
C_gdl_St=CCT_S-(Iy.*(100.^2).*lgdl./(4.*F.*Dg));
% Oxygen transport resistance, Eq. 5.42, 10^-3 ohms
Rgdl=(1000./Ac).*((100.^2).*(b.*lgdl)./(4.*96485.*Dg.*
  (C_gdl_St)));
% Local GDL-channel impedance, Eq. 5.41, 10^-3 ohms
Zgd=((Rgdl).*(tanh((sqrt(-1).*2.*pi.*f.*T_gdl).^0.5))./((sqrt
  (-1).*2.*pi.*f.*T_gdl).^0.5));
% Argument of local GDL-channel impedance
argm=abs(Zgd);
% Phase angle of local GDL-channel impedance
angm=angle(Zgd);
% Real component of local GDL-channel impedance
modm=argm.*cos(angm);
% Imaginary component of local GDL-channel impedance
Imgem=argm.*sin(angm);
```

```
%%% Estimation of real and imaginary residuals
for n=1:numel(f)
% Residual of real component
Jpr=abs((modw(n)-(modm(n)))./argw(n));
KpR=(Res_r+Jpr);
Res_r=Jpr;
kpr(n,:)=KpR;
% Residual of imaginary component
Jpi=abs((Imgew(n)-(Imgem(n)))./argw(n));
KpI=(Res_i+Jpi);
Res_i=Jpi;
kpi(n,:)=KpI;
end
% Average sum of real and imaginary residuals
Res=(mean(kpi+kpr));
% Number of iterations
Iteration=Iteration+1;
end
% Complex-impedance plot
plot(modw,-Imgew,'ok','LineWidth',2)
hold on
plot(modm,-Imgem,'r','LineWidth',2);
% Axis labels
xlabel('Z_r / 10^-^3 \Omega','FontSize',12,'FontWeight','bold',
  'Color','k')
ylabel('- Z_j / 10^-^3 \Omega','FontSize',12,'FontWeight',
  'bold','Color','k')
hleg1 = legend('Warburg impedance, Eq. 5.1','Local GDL-channel
  impedance, Eq. 5.41');
hleg1=gca;
hleg1.FontSize = 16;
hleg1.FontWeight = 'bold';
axis([0 8 0 6])
strg = {strcat('Dgdl =' , num2str(Dg,'%.4e'),' m^2/s')};
strl = {strcat('lgdl =' , num2str(lgdl,'%.4e'),' m')};
title([strg strl]);
grid;
clear all
```

See Fig. A.5.

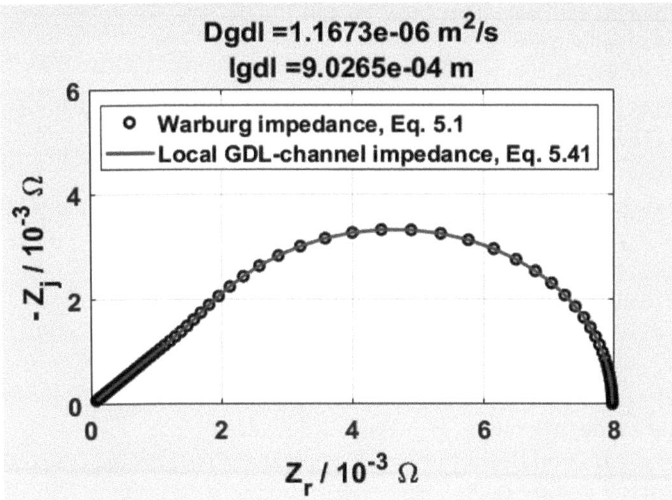

FIG. A.5 Plot in MATLAB resulting from the execution of script A.4.1.

A.4.2 Script to reproduce Fig. 5.15

```
% Frequency range 1000 Hz - 0.001 Hz
Frequency=logspace(-3,3,100);
f=(sort(Frequency,'descend'))';
% Channel inlet temperature
T=273.15+20;
% Channel inlet pressure
p=101325*1;
% Air relative humidity
RH=0;
% Oxygen Stoichiometry
so2=2.3;
% Channel depth, m
hc=0.001;
% Current density, A/cm^2
I=0.125;
% Cell active area, cm^2
Ac=100;
% Tafel slope, V
b=0.165;
% Oxygen diffusion coefficient in GDL, m^2/s
Dg=1.1673e-06;
```

```
% Oxygen diffusion distance in GDL, m
lgdl=9.0265e-04;
% Gas constant
R=8.3145;
% Faraday constant
F=96485;
% Coefficients for calculation of vapour saturation pressure
aa=-5800.2206;
bb=1.3914993;
cc=-0.048640239;
dd=0.41764768E-04;
ee=-0.14452093E-07;
ff=6.5459673;
% Saturation pressure of vapour, Eq. 5.23
Pvs=exp((aa/T)+bb+(cc*T)+(dd*T^2)+(ee*T^3)+ff*log(T));
% Humidity molar ratio, Eq. 5.22
H=RH*Pvs/(p-RH*Pvs);
% Channel length, it can be any value
Lc=1;
% Symbolic value of local channel length
syms yy
% Oxygen diffusion time constant in GDL, Eq. 5.2
T_GDL=(lgdl^2)/Dg;
% Calculation of inlet current density in channel
xdata=0:0.001:1;
x0=I;
IY_S=@(x,xdata)(x(1).*exp(-xdata.*x(1)./(I.*so2)));
I_in=@(x,xdata)(mean(IY_S(x,xdata)));
% Inlet current density in channel, A/cm^2
jo=lsqcurvefit(I_in,x0,xdata,I);
% Oxygen concentration at inlet of channel, mol/m^3
c_ref=p./(5.*(H+1).*R.*T);
% Velocity in channel, Eq. 5.17, m/s
Vc=so2.*Lc.*(I.*100.^2)./(4.*F.*hc.*c_ref);
% Inlet stoichiometry ratio
Lpe=4.*F.*c_ref.*Vc.*hc./(Lc.*jo*100.^2);
% Current density distribution, Eq. 5.26, A/cm^2
Iy=jo.*exp(-yy./(Lc.*Lpe));
%%% Impedance model of GDL-channel considering oscillating oxygen
  concentration
% Laplace domain to frequency domain
s=2.*pi.*f.*sqrt(-1);
```

```
% Eq. 5.20
beta=s+(sqrt(s.*Dg).*tanh(sqrt(s.*T_GDL))./hc);
% Eq. 5.21
gamma=exp(yy.*beta./Vc);
% Parameter lambda defined in Eq. 5.11
Lambda=sqrt(s./Dg);
% Eq. 5.38
MCC_d=1+((Vc)./(cosh(sqrt(s.*T_GDL)).*(beta.*Lc.*Lpe-Vc))).*
  ((1./gamma)-exp(-yy./(Lc.*Lpe)));
% Total molar oxygen concentration in wet air
M_CCT=MCC_d./(H+0.8+(MCC_d./5));
% Oscillating oxygen concentration in the channel, Eq. 5.37
CCT=p.*M_CCT./(5.*R.*T);
% Current density considering amplitude of sinusoidal perturba-
  tion, the value 0.1 is the amplitude value of 10% of the DC cur-
  rent density, see Figure 5.7
IT=Iy+(Iy.*0.1);
% Oxygen concentration in CCL/GDL interface, Eq. 5.36
C_gdl=(CCT)-((IT.*100.^2).*tanh(sqrt(s.*T_GDL))./(4.*F.*Dg.
  *Lambda));
% Local oxygen transport resistance, Eq. 5.35, 10^-3 ohms
Rgdl_ac=((100.^2).*(b.*lgdl)./(4.*96485.*Dg.*(C_gdl))).*(1000./
  Ac);
% Local GDL-channel impedance, Eq. 5.34, 10^-3 ohms
Z_ac=(Rgdl_ac.*(tanh((sqrt(-1).*2.*pi.*f.*T_GDL).^0.5))./((sqrt
  (-1).*2.*pi.*f.*T_GDL).^0.5));
% Estimation of total GDL-channel impedance, Eq. 5.48, 10^-3 ohms
Zt_ac=int(1./Z_ac,yy,[0 Lc]);
ZT_ac=double(Lc./Zt_ac);
% Argument of total GDL-channel impedance
ac_arg=abs(ZT_ac);
% Phase angle of total GDL-channel impedance
ac_ang=angle(ZT_ac);
% Real component of GDL-channel impedance
ac_mod=ac_arg.*cos(ac_ang);
% Imaginary component of GDL-channel impedance
ac_Imge=-ac_arg.*sin(ac_ang);
%%% Impedance model of GDL-channel considering steady-state
  oxygen concentration
% Molar oxygen concentration, Eq. 5.28
Ms_A=exp(-yy./(Lpe.*Lc));
MCCT_s=Ms_A./((Ms_A./5)+0.8+H);
```

```
% Oxygen concentration in channel, Eq. 5.27
CCT_S=(p.*MCCT_s./(5.*R.*T));
% Oxygen concentration at GDL/CCL interface, Eq. 5.43
C_gdl_S=CCT_S-(Iy.*(100.^2).*lgdl./(4.*F.*Dg));
% Local oxygen transport resistance, Eq. 5.42, 10^-3 ohms
Rgdl_dc=((100.^2).*(b.*lgdl)./(4.*96485.*Dg.*(C_gdl_S))).*
  (1000./Ac);
% Local GDL-channel impedance, Eq. 5.41, 10^-3 ohms
Z_dc=((Rgdl_dc).*(tanh((sqrt(-1).*2.*pi.*f.*T_GDL).^0.5))./
  ((sqrt(-1).*2.*pi.*f.*T_GDL).^0.5));
% Estimation of total GDL-channel impedance, Eq. 5.47, 10^-3 ohms
Zt_dc=int(1./Z_dc,yy,[0 Lc]);
ZT_dc=double(Lc./Zt_dc);
% Argument of total GDL-channel impedance
dc_arg=abs(ZT_dc);
% Phase angle of total GDL-channel impedance
dc_ang=angle(ZT_dc);
% Real component of GDL-channel impedance
dc_mod=dc_arg.*cos(dc_ang);
% Imaginary component of GDL-channel impedance
dc_Imge=-dc_arg.*sin(dc_ang);
% Complex impedance plot
plot(ac_mod,ac_Imge,'-dk','LineWidth',1)
hold on
plot(dc_mod,dc_Imge,'-or','LineWidth',1)
% Axis labels
xlabel('Z_r / 10^-^3 \Omega','FontSize',12,'FontWeight','bold',
  'Color','k')
ylabel('- Z_j / 10^-^3 \Omega','FontSize',12,'FontWeight',
  'bold','Color','k')
axis([0 10 0 5])
hleg1 =legend('Eq. 5.48','Eq. 5.47', 'northwest');
hleg1=gca;
hleg1.FontSize = 16;
hleg1.FontWeight = 'bold';
grid;
clear all
```

See Fig. A.6.

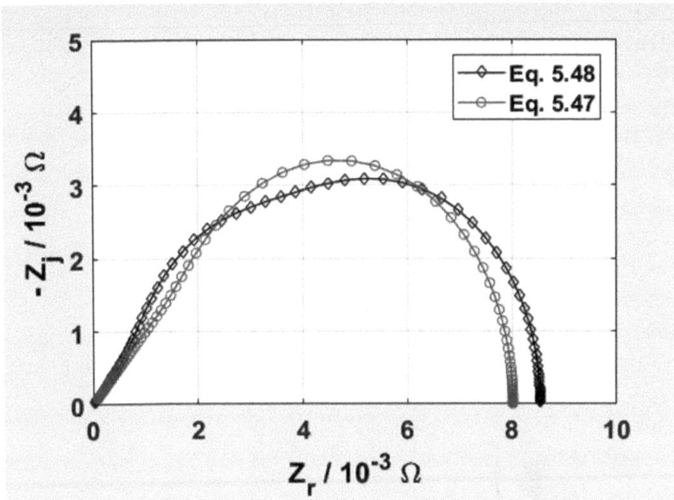

FIG. A.6 Plot in MATLAB resulting from the execution of script A.4.2.

A.4.3 Script to reproduce Fig. 5.24

```
% Frequency range 10000 Hz - 0.15 Hz
Frequency=logspace(-1,4,200);
f=(sort(Frequency,'descend'))';
f(f<0.15 | f>10000) = [];
% Channel inlet temperature
T=273.15+20;
% Channel inlet pressure
p=101325*1;
% Air relative humidity
RH=0;
% Oxygen Stoichiometry
so2=2.78;
% Channel depth, m
hc=0.001;
% Current density, A/cm^2
I=0.125;
% Cell active area, cm^2
Ac=100;
% Warburg resistance, 10^-3 ohms
Rsg=5.769;
% Warburg time constant, s
Tsg=0.751;
% Charge transfer resistance, 10^-3 ohms
```

```
Rc=10.6;
% Double-layer capacitance, F
C=0.253;
% Ohmic resistance, 10^-3 ohms
Rohm=2.562;
% Tafel slope, V
b=0.1325;
% Gas constant
R=8.3145;
% Faraday constant
F=96485;
% Coefficients for calculation of vapour saturation pressure
aa=-5800.2206;
bb=1.3914993;
cc=-0.048640239;
dd=0.41764768E-04;
ee=-0.14452093E-07;
ff=6.5459673;
% Saturation pressure of vapour, Eq. 5.23
Pvs=exp((aa/T)+bb+(cc*T)+(dd*T^2)+(ee*T^3)+ff*log(T));
% Humidity molar ratio, Eq. 5.22
H=RH*Pvs/(p-RH*Pvs);
% Channel length, it can be any value
Lc=100;
% Calculation of inlet current density in channel
ydata=0:0.001:1;
y0=I;
IY_S=@(y,ydata)(y(1).*exp(-ydata.*y(1)./(I.*so2)));
I_in=@(y,ydata)(mean(IY_S(y,ydata)));
% Inlet current density in channel, A/cm^2
jo=lsqcurvefit(I_in,y0,ydata,I);
% Fraction of channel length for SO2=2.78 icell=0.125 A/cm^2
y=0.481*Lc;
% Oxygen concentration at inlet of channel, mol/m^3
c_ref=p./(5.*(H+1).*R.*T);
% Velocity of oxygen in channel, Eq. 5.17, m/s
Vc=so2.*(I.*100.^2).*Lc./(4.*F.*hc.*c_ref);
% Inlet stoichiometry ratio
Lpe=4.*F.*c_ref.*Vc.*hc./(Lc.*jo*100.^2);
% Current density distribution, Eq. 5.26, A/cm^2
Iy=jo.*exp(-y./(Lc.*Lpe));
% Current density considering amplitude of sinusoidal perturba-
  tion, the value 0.1 is the amplitude value of 10% of the DC cur-
  rent density, see Figure 5.7
```

```
IT=Iy+(Iy.*0.1);
% Initial conditions for estimation of diffusion parameters
x0=[5e-04 2e-06];
xdata=f;
% Initial conditions for estimation of real and imaginary
  residuals
Res=2;
Res_r=0;
Res_i=0;
Iteration=0;
% Maximum number of iterations
limit=100;
% Conditional statement for real and imaginary residuals
while Res > 0.0001
if Iteration==limit
break
end
% Laplace domain to frequency domain
s=2.*pi.*f.*sqrt(-1);
% Function representing oxygen diffusion time constant in GDL,
  Eq. 5.2
T_GDL_f=@(x,xdata)(x(1).*x(1)./x(2));
% Function representing Eq. 5.20
beta_f=@(x,xdata)(s+(sqrt(s.*x(2)).*tanh(sqrt(s.*T_GDL_f(x,
  xdata)))./hc));
% Function representing parameter lambda defined in Eq. 5.11
Lambda_f=@(x,xdata)(sqrt(s./x(2)));
% Function representing Eq. 5.21
gamma_f=@(x,xdata)(exp(y.*beta_f(x,xdata)./(Vc)));
% Function representing Eq. 5.38
MCC_d_f=@(x,xdata)(1+((Vc)./(cosh(sqrt(s.*T_GDL_f(x,xdata))).*
  (beta_f(x,xdata).*Lc.*Lpe-Vc))).*((1./gamma_f(x,xdata))-exp
  (-y./(Lc.*Lpe))));
% Function representing Eq. 5.37
M_CCT_f=@(x,xdata)(MCC_d_f(x,xdata)./(H+0.8+(MCC_d_f(x,
  xdata)./5)));
CCT_f=@(x,xdata)(p.*M_CCT_f(x,xdata)./(5.*R.*T));
% Function representing oxygen concentration in CCL/GDL inter-
  face, Eq. 5.36
C_gdl_f=@(x,xdata)((CCT_f(x,xdata))-((IT.*100.^2).*tanh(sqrt
  (s.*T_GDL_f(x,xdata)))./(4.*F.*x(2).*Lambda_f(x,xdata))));
% Function representing oxygen transport resistance, Eq. 5.35,
  10^-3 ohms
Rgdl_ac_f=@(x,xdata)(((100.^2).*(b.*x(1))./(4.*96485.*x(2).*
  (C_gdl_f(x,xdata)))).*(1000./Ac));
```

```
% Function representing local impedance of GDL-channel, Eq. 5.34,
  10^-3 ohms
Z_ac_f=@(x,xdata)((Rgdl_ac_f(x,xdata).*(tanh((sqrt(-1).*2.*pi.
  *f.*T_GDL_f(x,xdata)).^0.5))./((sqrt(-1).*2.*pi.*f.*T_GDL_f
  (x,xdata)).^0.5)));
% Function representing Randles circuit considering GDL-channel
  impedance model, Eq. 5.50 with Eq. 5.34, 10^-3 ohms
ZT_f=@(x,xdata)(((Rohm./1000)+((Rc+Z_ac_f(x,xdata))./1000)./
  (1+((Rc+Z_ac_f(x,xdata))./1000).*C.*(sqrt(-1).*2.*pi.*f))).
  *1000);
% Function representing argument from Randles circuit with
  GDL-channel model
arg_m=@(x,xdata)abs(ZT_f(x,xdata));
% Function representing phase angle from Randles circuit with
  GDL-channel model
ang_m=@(x,xdata)angle(ZT_f(x,xdata));
% Function representing Real component from Randles circuit with
  GDL channel model
real_m=@(x,xdata)arg_m(x,xdata).*cos(ang_m(x,xdata));
% Function representing Imaginary component from Randles circuit
  with GDL channel model
img_m=@(x,xdata)arg_m(x,xdata).*sin(ang_m(x,xdata));
% Finite-length Warburg impedance, Eq. 5.1, 10^-3 ohms
Zw=((Rsg.*tanh((sqrt(-1).*2.*pi.*f.*Tsg).^0.5))./((sqrt(-1).*2.
  *pi.*f.*Tsg).^0.5));
% Impedance of Randles circuit considering finite-length Warburg
  impedance, Eq. 5.50 with Eq. 5.1, 10^-3 ohms
ZRan=((Rohm./1000)+((Rc+Zw)./1000)./(1+((Rc+Zw)./1000).*0.253.*
  (sqrt(-1).*2.*pi.*f))).*1000;
% Argument from Randles circuit with finite-length Warburg
  impedance
argw=abs(ZRan);
% Phase angle from Randles circuit with finite-length Warburg
  impedance
angw=angle(ZRan);
% Real component from Randles circuit with finite-length Warburg
  impedance
modw=argw.*cos(angw);
% Imaginary component from Randles circuit with finite-length
  Warburg impedance
Imgew=argw.*sin(angw);
% Non-linear regression
options=optimoptions('lsqcurvefit','Algorithm',
  'levenberg-marquardt');
```

```
lb = [];
ub = [];
xr=lsqcurvefit(real_m,x0,xdata,modw,lb,ub,options);
xim=lsqcurvefit(img_m,x0,xdata,Imgew,lb,ub,options);
% Estimated GDL diffusion distance, m
lgdl=(abs(xr(1))+abs(xim(1)))/2;
% Estimated GDL diffusion coefficient, m^2/s
Dg=(abs(xr(2))+abs(xim(2)))/2;
% Vector of diffusion parameters as initial condition for next
  iteration
x0=[lgdl Dg];
%%% Simulation of Randles circuit with GDL-channel impedance
  model considering oscillating oxygen concentration in channel
% Oxygen diffusion time constant, Eq. 5.2, s
T_GDL=(lgdl^2)/Dg;
% Eq. 5.20
beta=s+(sqrt(s.*Dg).*tanh(sqrt(s.*T_GDL))./hc);
% Eq. 5.21
gamma=exp(y.*beta./Vc);
% Parameter lambda defined in Eq. 5.11
Lambda=sqrt(s./Dg);
% Molar concentration of oxygen in the channel, Eq. 5.38
MCC_d=1+((Vc)./(cosh(sqrt(s.*T_GDL)).*(beta.*Lc.*Lpe-Vc))).*
  ((1./gamma)-exp(-y./(Lc.*Lpe)));
% Total molar oxygen concentration in wet air
M_CCT=MCC_d./(H+0.8+(MCC_d./5));
% Oxygen concentration in the channel, Eq. 5.37
CCT=p.*M_CCT./(5.*R.*T);
% Oxygen concentration in CCL/GDL interface, Eq. 5.36
C_gdl=(CCT)-((IT.*100.^2).*tanh(sqrt(s.*T_GDL))./(4.*F.*Dg.
  *Lambda));
% Oxygen transport resistance, Eq. 5.35, 10^-3 ohms
Rgdl_ac=((100.^2).*(b.*lgdl)./(4.*96485.*Dg.*(C_gdl))).*(1000./
  Ac);
% Impedance of GDL-channel, Eq. 5.34, 10^-3 ohms
Z_ac=(Rgdl_ac.*(tanh((sqrt(-1).*2.*pi.*f.*T_GDL).^0.5))./((sqrt
  (-1).*2.*pi.*f.*T_GDL).^0.5));
% Impedance of Randles circuit considering GDL-channel impedance,
  Eq. 5.50 with Eq. 5.34, 10^-3 ohms
Z_ac_Ran=((Rohm./1000)+((Rc+Z_ac)./1000)./(1+((Rc+Z_ac)./1000).
  *C.*(sqrt(-1).*2.*pi.*f))).*1000;
% Argument from Randles circuit with GDL-channel model
argm=abs(Z_ac_Ran);
% Phase angle from Randles circuit with GDL-channel model
```

```
angm=angle(Z_ac_Ran);
% Real component from Randles circuit with GDL-channel model
modm=argm.*cos(angm);
% Imaginary component from Randles circuit with GDL-channel model
Imgem=argm.*sin(angm);
% Estimation of real and imaginary residuals
for n=1:numel(f)
% Residual of real component
Jpr=abs((modw(n)-(modm(n)))./argw(n));
KpR=(Res_r+Jpr);
Res_r=Jpr;
kpr(n,:)=KpR;
% Residual of imaginary component
Jpi=abs((Imgew(n)-(Imgem(n)))./argw(n));
KpI=(Res_i+Jpi);
Res_i=Jpi;
kpi(n,:)=KpI;
end
% Average sum of real and imaginary residuals
Res=(mean(kpi+kpr));
% Number of iterations
Iteration=Iteration+1;
end
% Complex-impedance plot
plot(modw,-Imgew,'-^k','MarkerSize',5,'LineWidth',1)
hold on
plot(modm,-Imgem,'-sr','MarkerSize',5,'LineWidth',1);
% Axis labels
xlabel('Z_r / 10^-^3 \Omega','FontSize',12,'FontWeight','bold',
  'Color','k')
ylabel('- Z_j / 10^-^3 \Omega','FontSize',12,'FontWeight',
  'bold','Color','k')
hleg1 =legend('Randles with Eq. 5.1','Randles with Eq. 5.34',
  'northwest');
hleg1=gca;
hleg1.FontSize = 16;
hleg1.FontWeight = 'bold';
strg = {strcat('Dgdl = ' , num2str(Dg,'%.4e'),' m^2/s')};
strl = {strcat('lgdl = ' , num2str(lgdl,'%.4e'),' m')};
title([strg strl]);
axis([0 20 0 8])
grid;
clear all
```

See Fig. A.7.

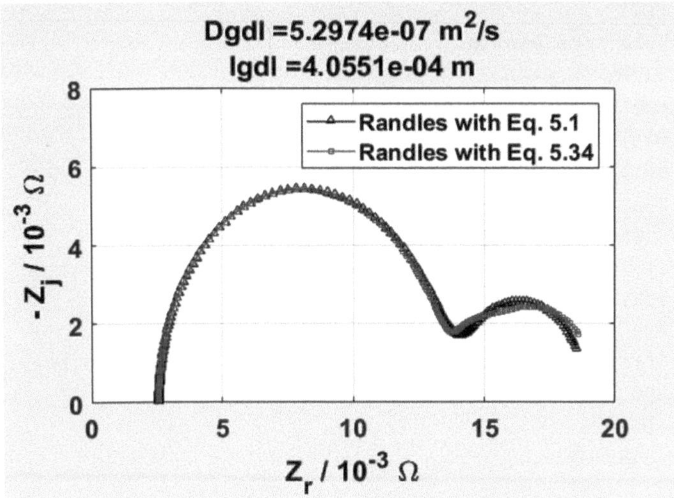

FIG. A.7 Plot in MATLAB resulting from the execution of script A.4.3.

A.5 Chapter 6

A.5.1 Script to estimate diffusion parameters defined in Table 6.4

```
% Frequency range 1000 Hz - 0.01 Hz
Frequency=logspace(-2,3,100);
f=(sort(Frequency,'descend'))';
% Channel inlet temperature
T=273.15+37;
% Temperature of PEMFC
TF=273.15+50;
% Channel inlet pressure
p=101325*1.7;
% Air relative humidity
RH=0.5;
% Oxygen Stoichiometry
so2=2.5;
% Channel depth, m
hc=0.001;
% Current density, A/cm^2
I=0.11;
% Warburg resistance in GDL, ohm.cm^2
Rsg=0.2509;
% Warburg time constant in GDL, s
Tsg=0.0662;
% Warburg resistance in CCL, ohm.cm^2
```

```
Rs=0.2275;
% Warburg time constant in CCL, s
Ts=0.3178;
% Charge transfer resistance, ohm.cm^2
Rc=0.8487;
% Tafel slope, V
b=Rc.*I;
% Gas constant
R=8.3145;
% Faraday constant
F=96485;
% Coefficients for calculation of vapour saturation pressure
aa=-5800.2206;
bb=1.3914993;
cc=-0.048640239;
dd=0.41764768E-04;
ee=-0.14452093E-07;
ff=6.5459673;
% Saturation pressure of vapour, Eq. 5.23
Pvs=exp((aa/T)+bb+(cc*T)+(dd*T^2)+(ee*T^3)+ff*log(T));
% Humidity molar ratio, Eq. 5.22
H=RH*Pvs/(p-RH*Pvs);
% Channel length, it can be any value
Lc=1;
% Losses related to pressure drop and water produced by oxygen
  reaction
wf=0.6;
% Calculation of inlet current density in channel
ydata=0:0.01:1;
y0=I;
IY_S=@(y,ydata)(y(1).*exp(-ydata.*y(1)./(I.*so2)));
I_in=@(y,ydata)(mean(IY_S(y,ydata)));
% Inlet current density in channel, A/cm^2
jo=lsqcurvefit(I_in,y0,ydata,I);
% Fraction of channel length for SO2=2.5 and icell=0.11 A/cm^2
y=0.48*Lc;
% Oxygen concentration at inlet of channel, mol/m^3
c_ref=p./((1+H).*5.*R.*T);
% Velocity of oxygen in channel, Eq. 5.17, m/s
Vc=so2.*(I.*100.^2).*Lc./(4.*F.*hc.*c_ref);
% Inlet stoichiometry ratio
Lpe=4.*F.*c_ref.*Vc.*hc./(Lc.*jo*100.^2);
% Current density distribution, Eq. 5.26, A/cm^2
Iy=jo.*exp(-y.*jo/(Lc.*I.*so2));
```

```
% Molar oxygen concentration, Eq. 5.28
Ms_A=exp(-y./(Lpe.*Lc));
% Oxygen concentration in channel, Eq. 5.27
MCCT_s=Ms_A./((Ms_A./5)+0.8+H);
CCT_S=(p.*MCCT_s./(5.*R.*T));
% Initial conditions for estimation of diffusion parameters in
  GDL
x0=[500e-06 2e-06];
xdata=f;
% Initial conditions for estimation of real and imaginary
  residuals
Res_gdl=2;
Resr_gdl=0;
Resi_gdl=0;
Iteration_gdl=0;
% Maximum number of iterations
limit=100;
% Conditional statement for real and imaginary residuals for GDL
  impedance
while Res_gdl > 0.0001
if Iteration_gdl==limit
break
end
% Function representing oxygen concentration at CCL/GDL inter-
  face, Eq. 5.43
GC=@(x,xdata)((Iy.*(100.^2).*x(1)./(4.*F.*x(2))));
C_gdl_S=@(x,xdata)(wf.*(CCT_S-GC(x,xdata)));
% Function representing oxygen transport resistance in
  GDL-channel, Eq. 6.2
Rgdl_dc=@(x,xdata)((100.^2).*(b.*x(1))./(4.*96485.*x(2).*
  (C_gdl_S(x,xdata))));
% Function representing oxygen diffusion time constant in GDL,
  Eq. 5.2
T_GDL=@(x,xdata)(x(1).*x(1)./(x(2)));
% Function representing local GDL-channel diffusion impedance,
  Eq. 6.1
Z_dc=@(x,xdata)((Rgdl_dc(x,xdata)).*(tanh((sqrt(-1).*2.*pi.
  *xdata.*T_GDL(x,xdata)).^0.5))./((sqrt(-1).*2.*pi.*xdata.
  *T_GDL(x,xdata)).^0.5));
% Function representing argument from GDL-channel model
arg_m=@(x,xdata)abs(Z_dc(x,xdata));
% Function representing phase angle from GDL-channel model
ang_m=@(x,xdata)angle(Z_dc(x,xdata));
% Function representing real component from GDL-channel model
```

```
real_m=@(x,xdata)arg_m(x,xdata).*cos(ang_m(x,xdata));
% Function representing imaginary component from GDL-channel
  model
img_m=@(x,xdata)arg_m(x,xdata).*sin(ang_m(x,xdata));
% Finite-length Warburg impedance for GDL, Eq. 6.23
Zw=((Rsg.*tanh((sqrt(-1).*2.*pi.*f.*Tsg).^0.5))./((sqrt(-1).*2.
  *pi.*f.*Tsg).^0.5));
% Argument of finite-length Warburg impedance for GDL
argw=abs(Zw);
% Phase angle of finite-length Warburg impedance for GDL
angw=angle(Zw);
% Real component of finite-length Warburg impedance for GDL
modw=argw.*cos(angw);
% Imaginary component of finite-length Warburg impedance for GDL
Imgew=argw.*sin(angw);
% Non-linear regression
options=optimoptions('lsqcurvefit','Algorithm',
  'levenberg-marquardt');
lb = [];
ub = [];
xr=lsqcurvefit(real_m,x0,xdata,modw,lb,ub,options);
xim=lsqcurvefit(img_m,x0,xdata,Imgew,lb,ub,options);
% Estimated GDL diffusion distance, m
lgdl=(abs(xr(1))+abs(xim(1)))/2;
% Estimated GDL Diffusion coefficient, m^2/s
Dg=(abs(xr(2))+abs(xim(2)))/2;
% Vector of Diffusion parameters as initial condition for next
  iteration
x0=[lgdl Dg];
%%% Simulation of GDL-channel impedance model considering
  steady-state oxygen concentration in channel
% Oxygen diffusion time constant in GDL, Eq. 5.2
T_gdl=(lgdl^2)/(Dg);
% Oxygen concentration at GDL/CCL interface, Eq. 5.43
C_gdl_St=wf.*(CCT_S-(Iy.*(100.^2).*lgdl/(4.*F.*Dg)));
% Oxygen transport resistance in GDL-channel, Eq. 6.2
Rgdl=((100.^2).*(b.*lgdl)./(4.*96485.*Dg.*(C_gdl_St)));
% Local GDL-channel impedance, Eq. 6.1
Zgd=((Rgdl).*(tanh((sqrt(-1).*2.*pi.*f.*T_gdl).^0.5))./((sqrt
  (-1).*2.*pi.*f.*T_gdl).^0.5));
% Argument of GDL-channel impedance
argm=abs(Zgd);
% Phase angle of GDL-channel impedance
angm=angle(Zgd);
```

```
% Real component of GDL-channel impedance
modm=argm.*cos(angm);
% Imaginary component of GDL-channel impedance
Imgem=argm.*sin(angm);
% Estimation of real and imaginary residuals
for n=1:numel(f)
% Residual of real component
R_gdl=abs((modw(n)-(modm(n)))./argw(n));
Kpr_gdl=(Resr_gdl+R_gdl);
Resr_gdl=R_gdl;
kpr_gdl(n,:)=Kpr_gdl;
% Residual of imaginary component
I_gdl=abs((Imgew(n)-(Imgem(n)))./argw(n));
Kpi_gdl=(Resi_gdl+I_gdl);
Resi_gdl=I_gdl;
kpi_gdl(n,:)=Kpi_gdl;
end
% Average sum of real and imaginary residuals
Res_gdl=(mean(kpi_gdl+kpr_gdl));
% Number of iterations
Iteration_gdl=Iteration_gdl+1;
end
% Initial conditions for estimation of diffusion parameters in
  CCL
n0=[20e-06 2e-07];
ndata=f;
% Initial conditions for estimation of residuals for CCL
  impedance
Resr_ccl=0;
Resi_ccl=0;
Res_ccl=2;
Iteration_ccl=0;
while Res_ccl > 0.0001
if Iteration_ccl==limit
break
end
% Function representing oxygen diffusion time constant in CCL,
  Eq. 4.98
T_CCL=@(n,ndata)(n(1).*n(1)./n(2));
% Function representing oxygen transport resistance in CCL,
  Eq. 6.10
Rccl=@(n,ndata)((100.^2).*R.*TF.*n(1)./(16.*F.*F.*n(2).
  *C_gdl_St));
% Function representing CCL diffusion impedance, Eq. 6.9
```

```
Zwc=@(n,xdata)((Rccl(n,ndata).*tanh((sqrt(-1).*2.*pi.*ndata.
  *T_CCL(n,ndata)).^0.5))./((sqrt(-1).*2.*pi.*ndata.*T_CCL(n,
  ndata)).^0.5));
% Function representing argument from CCL diffusion impedance
arg_mc=@(n,ndata)abs(Zwc(n,ndata));
% Function representing phase angle from CCL diffusion impedance
ang_mc=@(n,ndata)angle(Zwc(n,ndata));
% Function representing real component CCL diffusion impedance
real_mc=@(n,ndata)arg_mc(n,ndata).*cos(ang_mc(n,ndata));
% Function representing imaginary component CCL diffusion
  impedance
img_mc=@(n,ndata)arg_mc(n,ndata).*sin(ang_mc(n,ndata));
% Finite-length Warburg impedance for CCL, Eq. 6.21
Zwccl=((Rs.*tanh((sqrt(-1).*2.*pi.*f.*Ts).^0.5))./((sqrt(-1).
  *2.*pi.*f.*Ts).^0.5));
% Argument of finite-length Warburg impedance for CCL
argwc=abs(Zwccl);
% Phase angle of finite-length Warburg impedance for CCL
angwc=angle(Zwccl);
% Real component of finite-length Warburg impedance for CCL
modwc=argwc.*cos(angwc);
% Imaginary component of finite-length Warburg impedance for CCL
Imgewc=argwc.*sin(angwc);
% Non-linear regression
options=optimoptions('lsqcurvefit','Algorithm',
  'levenberg-marquardt');
lb = [];
ub = [];
nr=lsqcurvefit(real_mc,n0,ndata,modwc,lb,ub,options);
nim=lsqcurvefit(img_mc,n0,ndata,Imgewc,lb,ub,options);
% Estimated CCL diffusion distance, m
Lccl=(abs(nr(1))+abs(nim(1)))/2;
% Estimated CCL Diffusion coefficient, m^2/s
D_ccl=(abs(nr(2))+abs(nim(2)))/2;
% Vector of Diffusion parameters as initial condition for next
  iteration
n0=[Lccl D_ccl];
%%% Simulation of diffusion impedance model of the CCL
% Oxygen diffusion time constant in CCL, Eq. 4.98
T_CCL=(Lccl^2)/D_ccl;
% Oxygen transport resistance in CCL, Eq. 6.10
Rccl=(100.^2).*R.*TF.*Lccl./(16.*F.*F.*D_ccl.*C_gdl_St);
% Diffusion impedance of CCL, Eq. 6.9
```

```
Zccl=((Rccl.*tanh((sqrt(-1).*2.*pi.*f.*T_CCL).^0.5))./((sqrt
  (-1).*2.*pi.*f.*T_CCL).^0.5));
% Argument of diffusion impedance of CCL
argmc=abs(Zccl);
% Phase angle of diffusion impedance of CCL
angmc=angle(Zccl);
% Real component of diffusion impedance of CCL
modmc=argmc.*cos(angmc);
% Imaginary component of diffusion impedance of CCL
Imgemc=argmc.*sin(angmc);
% Estimation of real and imaginary residuals
for n=1:numel(f)
% Residual of real component
R_ccl=abs((modwc(n)-(modmc(n)))./argwc(n));
Kpr_ccl=(Resr_ccl+R_ccl);
Resr_ccl=R_ccl;
kpr_ccl(n,:)=Kpr_ccl;
% Residual of imaginary component
I_ccl=abs((Imgewc(n)-(Imgemc(n)))./argwc(n));
Kpi_ccl=(Resi_ccl+I_ccl);
Resi_ccl=I_ccl;
kpi_ccl(n,:)=Kpi_ccl;
end
% Average sum of real and imaginary residuals
Res_ccl=(mean(kpi_ccl+kpr_ccl));
% Number of iterations
Iteration_ccl=Iteration_ccl+1;
end
% Complex-impedance plot
plot(modw,-Imgew,'ko','LineWidth',2)
hold on
plot(modm,-Imgem,'r','LineWidth',2);
hold on
plot(modwc,-Imgewc,'bs','LineWidth',2)
hold on
plot(modmc,-Imgemc,'m','LineWidth',2);
xlabel('Z_r / \Omega cm^2','FontSize',16,'FontWeight','bold',
  'Color','k')
ylabel('- Z_j / \Omega cm^2','FontSize',16,'FontWeight','bold',
  'Color','k')
hlegl=legend('Eq.  6.23','Eq.  6.1  (y*)','Eq.  6.21','Eq.  6.9
  (y*)', 'northwest');
hlegl=gca;
```

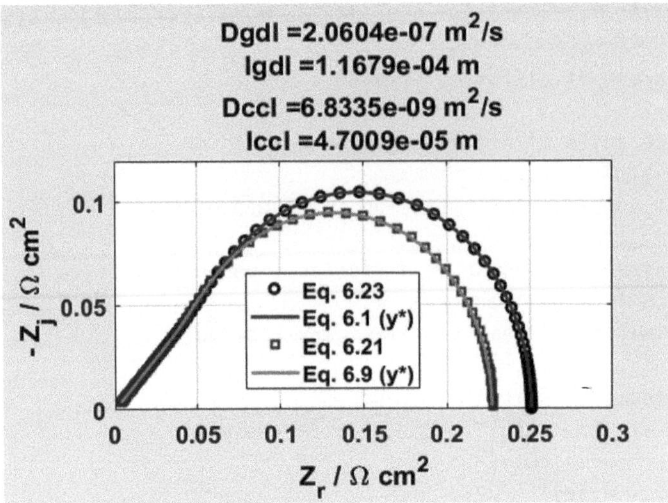

FIG. A.8 Plot in MATLAB resulting from the execution of script A.5.1.

```
hlegl.FontSize = 15;
hlegl.FontWeight = 'bold';
strg = {strcat('Dgdl =' , num2str(Dg,'%.4e'),' m^2/s')};
strgl = {strcat('lgdl =' , num2str(lgdl,'%.4e'),' m')};
strc = {strcat('Dccl =' , num2str(D_ccl,'%.4e'),' m^2/s')};
strcl = {strcat('lccl =' , num2str(Lccl,'%.4e'),' m')};
title([strg strgl strc strcl]);
grid;
clear all
```

See Fig. A.8.

A.5.2 Script to reproduce Fig. 6.11

```
% Frequency range 10000 Hz - 0.01 Hz
Frequency=logspace(-2,4,200);
f=(sort(Frequency,'descend'))';
% Channel inlet temperature
T=273.15+37;
% Temperature of PEMFC
TF=273.15+50;
% Channel inlet pressure
p=101325*1.7;
% Air relative humidity
RH=0.5;
% Oxygen Stoichiometry
so2=2.5;
```

```
% Channel depth, m
hc=0.001;
% Current density, A/cm^2
I=0.11;
% Charge transfer resistance, ohm.cm^2
Rc=0.8487;
% Tafel slope, V
b=0.0934;
% Gas constant
R=8.3145;
% Faraday constant
F=96485;
% Oxygen diffusion distance in CCL, m
Lccl=47e-06;
% Oxygen diffusion coefficient in CCL, m^2/s
D_ccl=6.8335e-09;
% Diffusion distance in GDL, m
lgdl=116.8e-06;
% Empirical oxygen diffusion coefficient in GDL, Eq. 6.28, m^2/s
Dr=2.9443e-07;
ws=0.0102;
Ep=0.646;
Dg=(Dr*(((Ep-0.11)/(1-0.11))^0.785)*((1-ws)^2)*((TF)/(T))^1.5);
%%% Parameters from Lumped model, Eq. 6.18
% Inductance from electrical cables
L=9.3394e-07;
% Ohmic resistance
Re=0.2398;
% Proton resistance
Rp=0.1294;
% Parameters from constant phase element
Y=0.0366;
P=0.8999;
% Coefficients for calculation of vapour saturation pressure
aa=-5800.2206;
bb=1.3914993;
cc=-0.048640239;
dd=0.41764768E-04;
ee=-0.14452093E-07;
ff=6.5459673;
% Saturation pressure of vapour, Eq. 5.23
Pvs=exp((aa/T)+bb+(cc*T)+(dd*T^2)+(ee*T^3)+ff*log(T));
% Humidity molar ratio, Eq. 5.22
H=RH*Pvs/(p-RH*Pvs);
% Channel length, it can be any value
```

```
Lc=1;
% Losses related to pressure drop and water produced by oxygen
  reaction
wf=0.6;
% Calculation of inlet current density in channel
xdata=0:0.01:1;
x0=I;
IY_S=@(x,xdata)(x(1).*exp(-xdata.*x(1)./(I.*so2)));
I_in=@(x,xdata)(mean(IY_S(x,xdata)));
% Inlet current density in channel, A/cm^2
jo=lsqcurvefit(I_in,x0,xdata,I);
% Fraction of channel length for SO2=2.5 and icell=0.11 A/cm^2
yy=0.48*Lc;
% Oxygen concentration at inlet of channel, mol/m^3
c_ref=p./((1+H).*5.*R.*T);
% Velocity of oxygen in channel, Eq. 5.17, m/s
Vc=so2.*Lc.*(I.*100.^2)./(4.*F.*hc.*c_ref);
% Inlet stoichiometry ratio
Lpe=4.*F.*c_ref.*Vc.*hc./(Lc.*jo*100.^2);
% Current density distribution, Eq. 5.26, A/cm^2
Iy=jo.*exp(-yy./(Lc.*Lpe));
% Molar oxygen concentration, Eq. 5.28
Ms_A=exp(-yy./(Lpe.*Lc));
% Oxygen concentration in channel, Eq. 5.27
MCCT_s=Ms_A./((Ms_A./5)+0.8+H);
CCT_S=(p.*MCCT_s./(5.*R.*T));
% Oxygen concentration at GDL/CCL interface, Eq. 5.43
GC=(Iy.*(100.^2).*lgdl./(4.*F.*Dg));
Cfi=c_ref-GC;
% Parameter defined in numerator of Eq. 6.25
alph_s=((R.*TF./(4.*F.*(Rc+0.2509).*Iy))+(lgdl./(4.*F.*Dg)).*
  (R.*TF./(4.*F.*(Rc+0.2509).*Cfi./100^2)));
%%% Oxygen concentration at GDL/CCL interface, Eq. 5.43 at nomi-
  nal conditions with GDL diffusion coefficient at nominal condi-
  tions Dgdl=2.0604e-07 m^2/s and humidity molar ratio at nominal
  conditions H=0.0186;
GCn=(Iy.*(100.^2).*lgdl./(4.*F.*2.0604e-07));
c_refn=(101325*1.7)./((1+0.0186).*5.*R.*(273.15+37));
Cfin=c_refn-GCn;
% Parameter defined in denominator of Eq. 6.25
alph_n=((R.*323.15./(4.*F.*(Rc+0.2509).*Iy))+(lgdl./(4.*F.
  *2.0604e-07)).*(R.*323.15./(4.*F.*(Rc+0.2509).*Cfin./100^2)));
% Factor related to the deviation of the charge transfer coeffi-
  cient from nominal conditions, Eq. 6.25
Alp=alph_s./alph_n;
```

```
% Recalculation of charge transfer resistance
Rc=b./(Iy.*Alp);
% Recalculation of Tafel slope
b=Rc.*Iy;
% Oxygen transport resistance, Eq. 6.2
C_gdl_S=wf.*(CCT_S-GC);
Rgdl_dc=(100.^2).*(b.*lgdl)./(4.*96485.*Dg.*(C_gdl_S));
% Oxygen diffusion time constant in GDL, Eq. 5.2
T_GDL=(lgdl^2)/Dg;
% GDL-channel impedance with steady-state oxygen concentration in
  channel, Eq. 6.1
Z_dc=((Rgdl_dc).*(tanh((sqrt(-1).*2.*pi.*f.*T_GDL).^0.5))./
  ((sqrt(-1).*2.*pi.*f.*T_GDL).^0.5));
% GDL-channel impedance with steady-state oxygen concentration in
  channel and capacitance effect from CCL, Eq. 6.13
ZKdc=Z_dc./(1+(Rc.*Y.*(sqrt(-1).*2.*pi.*f).^P));
% Laplace domain to frequency domain
s=2.*pi.*f.*sqrt(-1);
% Parameter lambda defined in Eq. 5.11
Lambda=sqrt(s./(Dg));
% Eq. 5.20
beta=s+(sqrt(s.*Dg).*tanh(sqrt(s.*T_GDL))./hc);
% Eq. 5.21
gamma=exp(yy.*beta./Vc);
% Molar concentration of oxygen in the channel, Eq. 5.38
MCC_d=1+((Vc)./(cosh(sqrt(s.*T_GDL)).*(beta.*Lc.*Lpe-Vc))).*
  ((1./gamma)-exp(-yy./(Lc.*Lpe)));
% Total molar Oxygen concentration in wet air
M_CCT=MCC_d./(H+0.8+(MCC_d./5));
% Oxygen concentration in the channel, Eq. 5.37
CCT=p.*M_CCT./(5.*R.*T);
% Current density considering amplitude of sinusoidal perturba-
  tion, the value 0.05 is the amplitude value of 5% of the DC
  current density, see Figure 5.7
IT=Iy+(Iy.*0.05);
% Oxygen transport resistance with oscillating oxygen concentra-
  tion in channel, Eq. 6.4
C_gdl=wf.*((CCT)-((IT.*100.^2).*tanh(sqrt(s.*T_GDL))./(4.*F.
  *Dg.*Lambda)));
Rgdl_ac=(100.^2).*(b.*lgdl)./(4.*96485.*Dg.*C_gdl);
% GDL impedance with oscillating oxygen concentration in channel,
  Eq. 6.3
Z_ac=(Rgdl_ac.*(tanh((sqrt(-1).*2.*pi.*f.*T_GDL).^0.5))./((sqrt
  (-1).*2.*pi.*f.*T_GDL).^0.5));
```

```
% GDL-channel impedance with oscillating oxygen concentration in
  channel and capacitance effect from CCL, Eq. 6.13
ZKac=Z_ac./(1+(Rc.*Y.*(sqrt(-1).*2.*pi.*f).^P));
% Oxygen diffusion time constant in CCL, Eq. 4.98
T_CCL=(Lccl^2)/D_ccl;
% Oxygen transport resistance in CCL, Eq. 6.10
Rccl=(100.^2).*R.*TF.*Lccl./(16.*F.*F.*D_ccl.*C_gdl_S);
% Diffusion impedance of CCL, Eq. 6.9
ZWccl=((Rccl.*tanh((sqrt(-1).*2.*pi.*f.*T_CCL).^0.5))./((sqrt
  (-1).*2.*pi.*f.*T_CCL).^0.5));
% Diffusion impedance in CCL neglecting oxygen transport
  limitations
lambda_ccl=sqrt((Rp.*Y.*((sqrt(-1).*2.*pi.*f).^P))+(Rp./(Rc)));
Z_CCL=(Rp.*coth(lambda_ccl)./lambda_ccl);
% Cathode impedance
Lambda=sqrt((Rp.*Y.*((sqrt(-1).*2.*pi.*f).^P))+(Rp./(Rc
  +ZWccl)));
Z_cathode=(L.*(sqrt(-1).*2.*pi.*f))+Re+ZKdc+(Rp.*coth(Lambda)./
  Lambda);
% Plotting cathodic impedance response
Zcat_arg=abs(Z_cathode);
Zcat_ang=angle(Z_cathode);
Zcat_re=Zcat_arg.*cos(Zcat_ang);
Zcat_Img=-Zcat_arg.*sin(Zcat_ang);
plot(Zcat_re,Zcat_Img,'-or')
hold on
% Plotting impedance of the CCL
Zccl_arg=abs(Z_CCL);
Zccl_ang=angle(Z_CCL);
Zccl_re=Re+Zccl_arg.*cos(Zccl_ang);
Zccl_Img=-Zccl_arg.*sin(Zccl_ang);
plot(Zccl_re,Zccl_Img,'-dk')
hold on
% Plotting oxygen diffusion impedance of CCL
Zwccl_arg=abs(ZWccl);
Zwccl_ang=angle(ZWccl);
Zwccl_re=Zccl_re(end)+Zwccl_arg.*cos(Zwccl_ang);
Zwccl_Img=-Zwccl_arg.*sin(Zwccl_ang);
plot(Zwccl_re,Zwccl_Img,'-sg')
hold on
% Plotting impedance of GDL-channel with steady-state oxygen in
  channel
Kdc_arg=abs(ZKdc);
```

```
Kdc_ang=angle(ZKdc);
Kdc_re=Zwccl_re(end)+Kdc_arg.*cos(Kdc_ang);
Kdc_Img=-Kdc_arg.*sin(Kdc_ang);
plot(Kdc_re,Kdc_Img,'-^b')
hold on
% Plotting impedance of GDL-channel with oscillating oxygen in
  channel
Kac_arg=abs(ZKac);
Kac_ang=angle(ZKac);
Kac_re=Zwccl_re(end)+Kac_arg.*cos(Kac_ang);
Kac_Img=-Kac_arg.*sin(Kac_ang);
plot(Kac_re,Kac_Img,'-vm')
axis([0 1.8 0 0.8])
xlabel('Z_r / \Omega cm^2','FontSize',16,'FontWeight','bold',
  'Color','k')
ylabel('- Z_j / \Omega cm^2','FontSize',16,'FontWeight','bold',
  'Color','k')
hleg1=legend('Cathode','CCL no O_2 loss','O_2 loss in CCL',
  'GDL-channel_s_s', 'GDL-channel_a_c')
hleg1=gca;
hleg1.FontSize = 12;
hleg1.FontWeight = 'bold';
grid;
clear all;
```

See Fig. A.9.

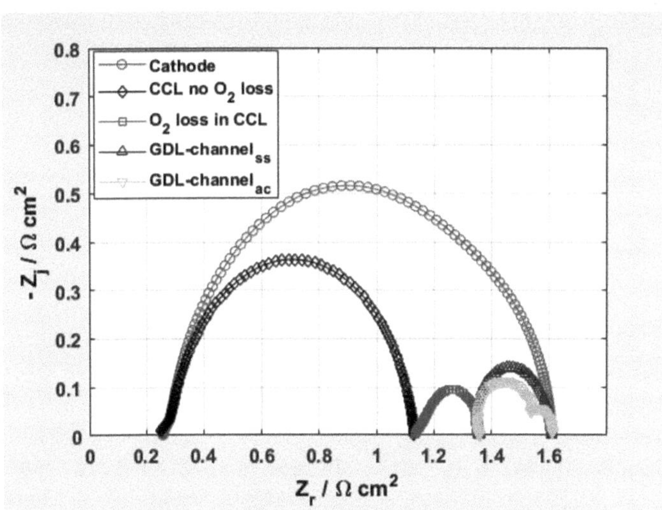

FIG. A.9 Plot in MATLAB resulting from the execution of script A.5.2.

A.6 Chapter 7

A.6.1 Script to reproduce Fig. 7.11

```
%%% EIS measurements of cell 1 at 0.1875 A/cm^2 neglecting induc-
  tive loop
% Data of frequencies
f=[9999.999997  8858.6679  7847.599693  6951.927958  6158.482108
  5455.594777  4832.930239  4281.332394  3792.690188  3359.818282
  2976.351442  2636.650896  2335.721467  2069.13808  1832.980709
  1623.776737  1438.449887  1274.274986  1128.837891  999.999999
  885.86679 784.759969 695.192796 615.848211 545.559477 483.293024
  428.133239 379.269019 335.981829 297.635144 263.66509 233.572147
  206.913808 183.298071 162.377674 143.844989 127.427499 112.883789
  100  88.586679  78.475997  69.51928  61.584821  54.555948  48.329302
  42.813324  37.926902  33.598183  29.763514  26.366509  23.357215
  20.691381  18.329807  16.237767  14.384499  12.74275  11.288379  10
  8.858668 7.8476 6.951928 6.158482 5.455595 4.83293 4.281332 3.79269
  3.359818  2.976351  2.636651  2.335721  2.069138  1.832981  1.623777
  1.43845 1.274275 1.128838 1 0.885867 0.78476]';
% Real components of EIS measurements neglecting inductive loop
Z_r=[0.235296 0.216288 0.206208 0.218 0.215056 0.220048 0.210496
  0.224352  0.219712  0.2248  0.22624  0.227424  0.232976  0.228944
  0.233136  0.23576  0.238448  0.240816  0.242656  0.245872  0.249168
  0.25192  0.254688  0.258544  0.261152  0.26536  0.269216  0.273168
  0.277776  0.282592  0.287664  0.29384  0.30032  0.306912  0.314624
  0.322976  0.332624  0.34344  0.356272  0.370128  0.385744  0.403152
  0.422272  0.44376  0.468192  0.49304  0.519648  0.548304  0.577408
  0.60656  0.636656  0.665552  0.693968  0.721232  0.7472  0.771744
  0.7944  0.816384  0.83736  0.856432  0.8748  0.891424  0.907296
  0.921456  0.933424  0.943552  0.953552  0.960672  0.966448  0.97056
  0.97384  0.974384  0.97688  0.9756  0.975424  0.973936  0.970768
  0.97016  0.966944]';
% Imaginary components of EIS measurements neglecting inductive
  loop
Z_i=[-0.003408 -0.018128 -0.025424 -0.026528 -0.024192 -0.03784
  -0.031008 -0.032816 -0.029872 -0.031232 -0.032272 -0.034928
  -0.036512 -0.035712 -0.040368 -0.040928 -0.043984 -0.045936
  -0.048832 -0.050752 -0.053552 -0.05664  -0.060832 -0.064016
  -0.068144 -0.072816 -0.076688 -0.082704 -0.088128 -0.094592
  -0.101616 -0.108672 -0.116864 -0.125888 -0.135616 -0.145984
  -0.156416 -0.167872 -0.179936 -0.19264  -0.206144 -0.219024
  -0.23128  -0.244256 -0.25432  -0.265072 -0.273472 -0.279648
  -0.284 -0.285984 -0.285472 -0.282496 -0.277984 -0.271808
```

```
  -0.263616 -0.254896 -0.244768 -0.233872 -0.223632 -0.210992
  -0.20008 -0.18712 -0.173376 -0.159904 -0.146064 -0.13224
  -0.117712 -0.10368 -0.090704 -0.07704 -0.06616 -0.055184
  -0.044768 -0.035248 -0.026064 -0.019024 -0.012784 -0.006944
  -0.001582608]';
% EIS measurements neglecting inductive loop in complex format
Z_data=complex(Z_r,Z_i);
% Argument of EIS measurements neglecting inductive loop
Zarg=abs(Z_data);
% Phase angle of EIS measurements neglecting inductive loop
Zang=angle(Z_data);
% Initial condition for estimation of parameters
x0=[0.1 0.01 0.001 0.9 0.01 0.1 0.1];
xdata=f;
% Initial conditions for estimation of real and imaginary
  residuals
Res_NL=2;
Resr_NL=0;
Resi_NL=0;
Iteration_NL=0;
% Maximum number of iterations
limit=100;
% Conditional statement for real and imaginary residuals for
  impedance
while Res_NL > 0.01
if Iteration_NL==limit
break
end
% Function representing finite-length Warburg impedance, Eq. 5.1
f_Zw=@(x,xdata)(x(1).*tanh((sqrt(-1).*2.*pi.*xdata.*x(2)).
  ^0.5))./((sqrt(-1).*2.*pi.*xdata.*x(2)).^0.5);
% Function representing constant phase element
f_Y=@(x,xdata)(x(3).*((sqrt(-1).*2.*pi.*xdata).^x(4)));
% Function representing parameter from transmission line model
f_lambda=@(x,xdata)(sqrt((x(5).*f_Y(x,xdata))+(x(5)./(x(6)+f_Zw
  (x,xdata)))));
% Function representing transmission line model (TLM), Eq. 7.4
  and neglecting Eq. 7.5
f_Z=@(x,xdata)(x(7)+(x(5).*coth(f_lambda(x,xdata))./f_lambda
  (x,xdata)));
% Function representing argument of TLM
arg_m=@(x,xdata)abs(f_Z(x,xdata));
% Function representing phase angle of TLM
```

```
ang_m=@(x,xdata)angle(f_Z(x,xdata));
% Function representing real component of TLM
real_m=@(x,xdata)arg_m(x,xdata).*cos(ang_m(x,xdata));
% Function representing imaginary component of TLM
img_m=@(x,xdata)arg_m(x,xdata).*sin(ang_m(x,xdata));
% Non-linear regression
options=optimoptions('lsqcurvefit','Algorithm',
  'levenberg-marquardt');
lb = [];
ub = [];
xr=lsqcurvefit(arg_m,x0,xdata,Zarg,lb,ub,options);
xim=lsqcurvefit(ang_m,x0,xdata,Zang,lb,ub,options);
% Estimated oxygen transport resistance
Rw=(abs(xr(1))+abs(xim(1)))/2;
% Estimated oxygen diffusion time constant
Tw=(abs(xr(2))+abs(xim(2)))/2;
% Estimated parameters of constant phase element
Y=(abs(xr(3))+abs(xim(3)))/2;
P=(abs(xr(4))+abs(xim(4)))/2;
% Estimated proton resistance
Rp=(abs(xr(5))+abs(xim(5)))/2;
% Estimated charge transfer resistance
Rc=(abs(xr(6))+abs(xim(6)))/2;
% Estimated ohmic resistance
Re=(abs(xr(7))+abs(xim(7)))/2;
% Impedance of finite-length Warburg impedance, Eq. 5.1
Zw=((Rw.*tanh((sqrt(-1).*2.*pi.*f.*Tw).^0.5))./((sqrt(-1).
  *2.*pi.*f.*Tw).^0.5));
% Parameter of transmission line model
lambda=sqrt((Rp.*Y.*((sqrt(-1).*2.*pi.*f).^P))+(Rp./(Rc+Zw)));
% Impedance of transmission line model, Eq. 7.4
Z_NL=Re+(Rp.*coth(lambda)./lambda);
% Argument of TLM
arg_NL=abs(Z_NL);
% Phase angle of TLM
ang_NL=angle(Z_NL);
% Real component of TLM
mod_NL=arg_NL.*cos(ang_NL);
% Imaginary component of TLM
Imge_NL=arg_NL.*sin(ang_NL);
% Vector of parameters as initial condition for next iteration
x0 =[Rw Tw Y P Rp Rc Re];
% Estimation of real and imaginary residuals
```

```
for n=1:numel(f)
% Residual of real component
R_NL=abs((Z_r(n)-(mod_NL(n)))./Zarg(n));
Kpr_NL=(Resr_NL+R_NL);
Resr_NL=R_NL;
kpr_NL(n,:)=Kpr_NL;
% Residual of imaginary component
I_NL=abs((Z_i(n)-(Imge_NL(n)))./Zarg(n));
Kpi_NL=(Resi_NL+I_NL);
Resi_NL=I_NL;
kpi_NL(n,:)=Kpi_NL;
end
% Average sum of real and imaginary residuals
Res_NL=(mean(kpi_NL+kpr_NL));
% Number of iterations
Iteration_NL=Iteration_NL+1;
end
%%% EIS measurements of cell 1 at 0.1875 A/cm^2 considering
  inductive loop
% Data of frequencies
f=[9999.999997 8858.6679 7847.599693    6951.927958    6158.482108
  5455.594777  4832.930239  4281.332394  3792.690188  3359.818282
  2976.351442   2636.650896   2335.721467   2069.13808 1832.980709
  1623.776737   1438.449887   1274.274986   1128.837891   999.999999
  885.86679 784.759969 695.192796 615.848211 545.559477 483.293024
  428.133239 379.269019 335.981829 297.635144 263.66509 233.572147
  206.913808 183.298071 162.377674 143.844989 127.427499 112.883789
  100   88.586679 78.475997 69.51928 61.584821 54.555948 48.329302
  42.813324 37.926902 33.598183 29.763514 26.366509 23.357215
  20.691381 18.329807 16.237767 14.384499 12.74275 11.288379
  10 8.858668 7.8476 6.951928 6.158482 5.455595 4.83293  4.281332
  3.79269  3.359818 2.976351 2.636651 2.335721 2.069138 1.832981
  1.623777 1.43845 1.274275 1.128838 1 0.885867 0.78476 0.695193
  0.615848 0.545559 0.483293 0.428133 0.379269 0.335982 0.297635
  0.263665 0.233572 0.206914 0.183298 0.162378 0.143845 0.127427
  0.112884 0.1]';
% Real components of EIS measurements considering inductive loop
Z_r=[0.235296 0.216288 0.206208 0.218 0.215056 0.220048
  0.210496 0.224352 0.219712 0.2248 0.22624 0.227424 0.232976
  0.228944 0.233136 0.23576 0.238448 0.240816 0.242656 0.245872
  0.249168 0.25192  0.254688 0.258544 0.261152 0.26536  0.269216
  0.273168 0.277776 0.282592 0.287664 0.29384  0.30032  0.306912
  0.314624 0.322976 0.332624 0.34344
```

```
0.356272 0.370128 0.385744 0.403152 0.422272 0.44376
0.468192 0.49304  0.519648 0.548304 0.577408 0.60656  0.636656
0.665552 0.693968 0.721232 0.7472 0.771744 0.7944 0.816384
0.83736    0.856432 0.8748 0.891424 0.907296 0.921456 0.933424
0.943552 0.953552 0.960672 0.966448 0.97056  0.97384  0.974384
0.97688   0.9756 0.975424 0.973936 0.970768 0.97016   0.966944
0.964448 0.962208 0.959504 0.95712  0.95384  0.950656 0.948256
0.945184 0.942896 0.940432 0.938144 0.935856 0.932672 0.929984
0.928112 0.926048 0.924976]';
% Imaginary components of EIS measurements considering inductive
  loop
Z_i=[-0.003408 -0.018128 -0.025424 -0.026528 -0.024192 -0.03784
  -0.031008 -0.032816 -0.029872 -0.031232 -0.032272 -0.034928
  -0.036512 -0.035712 -0.040368 -0.040928 -0.043984 -0.045936
  -0.048832 -0.050752 -0.053552 -0.05664  -0.060832 -0.064016
  -0.068144 -0.072816 -0.076688 -0.082704 -0.088128 -0.094592
  -0.101616 -0.108672 -0.116864 -0.125888 -0.135616 -0.145984
  -0.156416 -0.167872 -0.179936 -0.19264  -0.206144 -0.219024
  -0.23128  -0.244256 -0.25432  -0.265072 -0.273472 -0.279648
  -0.284 -0.285984 -0.285472 -0.282496 -0.277984 -0.271808
  -0.263616 -0.254896 -0.244768 -0.233872 -0.223632 -0.210992
  -0.20008  -0.18712  -0.173376  0.159904  0.146064 -0.13224
  -0.117712 -0.10368  -0.090704 -0.07704  -0.06616  -0.055184
  -0.044768 -0.035248 -0.026064 -0.019024 -0.012784 -0.006944
  -0.001582608 0.002272 0.006192 0.009536 0.012176 0.014624
  0.017952 0.01912 0.020848 0.021872 0.023008 0.023696 0.023808
  0.025152 0.026832 0.026336 0.02672 0.027328]';
% EIS measurements in complex format considering inductive loop
Z_data=complex(Z_r,Z_i);
% Argument of EIS measurements
Zarg=abs(Z_data);
% Initial condition
x0 =[Rw Tw Y P 0.1 0.1 0.8 Rp];
xdata=f;
% Initial conditions for estimation of residuals for impedance
Res_WL=2;
Resr_WL=0;
Resi_WL=0;
Iteration_WL=0;
% Conditional statement for real and imaginary residuals for
  impedance
while Res_WL > 0.01
if Iteration_WL==limit
```

```
break
end
% Function representing finite-length Warburg impedance, Eq. 5.1
f_Zw=@(x,xdata)(x(1).*tanh((sqrt(-1).*2.*pi.*xdata.*x(2)).
  ^0.5))./((sqrt(-1).*2.*pi.*xdata.*x(2)).^0.5);
% Function representing constant phase element
f_Y=@(x,xdata)(x(3).*((sqrt(-1).*2.*pi.*xdata).^x(4)));
% Function representing Randles circuit for simulation of induc-
  tive loop, Eq. 7.5
f_Randv=@(x,xdata)(x(5)./(1+(x(5).*x(6).*((sqrt(-1).*2.*pi.
  *xdata).^x(7)))));
% Function representing parameter from transmission line model
f_lambda=@(x,xdata)(sqrt((x(8).*f_Y(x,xdata))+(x(8)./(Rc
  +f_Randv(x,xdata)+f_Zw(x,xdata)))));
% Function representing transmission line model (TLM), Eq. 7.4
f_Z=@(x,xdata)(Re+(x(8).*coth(f_lambda(x,xdata))./f_lambda
  (x,xdata)));
% Function representing argument of TLM
arg_m=@(x,xdata)abs(f_Z(x,xdata));
% Function representing phase angle of TLM
ang_m=@(x,xdata)angle(f_Z(x,xdata));
% Function representing real component of TLM
real_m=@(x,xdata)arg_m(x,xdata).*cos(ang_m(x,xdata));
% Function representing imaginary component of TLM
img_m=@(x,xdata)arg_m(x,xdata).*sin(ang_m(x,xdata));
% Non-linear regression
options=optimoptions('lsqcurvefit','Algorithm',
  'levenberg-marquardt');
lb = [];
ub = [];
xr=lsqcurvefit(real_m,x0,xdata,Z_r,lb,ub,options);
xim=lsqcurvefit(img_m,x0,xdata,Z_i,lb,ub,options);
% Estimated oxygen transport resistance
Rw=(abs(xr(1))+abs(xim(1)))/2;
% Estimated oxygen diffusion time constant
Tw=(abs(xr(2))+abs(xim(2)))/2;
% Estimated parameters of constant phase element
Y=(abs(xr(3))+abs(xim(3)))/2;
P=(abs(xr(4))+abs(xim(4)))/2;
% Estimated parameters of impedance of Randles circuit, Eq. 7.5
Rad=-(abs(xr(5))+abs(xim(5)))/2;
Yad=-(abs(xr(6))+abs(xim(6)))/2;
Pad=(abs(xr(7))+abs(xim(7)))/2;
```

```
% Estimated proton resistance
Rp=(abs(xr(8))+abs(xim(8)))/2;
% Impedance of finite-length Warburg impedance, Eq. 5.1
Zw=((Rw.*tanh((sqrt(-1).*2.*pi.*f.*Tw).^0.5))./((sqrt(-1).*2.
  *pi.*f.*Tw).^0.5));
% Impedance of Randles circuit, Eq. 7.5
Rand=Rad./(1+(Rad.*Yad.*((sqrt(-1).*2.*pi.*f).^Pad)));
% Parameter of transmission line model
lambda=sqrt((Rp.*Y.*((sqrt(-1).*2.*pi.*f).^P))+(Rp./(Rc+Rand
  +Zw)));
% Impedance of transmission line model (TLM), Eq. 7.4
Z=Re+(Rp.*coth(lambda)./lambda);
% Argument of TLM
arg=abs(Z);
% Phase angle of TLM
ang=angle(Z);
% Real component of TLM
mod=arg.*cos(ang);
% Imaginary component of TLM
Imge=arg.*sin(ang);
% Vector of parameters as initial condition for next iteration
x0 =[Rw Tw Y P Rad Yad Pad Rp];
% Estimation of real and imaginary residuals
for n=1:numel(f)
% Residual of real component
R_WL=abs((Z_r(n)-(mod(n)))./Zarg(n));
Kpr_WL=(Resr_WL+R_WL);
Resr_WL=R_WL;
kpr_WL(n,:)=Kpr_WL;
% Residual of imaginary component
I_WL=abs((Z_i(n)-(Imge(n)))./Zarg(n));
Kpi_WL=(Resi_WL+I_WL);
Resi_WL=I_WL;
kpi_WL(n,:)=Kpi_WL;
end
% Average sum of real and imaginary residuals
Res_WL=(mean(kpi_WL+kpr_WL));
% Number of iterations
Iteration_WL=Iteration_WL+1;
end
% Complex-impedance plot
plot(Z_r,-Z_i,'-rd','LineWidth',2)
hold on
```

```
plot(mod_NL,-Imge_NL,'-sb','LineWidth',3)
hold on
plot(mod,-Imge,'-^k','LineWidth',2)
% Axis labels
xlabel('Z_r / \Omega cm^2','FontSize',16,'FontWeight','bold',
  'Color','k')
ylabel('- Z_j / \Omega cm^2','FontSize',16,'FontWeight','bold',
  'Color','k')
hlegl=legend('EIS  measurements','TLM  no  Inductive  loop','TLM
  with inductive loop','northwest');
hlegl=gca;
hlegl.FontSize = 16;
hlegl.FontWeight = 'bold';
grid;
clear all
```

See Fig. A.10.

B Alternative tests

If some parameters of an impedance model are estimated from alternative tests than that of electrochemical impedance spectroscopy (EIS), then the number of remaining parameters to be fitted in the impedance model with EIS measurements is reduced.

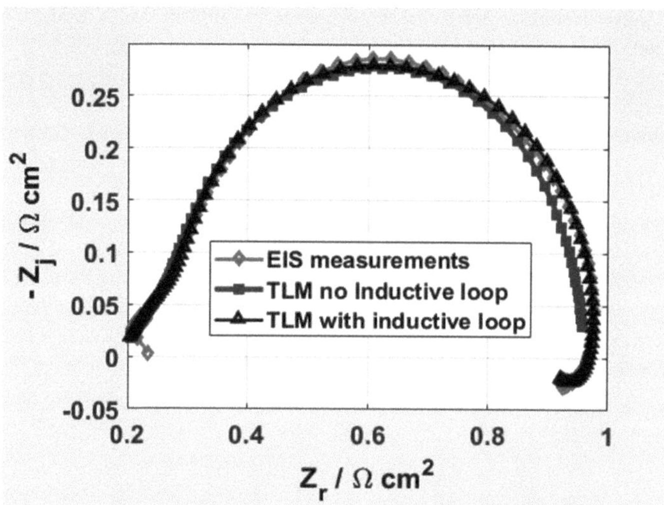

FIG. A.10 Plot in MATLAB resulting from the execution of script A.6.1.

TABLE B.1 Parameters of the cathodic impedance model that can be estimated from other experimental tests.

Parameter	Experimental test
Double-layer capacitance, C_{dl}	Cyclic voltammetry [1]
Electron resistance, R_e	In-plane and through-plane electronic conductivity tests [2]
Oxygen diffusion coefficient in CCL, D_{CCL}	Ex-situ diffusion test [3]
Proton resistance, R_P	EIS test in H_2/N_2 PEM fuel cell [4]
Tafel slope, b_{O_2}	cathodic current-potential curves [5]
Oxygen diffusion coefficient in GDL, D_{GDL}	Loschmidt cell technique [6]
Local current density in the channel, $i_F(y)$	Segmented cell system [7]

Some parameters of the cathodic impedance model defined in Chapter 6 can be estimated from alternative experimental tests. The estimation of the parameters is not limited to the experimental tests mentioned in (Table B.1).

References

[1] D.M. Morales, M. Risch, Seven steps to reliable cyclic voltammetry measurements for the determination of double layer capacitance, J. Phys. Energ. 3 (2021) 034013.

[2] M. Ahadi, M. Tam, J. Stumper, M. Bahrami, Electronic conductivity of catalyst layers of polymer electrolyte membrane fuel cells: through-plane vs. in-plane, Int. J. Hydrog. Energy 44 (2019) 3603–3614.

[3] A. Baricci, A. Casalegno, A novel (ex situ) method to quantify oxygen diffusion coefficient of polymer fuel cells backing and catalyst layers, J. Power Sources 325 (2016) 664–669.

[4] T. Gaumont, G. Maranzana, O. Lottin, J. Dillet, S. Didierjean, J. Pauchet, L. Guétaz, Measurement of protonic resistance of catalyst layers as a tool for degradation monitoring, Int. J. Hydrog. Energy 42 (2017) 1800–1812.

[5] D.A. Tryk, M.S. Lee, M. Uchida, H. Uchida, M. Watanabe, Tafel slope component analysis of polymer electrolyte fuel cell cathode current-potential behavior, ECS Trans. 35 (2011) 13–23.

[6] C. Chan, N. Zamel, X. Li, J. Shen, Experimental measurement of effective diffusion coefficient of gas diffusion layer/microporous layer in PEM fuel cells, Electrochim. Acta 65 (2012) 13–21.

[7] T. Reshetenko, A. Kulikovsky, On the distribution of local current density along a PEM fuel cell cathode channel, Electrochem. Commun. 101 (2019) 35–38.

Index

Note: Page numbers followed by *f* indicate figures and *t* indicate tables.

R

S

T

V

W

Z

FSC
www.fsc.org
MIX
Papier | Fördert gute Waldnutzung
FSC® C083411

Kolibri 360 GmbH
Ferdinand-Jühlke-Straße 7
99095 Erfurt, Deutschland
produktsicherheit@kolibri360.de